Vol. 29. **The Analytical Chemistry of Sulfur and Its Compounds** (*in three parts*). By J. H. Karchmer
Vol. 30. **Ultramicro Elemental Analysis.** By Günther Tölg
Vol. 31. **Photometric Organic Analysis** (*in two parts*). By Eugene Sawicki
Vol. 32. **Determination of Organic Compounds: Methods and Procedures.** By Frederick T. Weiss
Vol. 33. **Masking and Demasking of Chemical Reactions.** By D. D. Perrin
Vol. 34. **Neutron Activation Analysis.** By D. De Soete, R. Gijbels, and J. Hoste
Vol. 35. **Laser Raman Spectroscopy.** By Marvin C. Tobin
Vol. 36. **Emission Spectrochemical Analysis.** By Morris Slavin
Vol. 37. **Analytical Chemistry of Phosphorus Compounds.** Edited by M. Halmann
Vol. 38. **Luminescence Spectrometry in Analytical Chemistry.** By J. D. Winefordner, S. G. Schulman and T. C. O'Haver
Vol. 39. **Activation Analysis with Neutron Generators.** By Sam S. Nargolwalla and Edwin P. Przybylowicz
Vol. 40. **Determination of Gaseous Elements in Metals.** Edited by Lynn L. Lewis, Laben M. Melnick, and Ben D. Holt
Vol. 41. **Analysis of Silicones.** Edited by A. Lee Smith
Vol. 42. **Foundations of Ultracentrifugal Analysis.** By H. Fujita
Vol. 43. **Chemical Infrared Fourier Transform Spectroscopy.** By Peter R. Griffiths
Vol. 44. **Microscale Manipulations in Chemistry.** By T. S. Ma and V. Horak
Vol. 45. **Thermometric Titrations.** By J. Barthel
Vol. 46. **Trace Analysis: Spectroscopic Methods for Elements.** Edited by J. D. Winefordner
Vol. 47. **Contamination Control in Trace Element Analysis.** By Morris Zief and James W. Mitchell
Vol. 48. **Analytical Applications of NMR.** By D. E. Leyden and R. H. Cox
Vol. 49. **Measurement of Dissolved Oxygen.** By Michael L. Hitchman
Vol. 50. **Analytical Laser Spectroscopy.** Edited by Nicolo Omenetto
Vol. 51. **Trace Element Analysis of Geological Materials.** By Roger D. Reeves and Robert R. Brooks
Vol. 52. **Chemical Analysis by Microwave Rotational Spectroscopy.** By Ravi Varma and Lawrence W. Hrubesh
Vol. 53. **Information Theory As Applied to Chemical Analysis.** By Karel Eckschlager and Vladimír Štěpánek
Vol. 54. **Applied Infrared Spectroscopy: Fundamentals, Techniques, and Analytical Problem-solving.** By A. Lee Smith
Vol. 55. **Archaeological Chemistry.** By Zvi Goffer
Vol. 56. **Immobilized Enzymes in Analytical and Clinical Chemistry.** By P. W. Carr and L. D. Bowers
Vol. 57. **Photoacoustics and Photoacoustic Spectroscopy.** By Allan Rosencwaig
Vol. 58. **Analysis of Pesticide Residues.** Edited by H. Anson Moye
Vol. 59. **Affinity Chromatography.** By William H. Scouten
Vol. 60. **Quality Control in Analytical Chemistry.** By G. Kateman and F. W. Pijpers
Vol. 61. **Direct Characterization of Fineparticles.** By Brian H. Kaye
Vol. 62. **Flow Injection Analysis.** *Second Edition* By J. Ruzicka and E. H. Hansen

(*continued on back*)

Laser Microanalysis

# CHEMICAL ANALYSIS

A SERIES OF MONOGRAPHS ON ANALYTICAL CHEMISTRY
AND ITS APPLICATIONS



**VOLUME 105**

**A WILEY-INTERSCIENCE PUBLICATION**

**JOHN WILEY & SONS**

**New York / Chichester / Brisbane / Toronto / Singapore**

# Laser Microanalysis

**LIESELOTTE MOENKE-BLANKENBURG**

Martin-Luther-University Halle -Wittenberg
Department of Chemistry
Halle, German Democratic Republic

**A WILEY-INTERSCIENCE PUBLICATION**

**JOHN WILEY & SONS**

**New York / Chichester / Brisbane / Toronto / Singapore**

*To*

*Horst H. W.,*
*Jens U., and*
*Thurid C. Moenke*



*Library of Congress Cataloging in Publication Data:*

Moenke-Blankenburg, Lieselotte.
Laser microanalysis/Lieselotte Moenke-Blankenburg.
p. cm.—(Chemical analysis, ISSN 0069-2883; v. 105)
"A Wiley-Interscience publication."
Includes bibliographies and index.
ISBN 0-471-63707-6
1. Laser spectroscopy. I. Title. II. Series.
QD96.L3M64 1989 88-34686
543'.08584—dc19 CIP

Printed in the United States of America

10 9 8 7 6 5 4 3 2 1

# PREFACE

Since the publication of our previous books about laser microanalysis (LMA) in 1966, 1968, and 1973 and the reviews in 1984 and 1986, the techniques in this field have progressed and changed in two principal ways: After a steep ascent in the first 10 years, LMA based on optical emission spectrometry (OES) has reached an acceptable plateau and very little of moment is happening, whereas LMA based on mass spectrometry (MS) developed slowly in the beginning but has experienced a remarkable improvement in design and performance in the last 10 years. Recently, both methods have been utilized in interconnected systems, especially with inductively coupled plasma (ICP).

All methods of analytical application of lasers in OES, MS, atomic absorption spectrometry (AAS), atomic fluorescence spectrometry (AFS), and ICP-OES and ICP-MS that are described herein have a common background: the use of the laser as a radiation source, the effects of laser target interactions, microplasma generation, and the necessary presumptions for quantitative laser microanalysis of solids. Three chapters of the present book deal with these topics.

The second part of the book is devoted to special techniques, methodological treatments, and analytical features, and to applications in chemistry, mineralogy, metallurgy, biology and medicine, archaeology, environmental research, forensic science, and related areas.

I have tried to provide complete literature citations, especially for LM-OES, but LM-MS had to be treated selectively because of the large number of published papers.

The inspiration for writing this book came from Prof. Dr. J. D. Winefordner, editor of *Chemical Analysis*. Very sincere thanks are due to him and to Mr. J. L. Smith of John Wiley & Sons, also to the Martin-Luther-University Halle-Wittenberg for supporting this work.

My own experience and many of the laboratory results came from working closely with my former husband, Dr. H. H. W. Moenke, and our group in Jena, consisting of Mr. W. Quillfeldt, Mr. J. Mohr, Mr. M. Hüfner, Mr. D. Böwe, Mr. W. Maul, Mr. K. Berka, Mrs. U. Maier, Mrs. K. Müller, Mrs. C. Lindner, and Mrs. E. Böwe-Berlinghoff, during the period 1962 to 1980. Many thanks to all of them and to my co-workers at the University of Halle from 1982 to the present: Mrs. M. Birkenfeld, Dr. U. Zander, Mr. M. Gäckle, Mr. D. Günther, Mr. J. Kammel, Mrs. M. Kühnemund-Pacher, Mr. M. Landmann, and Miss K. Jahn.

I express my thanks to Mrs. A. Kuhr for typing the manuscript, and to Mrs. M. Holdefleiß and Mrs. E. Scheiner for preparing the illustrations and photographs. Special thanks are due to Thurid, my daughter, and Jens, my son, for manifold help.

I gratefully acknowledge permission to reproduce figures and tables from: Dr. J.-F. Eloy, Dr. G. Dimitrov, Dr. D. Simons, Dr. L. J. Radziemski, and Dr. H. S. Kwong; Pergamon Press Ltd., Akademische Verlagsgesellschaft Geest & Portig K. G., Springer-Verlag, VEB Gustav Fischer Verlag, VEB Deutscher Verlag der Wissenschaften, Elsevier Science Publishers, VEB Grundstoff-verlag; American Chemical Society, Optical Society of America, Society for Applied Spectroscopy, Royal Society of Chemistry; Leybold-Heraeus GmbH, B & M Spektronik GmbH & Co. K. G., Perkin-Elmer & Co. GmbH, Scanning Microscopy International, Cambridge Mass Spectrometry Ltd., and Spectro Analytical Instruments.

LIESELOTTE MOENKE-BLANKENBURG

*Halle, GDR*
*January 1989*

# FOREWORD

Since the 1950s, an astonishing variety of new analytical tools and techniques have been developed, for example, mass spectrometry, neutron activation analysis, electron microprobe analysis, scanning electron microscopy, ion microprobe analysis, analytical electron microscopy, and Mossbauer analysis. These tools and techniques have revolutionized the natural sciences and have significantly contributed to enormous progress in a wide range of scientific fields at a scale probably unprecedented in the history of science. Laser microprobe analysis, in the broadest sense, has contributed its share to this progress.

My own motivation and, I suppose, that of others in the 1960s, for an interest in applying laser microprobe techniques in materials analysis, was to find a way of doing with a laser beam what could not be done with an electron beam. Although the electron microprobe is a superb tool for the routine analysis of major, minor and trace elements down to about 100 ppm, it does not conveniently allow for measurement of lower concentrations nor for measurement of the isotopic compositions of materials. The application of the laser beam as a sampling device for providing a plasma that could be analyzed by sensitive devices such as mass spectrometers sparked considerable interest in this new device. Over the years, laser microprobes have therefore become very useful analytical tools that have opened up exciting new areas of research, such as the in situ dating of single mineral grains in meteorites and lunar samples.

Liselotte Moenke-Blankenburg (or "Lilo", as her fellow students used to call her) has made major contributions to the field of laser microprobe analysis for over 25 years, first at VEB Carl Zeiss, Jena, and later at Martin-Luther-University, Halle-Wittenberg. Her original contributions have been published in numerous articles and are devoted both to laser microprobe instrumentation and application, with emphasis on the analysis of geological materials. She has previously coauthored several books on this topic; the present volume is a culmination of her distinguished work in this field.

I am delighted and very honored to have been asked by Lilo to write this short Foreword. I have not seen her in some 30 years, but we have kept aware of each other's work through the literature, and of our personal lives through correspondence. I remember Lilo fondly and with great affection from our student days in Jena. She was a very dedicated, hard-working student, totally committed to her work. This love for her work has carried through her entire professional career and is quite apparent in the present book. Lilo has been an outstanding researcher, exceptional teacher and, last but not least, a very successful mother. I very much enjoyed reading this book, and I know her readers will find this marvelous account of laser microprobe analysis as exciting and stimulating as I have.

*Albuquerque, New Mexico*
*February 1989*

KLAUS KEIL

# FOREWORD

Man is a tool-using animal. Without tools he is nothing.

Since the very earliest times, man has sought to understand the world about him. His curiosity soon extended to all three physical states—solid, liquid and gas—of the matter present both on Earth and in the far-flung universe of the stars and planets. With the discovery, followed by the mastery, of fire, he encountered the fourth state of matter, which has only been characterized, in the 20th century. He has at all times applied his efforts to the discovery of a plan for observing and controlling these four states of matter. It became obvious that there was a need to record and classify—and thus to analyze—the various aspects of different materials.

Let us now consider the main features of the events that led to the invention and perfection of the tools, instruments, and methods used in the investigation of matter. These experimental methods are based on a knowledge of the fundamental relationships of the physical sciences. The progress of this science of methods has accompanied the discovery of the constituent particles of matter.

The usual and necessary approach of the physical chemist, or "*analyst*," involves enriching our knowledge not only in the field of chemical sciences, but also in the new field of experimental physical science known as the physics of particle–matter *interaction*. The *analyst's* first task is to record the main chemical and physical phenomena resulting either from chemical reactions or from particle–matter interactions. The second stage in his task involves promoting the development of the analytical method leading to the measurement of the amplitude of these phenomena, using appropriate detectors and methods of diagnostics in comparison with a standard reference, giving rise to the analyst's main problem: defining an accurate reference. Last, the analyst's task involves the study and interpretation of the results of his diagnostics, and reaching an assessment.

Numerous scientific developments took place thereafter until the emergence of the laser, a tool which made it possible to achieve further improvements in materials microanalysis using both optical emission spectroscopy and mass spectroscopy.

To understand how analytical instruments and methods evolved so prodigiously to the present day, it may be worthwhile to look briefly at the main historical and scientific landmarks.

Optical microscopy was the first method of analysis to play a leading role in the microscopic examination of materials, qualitatively and quantitatively. This was followed by:

- Metallography, using the optical microscope with photography, around 1850 after Daguerre invented the silver plate method (1839).

- Developments in optical microscopy as a result of Seidel's theory of lens aberration (1850).
- The spectroscope, invented in 1859 by Kirchhoff and Bunsen.
- First microscopic observation of heat-treated steel by Sorby (1864).
- Abbe's oil-immersion method (1880).
- The discovery of the electron by Helmholtz in 1881.
- The discovery of X-rays by Roentgen in 1895.
- The discovery of the properties of the first optical resonator by Perot and Fabry and its first application in spectroscopy in 1896.
- The discovery of the photon by Planck in 1900.
- The discovery of X-ray diffraction by M. von Laue and others in 1912.
- The first application of X-ray diffraction to determine the periodic structure of crystals by W. H. and W. L. Bragg in 1913.
- The discovery of the proton by Thomson in 1916.
- The discovery of isotopes by Dempster in 1918 and the construction of the first mass spectroscope by Aston in 1919.
- Auger's work on Auger electrons in 1925.
- The design of electric and magnetic field lenses for electron beams by Busch in 1926.
- The discovery of electron diffraction phenomena by Davisson and Germer in 1927.
- The discovery of the Raman photonic radiation lines in 1928.
- Knoll's discussion, in 1931, of the possibility of constructing an electron microscope.
- The discovery of the neutron by Chadwick in 1932.
- The construction of a transmission type electron microscope by Ruska and Borries in 1932.
- The first use of X-ray fluorescence analysis by von Hevesy in 1932, based on Moseley's earlier work.
- Building a scanning electron microscope (von Ardenne, 1938).
- The first microanalysis by electron spectroscopy by Hillier and Bakes in 1944, based on Rudberg's earlier work in 1930.
- The development of X-ray microanalysis by Hillier in 1947.
- The invention of the field-ion microscope by Müller in 1951.
- The construction of Castaing's famous X-ray microscope in 1952.
- Auger-electron spectroscopy based on the work of Auger in 1925 and proposed by Lander in 1953.
- Garwin's discussion in 1957 of the possibility of using muons as "a powerful tool for exploring magnetic fields in interatomic regions."

- Mössbauer spectroscopy, based on Mössbauer's work in 1956.
- The discovery in 1958 by Gordon Gould and also by Townes and Schawlow of the invaluable LASER.
- The invention of the first laser device by Malman in 1960.
- Building the ion microanalyzer (Castaing and Slodzian, 1962).
- The first laser microplasma experiment to produce photon emission, by Brech and Cross in 1962.
- The first laser mass spectrometry experiments, by Honig and Woolston, in 1963.
- The first time-resolved spectroscopy of laser-generated microplasmas by Archbold and others in 1964. (An excellent general description of other laser applications in microanalysis is given in this book.)
- The construction of the first nuclear microprobe, by Fergusson and Cookson in 1970.
- The Raman laser molecular microprobe, by Delhaye et al., in 1975.

Lasers were the product of fundamental research; their discovery was not the result of any direct practical motivation on the part of scientists, but rather the fruit of the natural curiosity of researchers. The laser was just what was needed to revitalize the use of microscopy and spectroscopy in both physics and chemistry.

*The spatial coherence* of the laser light source makes it possible to focus the laser beam on a very small area. The extremely high density of electromagnetic energy thus obtained (between $10^6$ and $10^{12}$ W/cm$^2$) gives rise to such intense heating of the irradiated matter that the laser interaction produces a change of state from solid to plasma within a very short time (between $10^{-9}$ and $10^{-8}$ sec). These successive physical phenomena give rise to the creation and emission of various particles such as excitons, polaritons, photons, phonons, excited molecules, neutral and excited atoms, ions, and also, under certain conditions, neutrons. These different kinds of particles may be used directly to study and characterize both the phenomena of the laser–material interaction and the material itself. Distinct physical effects such as optical emission, atomic absorption, atomic fluorescence, and photoionization make it possible to analyze the microplasma created by the laser–material interaction. The spatial coherence of the laser beam can be used to localize the volume of the interaction, and thus to obtain localized analytical information.

*The temporal coherence* of the laser beam, corresponding to its monochromaticity, determines new applications in spectroscopy. Laser radiation is used at a specific wavelength to fit the absorption lines of the particular molecular, atomic, and isotopic species. This specific absorption involves an increase in the energy of the excited particles, which gives rise to a disturbance of the particles' state of equilibrium; the return to equilibrium involves the restitution of part or all of the energy in one of the following energy forms: (1) Photonic: fluorescence, Raman effect, Rayleigh diffusion; (2) Electrical: optogalvanic effect, resonant photoionization; (3) Thermal: thermal lensing effect, optoacoustic effect; (4) Chemical: photochemistry.

These different effects of laser interaction have led to new developments in the various types of spectroscopy, including the Raman laser, the fluorescence laser, isotopic ionization, optoacoustics, thermal lensing, and laser plasma.

When lasers were first tried for analytical applications, Dr. L. Moenke-Blankenburg made a brilliant contribution to the field of laser optical emission spectroscopy with the original work she carried out at the Carl Zeiss Jena Company. The considerable experience with varied solid materials she has acquired over some twenty years of experiments in microanalysis and, more recently, her appointment as Professor in the Chemistry Department at Halle University provide an accurate idea of the extent to which she is qualified to describe and comment upon the field of laser microanalysis. The study and use of laser–matter interaction in microanalysis requires a thorough knowledge of both lasers and the physics of laser–matter interaction.

This book details the laser principle and the main properties of the laser beam, restricted to both types of solid laser used in microanalysis—in this case, rubies and neodymium garnet or glass lasers. There then follows a complete description of the main phenomena of laser–matter interactions, including the essential parameters and relationships required to understand the choice of device and the analytical conditions of each method. To show the difficulty of obtaining a quantitative analysis, the author provides some basic notions of statistics useful for determining the limits of accuracy and precision for each method of laser microanalysis. Next is a description of currently well known laser microanalysis techniques, with details of numerous investigations and instrumentations published in laser microanalysis applications in the last 30 years. Last, the author compares its main features with other techniques of solid-state microanalysis and surface analysis. This is a very well-documented book, which should be of great service to the students, analysts, and researchers wishing to acquire a wide and thorough knowledge of this original and promising experimental science.

*Grenoble, France*
*March 1989*

JEAN-FRANCOÏS ELOY

# CONTENTS

**CHAPTER 1 INTRODUCTION** 1

References 3

**CHAPTER 2 THE LASER AS A RADIATION SOURCE** 5

2.1. Principle 5
2.2. Solid-State Resonators 7
2.2.1. Ruby 7
2.2.2. Neodymium-Doped Glass and Garnet 8
2.3. Properties of Laser Radiation 9
2.3.1. Course of the Discharge in the Flashtube 10
2.3.2. Degree of Reflection of the Mirror Coating of the Reflector 10
2.3.3. Temperature of the Resonator Rod 10
2.3.4. Oscillation and Emission Behavior of the Laser 11
2.3.5. Stability of the Components of the Resonator 11
2.4. *Q*-Switching Technique 12
2.4.1. *Q*-Switching with the Aid of Electro-Optical Switches 12
2.4.2. *Q*-Switching with the Aid of Rotating Reflectors 13
2.4.3. *Q*-Switching with Saturable Absorbers 14
2.5. Mode of Action of an Energy and Spike-Number Measuring Accessory 15
References 19

**CHAPTER 3 LASER–TARGET INTERACTION: MICROPLASMA GENERATION** 21

3.1. Focusing of Laser Radiation 21
3.2. Technical Data of Normal-Mode Pulse and *Q*-Switched Solid-State Lasers Used for Laser Microanalysis 23

3.3. Vaporization of Solids 24
3.3.1. Vaporization by Normal Laser Pulses 25
3.3.2. Vaporization by Giant Pulse or Semi-$Q$-Switched Laser Pulse 29
3.3.3. Vapor Pressure and Temperature 35
3.3.4. Influences of Sample Parameters on Vaprization 37
*3.3.4.1. Thermal Properties of Solids* 37
*3.3.4.2. Reflectivity* 38
3.4. Microplasma Features for Analytical Use 39
3.4.1. Foundations 39
3.4.2. Local Thermodynamic Equilibrium 41
3.4.3. Energy Balance; Energy Losses in and Above the Target 43
3.5. Additional Excitation and Ionization of Laser-Beam-Produced Vapor by Spark Discharges and Other Sources 48
3.6. Vaporization and Excitation Under Variable Pressure and in Various Gas Media 57
3.7. Matrix Effects 64
References 65

**CHAPTER 4 PRESUMPTIONS FOR QUANTITATIVE LASER MICROANALYSIS OF SOLIDS: STATISTICS IN ANALYTICAL CHEMISTRY 75**

4.1. Accuracy and Precision of Laser Microanalysis 76
4.2. Homogeneity Tests of Reference Materials 77
4.3. Summary of Useful Statistics for Spectroanalytical Chemistry 84
4.3.1. Linear Regression 84
4.3.2. Linear Correlation 86
4.3.3. One-Way Analysis of Variance 86
4.3.4. Two-Way Analysis of Variance 88
References 90

**CHAPTER 5 LASER MICROANALYSIS BASED ON OPTICAL EMISSION SPECTROMETRY 93**

5.1. Instruments 93
5.1.1. Technical Details of Laser Microanalyzers 93
5.1.2. Construction and Mode of Action of Spectral Apparatus and Radiation Detectors 104

*5.1.2.1. Fundamental Concepts* 104
*5.1.2.2. Spectrographs with Photographic Recording* 105
*5.1.2.3. Photoelectric Recording* 106

5.2. Field of Application 120
5.2.1. Mineralogy, Geochemistry, and Cosmochemistry 120
5.2.2. Metallurgy and Related Fields 140
5.2.3. Silicate Technique, Crystal Synthesis, Chemistry, Forensic Science, Archaeology, Restoration of Art Objects, Medicine, and Biology 141
References 172

**CHAPTER 6 LASER MICROVAPORIZATION COMBINED WITH PLASMA EXCITATION BASED ON OPTICAL EMISSION SPECTROMETRY** **181**

6.1. ICP Spectrometry of Solids 181
6.1.1. Instrumentation of LM-ICP-OES 183
6.1.2. Theoretic and Methodic Investigations 188
6.1.3. Fields of Application 191

6.2. MIP Spectrometry of Solids 196
6.3. DCP Spectrometry of Solids 200
6.4. Comparison of LM-ICP-OES, LM-MIP-OES, and LM-DCP-OES 204
References 204

**CHAPTER 7 LASER MICROANALYSIS BASED ON ATOMIC ABSORPTION SPECTROMETRY** **207**

References 213

**CHAPTER 8 LASER MICROANALYSIS BASED ON ATOMIC FLUORESCENCE SPECTROMETRY** **215**

References 216

**CHAPTER 9 LASER MICROANALYSIS BASED ON MASS SPECTROMETRY** **219**

9.1. Foundation and Historical Development 219
9.2. Analytical Features 224
9.3. Instruments 227

9.3.1. Ion Sources 227
9.3.2. Analyzers 231
9.4. Fields of Application 236
9.4.1. Chemical Analysis of Inorganic Materials 236
9.4.2. Chemical Analysis of Organic Materials 259
References 259

**CHAPTER 10 LASER MICRO ICP MASS SPECTROMETRY 271**

References 273

**CHAPTER 11 COMPARISON WITH OTHER METHODS OF SOLID-STATE ANALYSIS 275**

References 278

**INDEX 281**

CHAPTER

1

# INTRODUCTION

Soon after the first report of laser action in ruby in 1960 [1] it was generally recognized [2–12] that the intense laser output beam could be used to excite material into a state of optical emission. From that time to the present, the field of laser micro analysis has been evolving and changing. Over the first 15 years in particular there has been an explosion of new ideas and developments in laser micro emission spectrometry and later in laser microanalysis based on absorption and fluorescence. Of late, the aura of glamor has supposely been transferred to laser micro ICP, MIP, and DCP spectrometry based on optical emission, and to laser micro mass spectrometry (Fig. 1).

Laser microanalysis is one of several physical methods for direct chemical analysis of solids. The mode of action is the following. A solid specimen is placed on the movable stage of a microscope. The surface of the specimen is investigated under the microscope and a region to be analyzed chemically is selected. The coherent radiation of a solid-state laser with an output energy of $\leqslant 1$ Joule (1 J) is focused by the optical system of the same microscope or, by a separate method, on the selected region of the specimen. It evaporates a picogram to microgram amount of substance. The microplasma so produced is imaged directly or after additional procedures into the entrance slit of a spectrometer. The spectrum is recorded photographically or by means of photoelectric detectors (Fig. 2). Laser microanalysis has three principal fields of application: as a microchemical method, as a method for local analysis, and as a method for distribution analysis (Table 1).

Microchemical analysis of nonconducting as well as conducting powders using laser radiation for vaporization, atomization, excitation, and ionization requires very small amounts of substance. Powders have to be applied quantitatively to an object disk with or without an adhesive or have to be pressed into pellets. Then, depending on parameters of the laser and the optics, the laser microanalysis based on emission or absorption spectrometry requires about 1 ng of substance with a laser focal spot of about 10 $\mu$m and about 100 $\mu$g with a larger focal spot of about 300 $\mu$m referred to a laser output energy of 1 J. The average consumption of material is $10^{-6}$ g.

Laser microanalysis based on mass spectrometry requires about 1 pg with a laser focal spot of $\leqslant 1$ $\mu$m and is therefore an ultramicro method according to

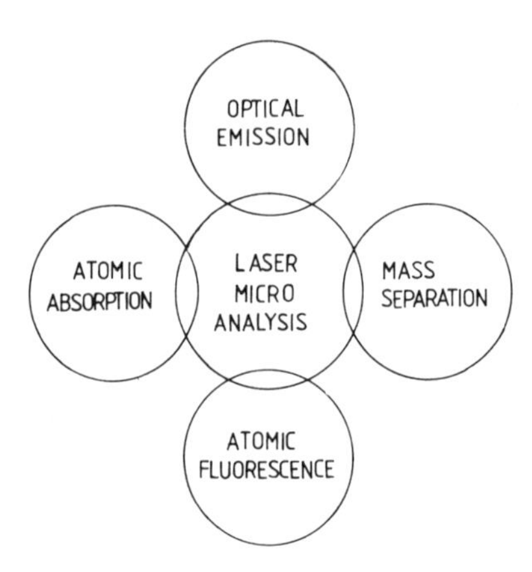


**Figure 1.** Principles of laser microanalysis. From Ch. 2., Ref. [20], reprinted with permission of Pergamon Journals Ltd., Oxford.

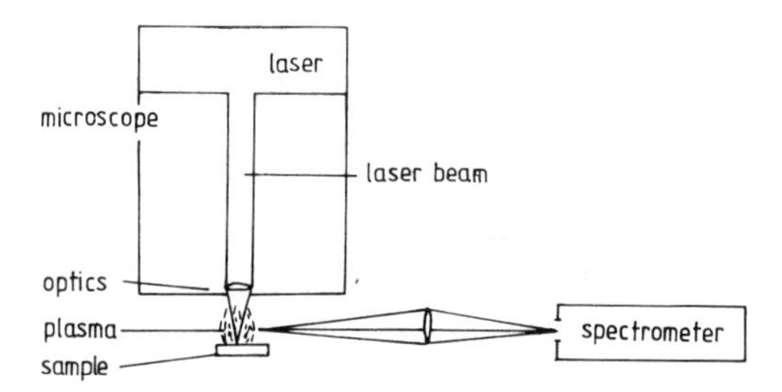


**Figure 2.** Schematic of laser microanalysis. From Ch. 2, Ref. [20], reprinted with permission of Pergamon Journals Ltd.

the IUPAC definition. Emission spectral analysis is an element-specific method that permits about 60 chemical elements to be detected simultaneously, depending on the spectral range covered by the spectrometric apparatus. The same is true of atomic absorption spectrometry, but without simultaneous detection. Laser micro mass analysis permits the detection of all elements, isotopes, molecules, and radicals sequentially and simultaneously. Laser microanalysis permits the qualitative and quantitative determination not only of microamounts but also of microregions, that is, microscopically small local inhomogeneities in solids without isolation of these inhomogeneities or heterogeneities from their matrices. There is a variable working range of $\leqslant 1$ to $\geqslant 300\,\mu$m.

In addition to the above-mentioned analytical aims, the method also provides information on the change in composition of a specimen as a function

**TABLE 1. Three Main Fields of Application of Laser Microanalysis[a]**

| | Microchemical Analysis | Local Analysis | Distribution Analysis |
|---|---|---|---|
| Amount of substance required | OES: μg–ng<br>MS: pg | | |
| Spatial resolving power | | OES: ⩽ 10–300 μm<br>MS: ⩽0.5–10 μm | |
| Length by line analysis | | | OES: μm–mm |
| Depth by layer analysis | | | OES: 1–100 μm |
| Area by area analysis | | | OES: approx. 50 $mm^2$ |

*Source:* Reprinted with permission from *Prog. Analyt. Spectrosc.* 9, L. Moenke, copyright 1986, Pergamon Journals Ltd.

[a] OES, optical emission spectometry; MS, mass spectrometry.

of the spatial coordinates, which means that distribution analysis is possible. A line analysis is achieved by a sequence of local analysis with a regulated stepwise movement of the specimen carriage after each laser shot. The length of the line that can be investigated depends on the apparatus and may range from some tenths of a micrometer to some tenths of a millimeter. A layer analysis is achieved by a vertical sequence of spot analysis after stepwise focusing of the radiation to a depth of from about 1 μm to some hundreds of micrometers. An area analysis is also possible, by a successive sequence of spot analyses—in this case in two dimensions. An area of about 50 $mm^2$ can be recorded. A volume analysis, that is, a bulk or macroanalysis, can be obtained by microanalysis summation. A volume of $10^{-2}$ to $10^{-1}$ $mm^3$ can be investigated.

## REFERENCES

1. T. H. Maiman, *Phys. Rev. Lett.*, **4**, 564 (1960); *Nature* (*London*), **187**, 493 (1960).
2. Anonymous, *Chem. Eng. News*, **40**, 52 (1962).
3. F. Brech and L. Cross, *Appl. Spectrosc.*, **16**, 59 (1962).
4. J. Debras-Guedon and N. Liodec, *C. R. Acad. Sci.*, **257**, 336 (1963).
5. E. F. Runge, F. R. Bryan, and R. W. Minck, *Can. J. Spectrosc.*, **9**, 5 (1964).
6. H. I. S. Ferguson, J. E. Mentall, and R. W. Nicholis, *Nature* (*London*), **204**, 1295 (1964).

7. M. Berndt, H. Krause, L. Moenke-Blankenburg, and H. Moenke, *Jenaer Jahrb.*, 45 (1965).
8. J. Debras-Guedon, N. Liodec, and J. Vilnat, *J. Math. Phys. Appl.*, **16**, 155 (1965).
9. W. D. Hagenah, *Z. Angew. Math. Phys.*, **16**, 130 (1965).
10. W. F. Jankewitsch and J. W. Besrutschko, *Zavod. Lab.*, **30**, 628 (1965).
11. A. V. Karyakin, M. V. Achmanova, and V. A. Kaigorodov, *Zh. Anal. Khim.*, **20**, 145 (1965).
12. H. Moenke and L. Moenke-Blankenburg, *Einführung in die Laser-Mikro-Emissionsspektralanalyse*, Akademische Verlagsgesellschaft Geest und Portig KG, Leipzig, German Democratic Republic, 1966.

CHAPTER

2

# THE LASER AS A RADIATION SOURCE

## 2.1. PRINCIPLE

The term "laser" is an acronym for "light amplification by stimulated emission of radiation." Since the device was based on the same principles as the maser, a microwave source developed in the 1950s, the term "optical maser" was also in use for a short time. The principle of the action of lasers is based on the induced, or stimulated, emission of radiation. In contrast to spontaneous emission, in which a quantum transition takes place from a higher-energy level to a lower level without this being caused by a radiation field, in stimulated emission this process is made to take place by light quanta.

The first prerequisite for optical amplification to be achieved in this way is inversion of the population densities of the energy levels. This is accomplished by "optical pumping" by means of a high-powered incoherent optical radiation by which the atoms are raised from the ground state into a higher-energy state. In the ruby laser the natural energy levels of the chromium ion are used for this purpose, and in glass and YAG lasers, those of the neodymium ion.

The second prerequisite for the induction of stimulated emission of radiation is inclusion of the active medium in a suitable resonator. For wavelengths for which the distance $d$ between the two surfaces is an integral multiple of a half-wavelength, that is,

$$d = n\frac{\lambda}{2} \tag{1}$$

a system of standing waves arises between the two surfaces. The light passes through the active medium between the surfaces many times because of reflection and can therefore effectively interact with atoms of the active substance. The coupling between radiation and matter is particularly strong for a beam of light that runs accurately parallel to the axis of the resonator. Beams of light the wavefronts of which are inclined to the bounding surfaces of the resonator and beams of other wavelengths leave the system after a few reflections. This means that the resonator has resonance properties only for monochromatic beams, the wavefronts of which run accurately parallel to its end surfaces and for which the condition above is fulfilled. Therefore, the light pulsations are a direct consequence of the process of amplification by stimulated emission. As the laser material is being excited, the population inversion between the levels of the laser transition reaches the oscillation threshold and the system becomes unstable. A photon, randomly produced by fluorescence, will stimulate the emission of more photons, and when this occurs along the optical axis of the resonance cavity, the electromagnetic field is amplified, resulting in an output pulse and a net decrease in the number of atoms in the excited (metastable) state. When the threshold value is reached, the field amplitude dies out and the process is repeated for the duration of the flashlamp discharge.

Solid-state lasers are generally used for laser microanalysis [1]. Suitable optical media are ruby ($Al_2O_3$, $Cr^{3+}$-doped), glass ($Nd^{3+}$-doped), and YAG (yttrium-aluminum garnet, $Y_3Al_5O_{12}$, $Nd^{3+}$-doped). The optical medium is contained in a resonator, which in the simplest case, consists of the two highly reflecting parallel surfaces, the distance between which is large compared with the cross-sectional area. In parallel with the resonator, there is a flashtube as an optical pump. Both units are surrounded by a reflector.

The radiation produced by lasers, such as ruby and neodymium-doped glass and garnet used in producing high-power effects, is generally used in pulsed operation. Widely differing time regimes are available from different methods of pulsing lasers. The effects produced by the laser depend strongly on the particular type of pulsing.

If the laser is simply pumped by a pulsed flashtube and the radiation allowed to emerge when the threshold conditions for laser operation are reached (i.e., when a sufficient population inversion is attained so that optical gain can exceed losses), one has what is generally termed a normal pulse laser. With this type of operation, pulse widths in the range 100 $\mu$s to 1 ms are typical. Usually, the emission is not uniform but consists of many microsecond-duration spikes called relaxation oscillations. The amplitude and spacings between the pulses are different. If the laser is operated near threshold, there may be only a small number of relaxation oscillations in the train. A laser operating in the relaxation oscillation type of output is sometimes said to be oscillating in the burst mode. Of course, the presence of the relaxation oscillations in a laser pulse will cause heating and cooling of an absorbing surface as the radiation strikes it, so that this behavior complicates definition of the effects.

Under some conditions the relaxation oscillations can become regular in amplitude and spacing. It is also possible to obtain quasi-continuous laser oscillation in the normal pulse mode so that relaxation oscillations are suppressed and the entire envelope is relatively smooth and lasts a time of the order of 1 ms. Analysis of the relaxation oscillations has been performed in terms of rate equations that describe the competition between a population of the upper laser level by the pumping source and emptying it through stimulated emission. The result is that the light is produced in short bursts [2–4]. The general features of the relaxation oscillations have been explained satisfactorily in this way.

Hellwarth [5] put forth a proposal to control the output of a laser by building up such a high overpopulation of excited atoms through a light seal on the active medium that upon its opening one or a few giant pulses of maximum power are obtained. This process is called *Q*-switching. By employing the *Q*-switching technique, one obtains pulses in the range 10 to 1000 ns. *Q*-switching involves changing the *Q* (or quality) of the laser resonant cavity. The quality factor of a cavity resonator is defined as the energy stored in the cavity divided by the energy lost from the cavity per cycle of the field. Thus a high-*Q* cavity will store energy well; a low-*Q* cavity will emit the stored energy rapidly.

Switching rapidly from a condition of high to low *Q* will result in rapid extraction at high power. This can be done in a variety of ways: for example, by rotating one of the laser mirrors or by inserting an absorbing element between the laser rod and one of the mirrors. There are many types of variable absorbing elements, such as Kerr cells or bleachable dyes. While the laser is being pumped by the flashtube light, the laser rod is optically cut off from the mirrors and there is no resonant cavity available. Therefore, laser operation is suppressed and the population inversion in the rod can be increased greatly over the normal threshold condition [6]. Then the *Q*-switching element is changed to a transparent condition so that the light from the rod can reach the mirrors. When the laser rod, now in a highly excited state, is finally optically coupled to the set of parallel mirrors, the laser operation will be emitted in a pulse of much higher power and much shorter duration than if *Q*-switching had not been employed. This is the technique by which the highest peak power have been produced.

## 2.2. SOLID-STATE RESONATORS

### 2.2.1. Ruby

The most familiar solid-state laser is the ruby laser, which contains 0.05% $Cr_2O_3$ in an $Al_2O_3$ lattice. The chromium ions are excited by the broadband emission of a flashlamp coiled around it or placed alongside it within an elliptical reflector. Ruby can be regarded as a three-level fluorescent solid, consisting of the ground state $E_0$, two broad pump bands, and one metastable state, which has a lifetime of approximately 3 ms. The ruby is pumped by optical excitation using wavelengths within the bands 500 to 600 nm and 320 to 440 nm to produce absorption. A spontaneous nonradiative transition then takes place at an intermediate level (a process known as phonon relaxation) into internal energy in the crystal in the form of heat. Ordinary fluorescence occurs when radiation is emitted, which is associated with spontaneous transition to the ground state. When this fluorescence travels through the crystal, additional transitions are stimulated.

When the crystal is placed within a resonant cavity, the radiation is reflected repeatedly in both directions through the crystal, inducing more transitions. Thus every passage through the crystal is accompanied by amplification of the radiation intensity, provided that the population of the radiation intensity, provided that the population of the $E_0$ level does not exceed that of the metastable level. This is ensured by continuous repumping of ground state ions. Laser output is obtained by partial transmittance through one of the resonator reflectors. The laser may thus be described as a high-gain amplifier with positive feedback. The wavelength of the ruby laser line is 694 nm.

One problem with the ruby laser, which is common to all lasers of this type, is the damage caused by the repeated cycle of heating and cooling associated with the generation of each pulse, which ultimately necessitates replacement of the ruby rod. To improve performance, the rod usually needs to be cooled by circulation of water in a jacket enclosing it.

### 2.2.2. Neodymium-Doped Glass and Garnet

The four-level system of neodymium-doped glass and YAG (garnet) (doped with approximately 3 wt% $Nd_2O_3$) is shown schematically in Fig. 3. In the neodymium system fluorescing does not occur directly to the ground state from the upper laser level as in the ruby system. The terminal level $E_3$ is an

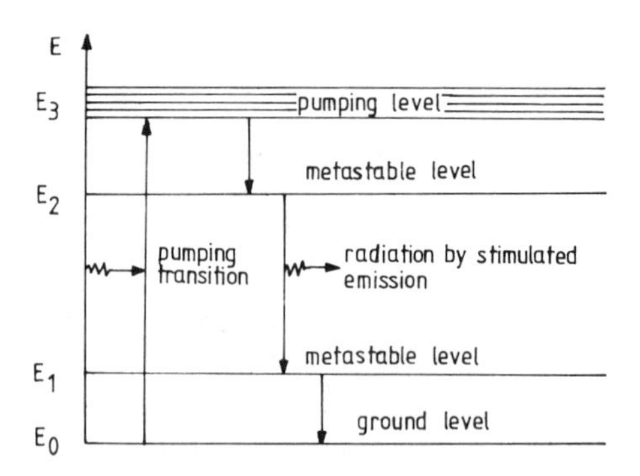


**Figure 3.** Energy-level diagram of a laser system. From [20], reprinted with permission of Pergamon Journals Ltd.

intermediate one located about 200 cm$^{-1}$ from the ground state $E_0$. The transition from this laser terminal level to the ground state occurs by phonon relaxation. Because the terminal level is essentially empty at room temperature, the population of the $E_1$ level can be increased by a relatively small amount of pump power above that of the $E_3$ level. This is unlike the ruby system, where more than half the chromium atoms must be raised to the fluorescence level before population inversion is achieved, and is the reason why the threshold pump power of the three-level ruby system is so much higher than that of the four-level neodymium system. The lifetime of the $E_1$ state of neodymium is approximately 0.3 ms, that is, about one-tenth that of the upper level of ruby. The wavelength of the laser transition of neodymium is 1.06 $\mu$m.

## 2.3. PROPERTIES OF LASER RADIATION

The properties of lasers that are useful for their employment as a source of energy in spectral analysis are thus first the monochromaticity and parallelism of the released beam and its consequent ideal focusability, and second, the output energy and output power, or the power density achievable at the focus that is optimal for the intended purpose [7–12].

The output beam of a laser is essentially monochromatic because of the well-defined upper and terminal energy levels of the leasing element. The exact degree of monochromaticity will become evident only after studying the mode structure. Furthermore, by virtue of the construction of the lasing element in the form of a rod, the device will emit a highly collimated beam. Waves propagating in off-axial directions will, after a number of oblique reflections, be reflected away from the system, so strong emission cannot occur except in the direction closely parallel to the rod axis.

The degree of collimation, or the angular divergence of the output beam, is theoretically of the order of $\lambda/D$, where $\lambda$ is the laser wavelength and $D$ the aperture of the device. It is, however, difficult in practice to attain this small value ($\approx 0.1$ mrad for a 1-cm-diameter neodymium rod), and divergences on the order of 5 mrad are usually achieved. Off-axis oscillations coupled with imperfect resonator reflectors are usually responsible for this increase in the divergence of the output beam above the theoretical value.

One of the most striking features of pulsed laser radiation, apart from those mentioned, is its very high intensity. By using a simple lens to focus the beam, it is possible to attain extremely high radiation densities. This important feature forms the basis of investigations concerned with the interaction between the laser beam and materials and the application of this interaction in analytical spectroscopy [13–22].

A ruby laser suitable for laser microanalysis on unswitched operation emits sequences of $\geqslant 100$ spikes with an average duration of 1 $\mu$s each. The output energy as a rule amounts to 1 J, and therefore the power is 10 kW, in accordance with the equation

$$W = ptN_{sp} \tag{2}$$

where $W$ = energy
$P$ = power
$t$ = spike duration
$N_{sp}$ = number of spikes

In $Q$-switched operation, the number of spikes decreases according to the cell segment used in six steps as far as single-spike operation. At the same time the peak powers rise to 2 MW at an energy of 0.1 J and a spike duration of 50 ns. The reproducibility of the output is of decisive importance for evaluating the suitability of the laser for quantitative spectrochemical analysis. Moenke et al. [23] studied the influencing factors in a ruby laser with the following results.

### 2.3.1. Course of the Discharge in the Flashtube

The emission and discharge processes in the flashtube deteriorate through damage to the tips of the electrodes, through impurities ejected from the material of the electrodes and the walls, and through consumption of the gas, which leads to a lower efficiency ratio between input energy and radiation output in the pumping band. In addition to this, there are absorption losses due to the darkened quartz wall of the flashlamp, which becomes coated internally with metal vapor. These effects are, apart from minor fluctuations, typical long-term changes. Short-term variations can be caused by a spatial change in the discharge channel. The working life of flashtubes can be increased further by systematic technological revision and improvement of the starting materials.

### 2.3.2. Degree of Reflection of the Mirror Coating of the Reflector

The wall of the elliptical-cylindrical reflector, which images the center of the flashtube on the axis of the resonator consists of highly polished aluminum and is subject to corrosion, contamination, and radiation damage by the ultraviolet (UV) light of the flashtube. This is a long-term effect.

### 2.3.3. Temperature of the Resonator Rod

The radiant energy converted into heat in the pumping of the resonator rod must be removed by a suitable cooling process. Only when the active medium has reached a constant temperature (in the case of air cooling, usually room temperature) can reproducible oscillation and emission behavior be expected. An adequate flow of air and sufficiently long pauses between the pumping of the rod, together with a constant frequency of the succession of shots, have a decisive effect on the reproducibility of the spectral line intensities.

### 2.3.4. Oscillation and Emission Behavior of the Laser

In addition to the effects of other components of the resonator, the oscillation and emission behavior of the laser is greatly affected by the temperature, illumination, and homogeneity (freedom from bubbles, streaks, and orientation defects, uniform satisfactory processing and doping) of the resonator rod. The influence of these factors is shown in the laser output parameters energy $W$, number of spikes $N$, half-value duration of the spikes $t_{1/2}$, energy per spike $P$, emission of radiation in the solid angle $\omega$, and radiation density distribution over the cross section of the beam $\theta$. The laser parameters can be optimized by stable construction, selection of the best-quality material for the resonator rod, and homogeneous illumination.

### 2.3.5. Stability of the Components of the Resonator

Because of the effects of radiation, all parts of a resonator through which radiation passes, especially the glass–air surfaces, show small disturbances after a certain time, which increase in frequency and finally, affect the performance of the laser. While disturbances at the dielectric coating of the output mirror and the plane surface of the resonator rod increase insignificantly over long operating periods (i.e., they are to be regarded as a long-term instability), the radiation effects and changes in the switching element (saturable absorber) and at the 100% reflector are more serious. A bleaching of the switching dyestuff leads, at the same input energy, to increased output energy, a higher number of spikes, and lower energy per spike. Switching dyestuffs that exhibit great stability in relation to laser radiation, the UV light of the flashtube, and daylight, and are also chemically stable, are therefore necessary.

A total-reflection prism is frequently used as the resonator closure on the 100% side, since dielectrically coated mirrors with such high reflecting powers and adequate stability are very expensive. The 90° edge of such a prism is subjected to high loads and splinters into small sections. The situation is restored by moving the prism in the direction of the 90° edge, thereby bringing an undamaged part into the path of the beam. More pronounced splintering, with the losses of reflection that they cause, produce typical jumplike instabilities. By selecting highly transparent silica glass, these deterioration phenomena can be reduced substantially.

## 2.4. *Q*-SWITCHING TECHNIQUE

Various arrangements have been developed for *Q*-switching (Fig. 4). In general, a distinction is made between passive and active optical switches. Passive switches employ absorbing material the transmission of which depends on the light flux in the resonator. Active switches require additional energy to operate the light seal, such as voltage for opening Kerr or Pockels cells or power for driving rotating prisms or mirrors [18].

### 2.4.1. *Q*-Switching with the Aid of Electro-optical Switches

Electro-optical switches with Pockels cells or Kerr cells are fast-active switches based on the polarization effect. Electric fields are used to alter the plane of polarization. In normal laser operation no voltage is applied to the Pockels cell. When a potential difference is applied, polarized light, produced by the appropriate orientation of the *C*-axis in the ruby or by an additional polarizer in the resonator space, is split in the electro-optical cell into two components vibrating in mutually perpendicular planes which are propagated with different velocities. The cell creates a phase difference that depends on the applied voltage and there is a corresponding quenching of the laser emission.

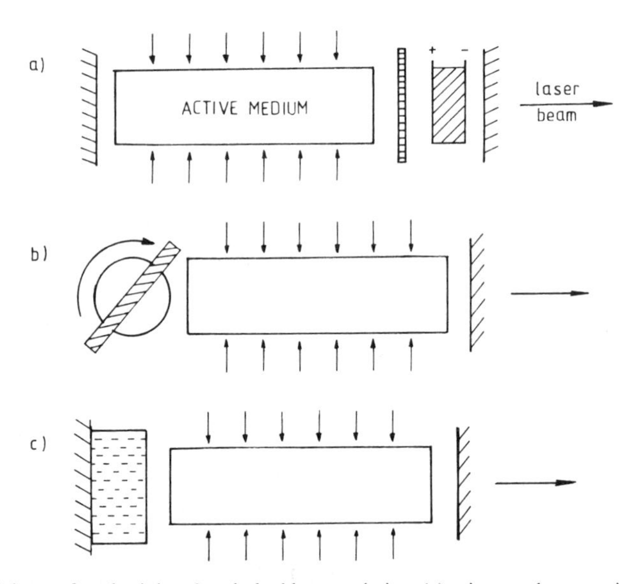


**Figure 4.** Schemes for obtaining *Q*-switched laser emission: (*a*) using an electrooptical switch; (*b*) using a rotating reflector; (*c*) using a saturable absorber. From Ref. [14], reprinted with permission by Springer-Verlag, Heidelberg.

Switching can be achieved in nanoseconds. However, the switching process requires considerable technical resources, since the high voltages necessary for operating the switch have to be applied in a short time.

Prisms of the Nicol, Wollaston, or Glan–Thompson types without cement can be used as polarizers, but only up to moderately high energies. For the highest pulse energies, stocks of plates at the Brewster angle are preferred. Systems for rotating the plane of polarization with electro-optical crystals such as potassium or ammonium dihydrogen phosphate (KDP or ADP) have the advantage that the control voltage for a 90° rotation in a double pass is relatively low. Depending on the arrangement of the electrodes and the special material, it amounts to about 5 to 12 kV with the usual longitudinal effect in which the control voltage is independent of the length of the crystal. For giant pulses, however, crystal cells are used only rarely, since they are destroyed by laser radiation.

Because of their greater load capacity, better optical isotropy and homogeneity, and simpler handling, cells making use of the quadratic Kerr effect in nitrobenzene are generally used for giant-pulse lasers with polarization switches. In this transverse effect, the control voltage is 20 to 30 kV, regardless of the aperture of the cell. Complete equipment, including pulse generators, is commercially available. Since the Kerr cell acts as a $\lambda/4$ switch, the control field is arranged at 45° to the direction of polarization. The nitrobenzene used must be extremely pure if it is desired to work with a bias voltage on the cell.

For opening, the voltage on the Kerr cell is briefly short circuited, for example, by means of a controlled spark gap or a thyratron. In practice, the bias voltage is fed as a pulse with approximately the same length as the pump pulse, so that loading of the cell due to the flow of current remains small. Laser pulses of 50 MW with a duration of 20 to 30 ns are to be expected.

### 2.4.2. $Q$-Switching with the Aid of Rotating Reflectors

One reflector of the resonator is made rotatable. A resonator with two plane mirrors requires accurate adjustment, for which reason a resonator formed of a 90° prism and a plane mirror is used exclusively. When the axis of rotation is perpendicular to the 90° edge of the prism, only a rough adjustment is necessary. To avoid a second beam, the hypotenuse surface should be provided with an antireflection coating.

With appropriate positioning of the reflector, the resonator oscillates when a sufficient overpopulation is present in the active material. The pumping process must therefore begin before the best $Q$-factor of the resonator has been reached. An electronic device synchronizes the firing of the flashtube with the parallel position of the prism in such a way that it takes place at the moment of maximum inversion. In the parallel position, the excited level is depopulated and the photon density in the resonator increases sharply. It reaches a maximum when the inversion has disappeared. As the prism continues to rotate, the $Q$-factor of the resonator deteriorates again and the continuing excitation process brings about the formation of a new overpopulation, but the threshold is not reached again. This situation applies only in the neighborhood of the threshold. With pronounced overpumping or when the speed of rotation of the reflector is too low, multiple pulses may occur. The delay between the firing of the flashtube and the parallel orientation of the reflector must be determined experimentally.

The optimum delay depends on the quality of the active material, the lifetime of the excited states, the change in the excitation energy with time, and the resonator geometry. The power consumption is low in comparison with that of electro-optical switching.

### 2.4.3. *Q*-Switching with Saturable Absorbers

As compared with the rotating-prism method, $Q$-switching with saturable absorbers is a simple and cheap process since it does not require a special motor with a precision supporting bracket, prism table, control system in the power supply unit, and a rearrangement of the triggering of the power supply unit.

Saturable absorbers are solutions of organic dyestuffs which, as passive reversible switches, exhibit no decomposition at high powers. For this purpose, the dyestuff is used in very low concentrations of $10^{-7}$ to $10^{-5}$ g/cm$^3$ in a cell in the resonator.

The transmission of the dyestuff must be selected in such a way that the population inversion in the active medium and its maximum just reaches the threshold of this complete system. Under the influence of the increasing number of light quanta, the absorption becomes smaller and smaller; within a few nanoseconds transmission of the dyestuff rises to practically 100% and the $Q$-factor of the resonator becomes better and better. This feedback effect gives an extremely fast switch.

With increasing concentration of dyestuff, the number of laser pulses falls off until finally only one pulse arises. Correspondingly, the threshold energy increases above that for an unswitched resonator. The width of the single-pulse region becomes greater with increasing concentration. If it is possible to adjust the excitation energy within this range, a more certain single-pulse operation is possible. The reproducibility of the pulse is not guaranteed if the width of the single-pulse region is less than the accuracy of measurement of the excitation energy. The half-value width decreases with increasing concentration of dyestuff.

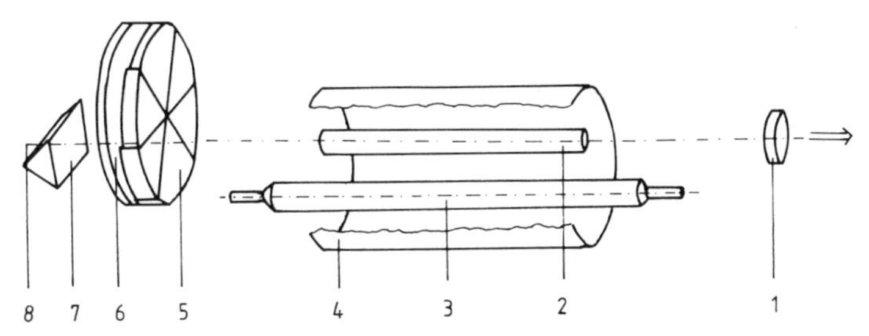


**Figure 5.** Optical schematic of a laser system with passive $Q$-switching. 1, Output mirror; 2, ruby laser rod; 3, flashtube; 4, elliptical = cylindrical reflector; 5, passive switch (stepped cell with switching liquid); 6, layer thickness of the step of the cell traversed by the beam; 7, roof prism; 8, resonator axis. From Ref. [20], reprinted with permission of Pergamon Journals Ltd.

Analytical results with a laser switched with bis(1-ethylquinolin-4-yl)trimethinecyanine iodide in methanol and, later, with vanadylphthalocyanine in nitrobenzene have been published by Moenke-Blankenburg and Mohr [24]. Bis(4-di-methylaminodithiobenzil)nickel in chlorobenzene has proved satisfactory for the switching of YAG lasers.

The arrangement of a ruby laser with a six-stage switching cell is shown in Fig. 5. The laser resonator consists of a roof prism [7] and a plane dielectric mirror for beam uncoupling. The cell with the dissolved switching dyestuff [5] is rotatable between the ruby [2] and the roof prism [7] and is easily removable. It contains six sectors with thickness differing in steps of two. This makes it possible to vary the laser output from a single giant pulse through several spikes (semi-$Q$-switched pulse) to unswitched emission (free-running mode). A stepwise change in transmission is obtained according to Lambert's law [25, 26],

$$T = T_0 e^{-kd} \tag{3}$$

where $T$ = transmitted light intensity
$T_0$ = incident light intensity
$k$ = absorption coefficient
$d$ = thickness of absorbing layer
by changing $k$ or, as in this case, the thickness $d$.

## 2.5. MODE OF ACTION OF AN ENERGY AND SPIKE-NUMBER-MEASURING ACCESSORY

Since the lasers developed for laser microanalysis have a variable energy output, it is desirable to measure the energy, at least in relative units, and the

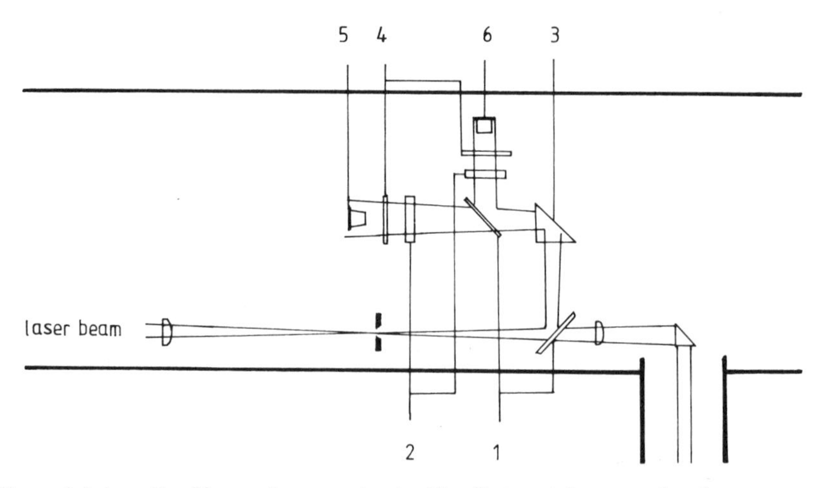


**Figure 6.** Schematic of laser microscope head with adjustment for measuring the energy and counting the spike number of laser output. 1, Semitransparent mirror; 2, diffusing screen; 3, prism; 4, filter; 5, photodiode for counting the number of spikes of each laser shot; 6, photodiode for measuring the relative output energy of each laser shot. From Ref. [27], reprinted with permission of Pergamon Journals Ltd.

number of spikes at each shot. A reliable connection exists between these parameters and the diameter and depth of the crater and therefore, indirectly, the amount of substance that is consumed in the analysis.

Figure 6 shows the scheme of a measuring adjustment proposed by Moenke-Blankenburg and Quillfeldt [27]. The principle is as follows: 1.5% of the total energy of the parallel bundle of laser rays is reflected away by means of a plane glass. The radiation is divided and is fed to a photodiode for the measurement of energy or the counting of spikes. The measuring process has four main purposes:

- Determination of the optimum laser parameters for quantitative analysis
- Adjustment of a reproducible laser energy and number of spikes for the vaporization and excitation of a definite amount of material before the beginning of the analysis
- Control of the constancy of the energy consumed and of the selected number of spikes during the analysis
- Ultilization of the values for correction procedures after the analysis

The determination of the optimum laser parameters for quantitative analysis is identical with the determination of the working point for the maximum intensity of the spectral lines and their reproducibility. It is known that the excitation function of any spectral line at a determined temperature of the plasma has an intensity maximum. The temperature of the laser microplasma is determined by the power of the laser. An approximate measure of this is the mean energy per spike. The spectral line intensity therefore forms a three-dimensional function with the energy and the number of spikes. The graphical representation of this function for a given class of spectral lines of a given matrix element is carried out by recording the spectra of $Q$-switched laser microanalysis in steps of 50 V each between the laser pulse voltages of 0.75 and 1.0 kV. From the intensities obtained from this, the relative energy, and the number of spikes it is possible to obtain both the two-dimensional relationships of intensity as a function of the number of spikes and intensity as a function of energy, and also the three-dimensional function of intensity as functions of energy and number of spikes (Fig. 7). The spectral line intensity reaches a maximum for each selected number of spikes as a function of the laser energy.

Particularly for comparative records of analytical samples and calibration samples differing in composition, structure, texture, and surface state, it is necessary to determine the dimensions of the crater corresponding to a given amount of substance and to relate these to the settings of the laser by determining the relative energy and the number of spikes. The relationships between number of spikes and depth of crater and between energy and diameter of crater can be shown graphically from the results of a relatively

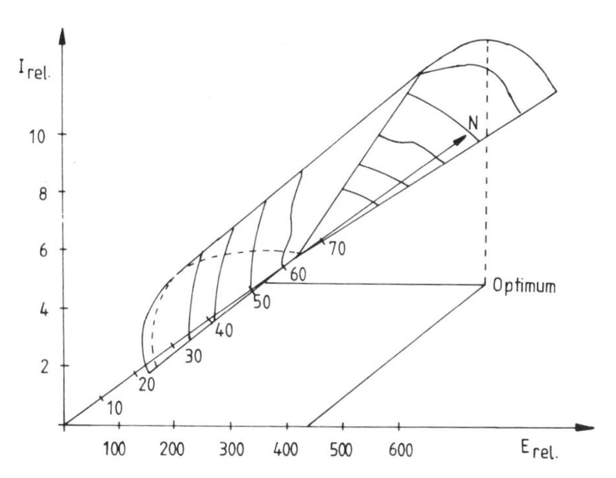


**Figure 7.** Relationship between laser output energy spike number, and spectral line intensity. $I_{rel}$ = relative intensity of a spectral line; $E_{rel}$ = relative measurement of laser output energy; $N$ = number of laser spikes. From Ref. [20], reprinted with permission of Pergamon Journals Ltd.

small number of shots. Once the relationships have been determined for calibration samples used frequently, they can be used repeatedly.

If it is possible to achieve a good reproducible intensity from the settings obtained as described with a known relationship between intensity, energy, and number of spikes, a correction of spectral line intensities can be carried out. As an example, suppose that the scatter of an uncorrected intensity ratio is between 0.86 and 0.99 and the percentage relative standard deviation is 5%. After correction of the intensities before the derivation of the ratio, the width of scatter decreases to 0.90 to 0.97 and a percentage relative standard deviation of 2.5% is obtained.

Here again, for correction it is necessary to show the relationship of the three magnitudes graphically. This can be done as a plot of intensity versus number of spikes, with the energy as parameter or as a plot of intensity versus energy. Then factors $a$ or $b$ are determined as the dependence of intensity on number of spikes or of energy:

$$a = \frac{\Delta I}{\Delta N} E \quad (4)$$

$$b = \frac{\Delta I}{\Delta E} N \quad (5)$$

where $\Delta I$ = difference between the mean value of the intensity and a measurement

$\Delta N$ = difference between the mean value of the number of spikes and a measurement

$\Delta E$ = difference between the mean value of the energy and a given measurement

After correction of the spectral line intensities or of the photographic blackenings directly for the lines of an analytical and a reference element, the derivation of a ratio can be carried out in the usual way. Because of the possibility of connecting a printer or a tape puncher to the indicating apparatus, electronic data processing is recommended for this correction.

The measurement of laser power and energy is an important topic in the subject of laser effects, because all quantitative characterizations of the effects produced by lasers specify the effects in terms of the laser energy or irradiance. The measurement of laser energy or laser power density is an area in which considerable improvement is needed. A compilation of techniques for measurement of laser parameters has been published by Heard [28].

## REFERENCES


1. H. Moenke and L. Moenke-Blankenburg, *Einführung in die Laser-Mikro-Emissionsspektralanalyse*, 2nd Ed., Akademische Verlagsgesellschaft Geest und Portig KG, Leipzig, German Democratic Republic, 1968; 3rd Ed., Izdatelstva, MIR, Moscow, 1968; 4th Ed., *Laser Microspectrochemical Analysis*, Adam Hilger Ltd., London, England, and Crane, Russak & Co., Inc., New York, 1973.
2. J. R. Singer, *Advances in Quantum Electronics*, Columbia University Press, New York, 1961.
3. A. Maitland and M. H. Dunn, *Laser Physics*, North-Holland Publishing Company, Amsterdam, 1969.
4. J. F. Ready, *Effects of High-Power Laser Radiation*, Academic Press, Inc., New York, 1971.
5. R. W. Hellwarth, in *Advances in Quantum Electronics*, J. R. Singer, Ed., Columbia University Press, New York, 1961, p. 334.
6. F. J. McClung and R. W. Hellwarth, *Proc. IEEE*, **51**, 46 (1963).
7. B. A. Langyel, *Laser. Physikalische Grundlagen und Anwendungsgebiete*, Berliner Union, Stuttgart, Federal Republic of Germany, 1967.
8. F. T. Arrechi and E. O. Schulz-Dubois, *Laser Handbook*, Vol. 1, North-Holland Publishing Company, Amsterdam, 1972.
9. W. Köchner, *Solid-State Laser Engineering*, in *Springer Series in Optical Sciences*, Vol. 1, Springer-Verlag, Berlin, 1976.
10. G. K. Grau, *Quantenelektronik, Optik und Laser*, Vieweg & Sohn, Wiesbaden, FRG, 1978.
11. K. Tradowsky, *Laser*, Vogel-Verlag, Würzburg, FRG, 1979.
12. W. Brunner and K. Junge, *Wissenspeicher Lasertechnik*, VEB Fachbuchverlag, Leipzig, GDR, 1982.
13. J.-F. Eloy, *Les Lasers de Puissance*, Masson Éditeur, Paris, 1985.
14. D. L. Andrews, *Lasers in Chemistry*, Akademie-Verlag, Berlin, 1987.
15. R. H. Scott and A. Strasheim, in *Applied Atomic Spectroscopy*, Vol. 1, E. L. Grove, Ed., Plenum Press, New York, 1978, pp. 73–118.
16. N. Omenetto, *Analytical Laser Spectroscopy*, Vol. 50 in *Chemical Analysis: A Series of Monographs on Analytical Chemistry and Its Applications*, P. J. Elving and J. D. Winefordner, Eds., John Wiley & Sons, Inc., New York, 1979.
17. K. Laqua, in *Analytical Laser Spectroscopy*, N. Omenetto, Ed., John Wiley & Sons, Inc., New York, 1979; see also in *Analytical Laser Spectroscopy*, S. Martellucci and A. N. Chester, Eds., Plenum Press, New York, 1985, p. 159.
18. L. Moenke-Blankenburg, in *Advances in Optical and Electron Microscopy*, Vol. 9, R. Barer and V. E. Cosslett, Eds., Academic Press, Inc., New York, 1984, pp. 243–322.
19. K. Dittrich and R. Wennrich, *Prog. Anal. At. Spectrosc.*, **7**, 139 (1984).

20. L. Moenke-Blankenburg, *Prog. Anal. Spectrosc.*, **9**, 335 (1986).
21. E. H. Piepmeier, *Analytical Applications of Lasers*, Vol. 87 in *Chemical Analysis: A Series of Monographs on Analytical Chemistry and Its Applications*, P. J. Elving and J. D. Winefordner, Eds., John Wiley & Sons, Inc., New York, 1986.
22. L. J. Radziemski, R. W. Solarz, and J. A. Paisner, *Laser Spectroscopy and Its Applications*, Marcel Dekker, Inc., New York, 1987.
23. H. Moenke, L. Moenke-Blankenburg, J. Mohr, and W. Quillfeldt, *Mikrochim. Acta*, **6**, 1154 (1970).
24. L. Moenke-Blankenburg and J. Mohr, *Jenaer Jahrb.*, 1970, p. 195.
25. J. Mohr, *Jena Rev.*, 1971.
26. J. Mohr and H. Schmidt, *Jena Rev.*, 1973, p. 93.
27. L. Moenke-Blankenburg and W. Quillfeldt, *Proc. Colloquium Spectroscopicum Internationale*, Prague, Czechoslovakia, 1977.
28. H. G. Heard, *Laser Parameter Measurement Handbook*, John Wiley & Sons, Inc., New York, 1968.

CHAPTER

3

# LASER–TARGET INTERACTION
## MICROPLASMA GENERATION

The interaction of laser light with solids, and subsequent development of an expanding radiant plasma, has been reviewed in detail by several authors [1–36].

### 3.1. FOCUSING OF LASER RADIATION

One property of laser radiation that is of interest in effects work is the directionality of the beam. Laser radiation is confined to a narrow cone of angles. Typically, for solid-state lasers, the spreading angle is on the order of a few milliradians. Because of the narrow divergence angle of laser radiation, it is easy to collect all the radiation with an optical lens system. The narrow beam angle also allows focusing of the laser light to a small spot. Therefore, the directionality of the radiation is an important factor in the ability of lasers to deliver high power densities to a target and therefore to produce interesting effects.

Classical optical theory predicts the spot radius at the focus of a simple lens, for uniform plane wavefronts, to be [16, 37–40]

$$r \approx 1.22\frac{\lambda f}{D} \tag{6}$$

where $\lambda$ = wavelength of the laser radiation
$f$ = focal length of the lens
$D$ = aperture of the laser device

The divergence angle of the laser beam $\sigma$ is

$$\sigma \approx 1.22\frac{\lambda}{d_e} \tag{7}$$

and

$$d_e = \frac{d}{n} \tag{8}$$

where $d_e$ = effective diameter of the resonator
$d$ = free diameter of the rod
$n$ = refraction index of the resonator material

Imperfections in the focusing optics will result in a spot size larger than the Airy disk. According to Scott and Strasheim [16], the important factor that contributes to a very much larger spot size than that established theoretically is the fact that the laser does not usually produce a uniform-plane wavefront. Although the output of a ruby or neodymium laser can be approximated to a plane wave by mode locking or by reducing the aperture of the system, this leads to an energy loss that cannot always be tolerated. The output beam normally has a divergence of over 1 mrad, resulting in a spot size one or two orders of magnitude greater than that predicted for plane waves.

A good approximation of the spot diameter can be derived from a geometrical construction (see [16]). If the divergence angle of the laser beam is $\sigma$, the spot diameter is given by

$$r \approx f \tan\frac{\sigma}{2} \tag{9}$$

where $f$ is the focal length of the lens at the laser wavelength. Angle $\sigma$ is generally very small, so $\tan(\sigma/2) \approx 2$, giving

$$r \approx \frac{\sigma}{2} f \tag{10}$$

The diameters of craters formed in metal samples by focused laser pulses are invariably greater than this value, owing to processes such as lateral heat diffusion or transfer of energy from the laser plasma to the sample.

Important parameters influencing the interaction between the laser beam and the surface are the energy and power density of the radiation falling onto the sample surface. To determine these parameters, the laser beam energy must first be measured calorimetrically. The power can be calculated once the time profile is known. In the case of a single pulse ($Q$-switched mode) the time profile must be recorded oscillographically. The half-width of the pulse is then measured and the mean power calculated. This is equal to the total pulse energy divided by the half-width. The mean power of a multipulse discharge such as a conventional-mode laser pulse train is determined by measuring the total energy of the pulse train and dividing this by the total duration of the train.

The mean energy and power densities produced by focusing the laser beam are averages not only in time but also in space, because of the transverse-mode structure of the beam. Thus the mean energy density is defined as the total energy deposited on the focal spot divided by the area of the spot, $\pi\sigma^2 f^2/4$. The mean power density of a focused $Q$-switched laser pulse of energy $E$ and half-width $\tau$ is given by

$$\phi = \frac{4E}{\pi\sigma^2 f^2 \tau} \tag{11}$$

## 3.2. TECHNICAL DATA ON NORMAL-MODE PULSE AND Q-SWITCHED SOLID-STATE LASERS USED FOR LASER MICROANALYSIS

The characteristics of the crater and of the laser cloud formed above the surface of the sample will depend on the nature of the laser pulse, the material of the analytical object, and the matrix around the analytical object. Given the two types of laser pulse, the normal long-pulsed mode and the $Q$-switched pulse, where the number of pulses is accurately known, one should consider some effects that result when these different types of laser pulses interact with samples under various conditions. With all types, when the laser beam strikes the surface of the sample, some of the energy is reflected, whereas the rest interacts with an area of the sample giving rise to a luminous vapor cloud.

The long-pulsed laser beam leaves deep craters with raised edges, and splattered molten particles surround the crater. The crater depth depends on the energy density of the beam, reflectivity, and matrix properties of the sample, as well as other sample properties. The $Q$-switched laser beam

**TABLE 2. Solid-State Laser Parameters for Laser Microanalysis**

| | Normal Pulse | Semi-$Q$-switched Pulse | $Q$-Switched Giant Pulse |
|---|---|---|---|
| Energy | 1 J | Continuously or stepwise | 0.1 J |
| Number of spikes | $\geqslant 100$ | Continuously or stepwise | 1 |
| Time duration | $\approx 1$ ms | Continuously or stepwise | $\approx 50$ ns |
| Power | $\approx 10$ kW | Continuously or stepwise | $\approx 2$ MW |
| Power density | | | |
| $r \approx 125\ \mu$m | $2 \times 10^7$ W cm$^{-2}$ | Continuously or stepwise | $4 \times 10^9$ W cm$^{-2}$ |
| $r \approx 5\ \mu$m | $10^{10}$ W cm$^{-2}$ | Continuously or stepwise | $3 \times 10^{12}$ W cm$^{-2}$ |

*Source*: Reprinted with permission from *Prog. Analyt. Spectrosc.* **9**, L. Moenke, copyright 1986, Pergamon Journals Ltd.

produces craters that are less dependent on sample properties, and therefore shallower than those mentioned above; there is only a minimum of molten particles.

In laser microanalysis, the following values are common, giving rise to plasma temperatures between 3000 and about 25,000 K (see Table 2): power densities in the range of approximately 10 MW cm$^{-2}$ to 10 GW cm$^{-2}$ for a normal-mode pulse of 1 J energy; ≈ 1 ms time duration; 10 kW power, focused to spot diameters of about 250 and 10 $\mu$m to approximately 1 GW cm$^{-2}$ to 1 TW cm$^{-2}$ for a *Q*-switched pulse of 0.1 J energy; 50 ns time duration of one giant pulse; 2 MW power, focused to the same spot diameters [36].

## 3.3. VAPORIZATION OF SOLIDS

Effects caused by absorption of focused laser radiation at the surfaces of solids are heating, melting, vaporization, atomization, excitation, and ionization (Fig. 8). Generation of free atoms by vaporization of solids is the effect used in laser microanalysis based on atomic absorption and atomic fluorescence spectrometry; generation of excited atoms and ions by excitation of the vapor is used in laser microanalysis based on atomic emission and mass spectrometry.

In times on the order of the duration of a laser pulse, the electrons, which absorb the photons, will produce many collisions, both among themselves and with lattice phonons. The energy absorbed by an electron will be distributed and passed on to the lattice. Ready [1] therefore regards the optical energy as

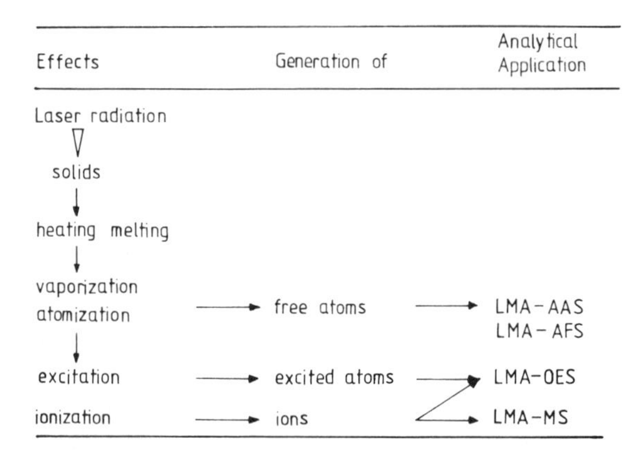


**Figure 8.** Effects caused by absorption of focused laser radiation at surfaces of solids. From Ref. [83], reprinted with permission of Pergamon Journals Ltd.

being turned into heat instantaneously at the point at which the light was absorbed. The intense local heating experienced by the target causes a rapid rise in the surface temperature of the material. Heat is conducted into the interior of the target and a thin molten layer forms below the surface [41]. As the thermal energy deposited at the surface increases, a point is reached where the deposited energy exceeds the latent heat of vaporization. When this happens, heat cannot be conducted away from the point of irradiation fast enough to prevent the surface from reaching its boiling temperature, and evaporation occurs from the surface. The distribution occurs so rapidly in the time scale of $Q$-switched and normal laser pulses that Ready [1] and other [33, 2–48] assume a local equilibrium to have been established during the pulse. This assumption may break down for the case of very short pulses.

The energy density deposited at the surface of the target by a power density $F$ is $Ft_e$; hence the average energy per unit mass acquired by the thin layer of molten metal is $Ft_e/d(at_e)^{1/2}$, where $d$ is the mass density of the target, $a$ the thermal diffusivity, and $t_e$ the duration of the laser pulse. For evaporation to occur, the energy deposited in this layer must exceed the latent heat of vaporization of the target, $L_v$. Thus the following threshold condition is obtained for the minimum absorbed power density ($F_{\min}$), below which no evaporation will occur [10]:

$$F_{\min} = dL_v a^{1/2} t_e^{-1/2} \tag{12}$$

### 3.3.1. Vaporization by Normal Laser Pulses

There is a great difference in the behavior of surfaces struck by laser pulses with millisecond duration compared to those in the nanosecond region. High-power short pulses do not produce much vaporization, but instead, remove only a small amount of material from the surface, whereas longer, higher-power pulses from a normal pulse laser produce deep holes in the target.

The spiking behavior of the laser will also influence the vaporization. In streak camera photographs of a surface being vaporized by a ruby laser exhibiting strong relaxation oscillations, one sees that the glowing vaporized material is emitted in pulsations because of the spiked nature of the laser output [40, 49–52]. Experimental data on the amount of material vaporized by a laser pulse exhibits considerable scatter, depending on the exact conditions [53–74]. Representative values of the amount of the material vaporized by a normal pulse laser are given in a number of sources [1, 75–80, 82].

For typical parameters, the surface of the material is raised to the vaporization temperature (boiling point) in a very short time, $t_v$, given by

Ready [1(1971)] as

$$t_v = \frac{\pi K d C(T_v - T_0)^2}{4F^2} \tag{13}$$

Here $K$, $C$, $d$, and $F$ are the thermal conductivity, heat capacity per unit mass, mass density, and power density, respectively, and $T_v$ and $T_0$ are the vaporization temperature and initial temperature, respectively. This equation is useful for rough estimates of the surface temperature, depth vaporized, and time taken to reach the boiling temperature when the irradiance is close to the threshold value.

Adrain and Watson [32] reported that for an iron target and an irradiance of 20 GW m$^{-2}$, the time predicted to reach the boiling temperature of 3000 K is about 4 $\mu$s, the steady-state velocity of the retreating molten layer is 0.3 m s$^{-1}$, and the depth vaporized by a 200-$\mu$s pulse is about 60 $\mu$m. In contrast, for a short laser pulse of 50 ns and an irradiance of about 1 TW m$^{-2}$, these predictions become 2 ns for $t_v$, 17 m s$^{-1}$ for $v$, and 0.8 $\mu$m for $x_t$. Ready [1(1971)], for example, measured the time to reach the vaporization temperature at a power density of $10^6$ W cm$^{-2}$ as 0.25 $\mu$s for Bi, 1.2 $\mu$s for Pb, 18.6 $\mu$s for Fe, and 104.6 $\mu$s for W.

Following the treatment of Ready [1 (1971)], for laser irradiation on a semi-infinite solid, the one-dimensional heat conduction equation is given by

$$K\frac{\partial^2 T}{\partial x^2} - \frac{K\partial T}{a\partial t} = -E\alpha \exp(-\alpha x) \tag{14}$$

where $T$ is the temperature as a function of time $t$ and depth $x$ in the solid. The thermal conductivity is denoted by $K$ and $\alpha$ is the absorption coefficient. For the simple case of a constant power density $F$ and a material with a large absorption coefficient, the temperature $T_s$ at the surface of the metal after a time $t_v$ is given by

$$T_s = \frac{2F}{K}\left(\frac{at_v}{\pi}\right)^{1/2} \tag{15}$$

When the material is exposed to a large constant flux and begins vaporizing after time $t_v$, the rate of material removal will approach a steady-state rate given by

$$v_{ss} = \frac{F}{d|L_v + C(T_v - T_0)|} \tag{16}$$

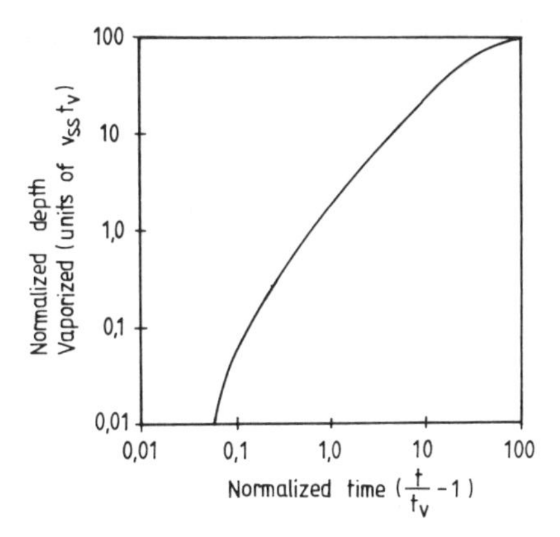


**Figure 9.** Normalized depth vaporized in typical metals as a function of normalized time for normal laser pulses according to a model that assumes continual removal of the vaporized material. Depth is in units of steady-state evaporation velocity $v_{ss} \times$ time $t_v$ to reach vaporization temperature. From Refs. [81, 83], reprinted with permission of Pergamon Journals Ltd.

The depth of vaporization may be expressed as a function of the time in units of $t_v$. As the time becomes very long, the velocity approaches the steady-state condition. Figure 9 gives results for the depth vaporized in typical metals by a laser pulse as a function of the normalized time, $(t/t_v)$ [81]. Hence the depth $x_t$ vaporized after a time $t$ is given by

$$x_t = v_{ss}(t - t_v) \tag{17}$$

where the time $t_v$ is that taken to reach the boiling temperature $T_v$ from the ambient temperature $T_0$ [see eq. (13)]. Equations (13), (16), and (17) are useful for rough estimates of the surface temperature, depth vaporized, and time taken to reach the boiling temperature when the power density is close to the threshold value.

Klocke [84] calculated the depth of a crater another way (Table 3). A satisfactory formula for calculating the depths of the crater can be written as

$$x = \frac{Ft_e(1-P)}{d[L_v + C(T_v - T_0)]} - \frac{K(T_v - T_0)}{F(1-P)} \tag{18}$$

where $F$ = power density ($\mathrm{W\,cm^{-2}}$)
$t_e$ = pulse duration (s)
$P$ = reflection factor
$d$ = mass density ($\mathrm{g\,cm^{-3}}$)
$C$ = heat capacity ($\mathrm{J^{-1}\,g^{-1}\,K^{-1}}$)

**TABLE 3. Comparison of Observed and Calculated Crater Depths Produced by Normal Laser Pulses[a]**

| Element | Observed Depth ($\mu$m) | Calculated Depth ($\mu$m) Reflection Taken into Account | Calculated Depth ($\mu$m) Reflection Neglected |
|---|---|---|---|
| Sn | 1586 | 2100 | 2400 |
| Bi | 1516 | 4700 | 5100 |
| Cd | 1648 | 4240 | 5100 |
| Pb | 1554 | 3810 | 4100 |
| Mg | 1676 | 3150 | 4150 |
| Ag | 960 | 925 | 1570 |
| Cu | 830 | 615 | 995 |
| Ti | 758 | — | 970 |
| Ni | 660 | — | 741 |
| Fe | 538 | 675 | 814 |
| Mo | 570 | 550 | 730 |

*Source*: Ref. 84; reprinted by permission from *Spectrochim. Acta.*, **24B**, H. Klocke, copyright 1969 Pergamon Press PLC.

[a] Additional work dealing with crater dimensions and behavior has been published [85–89].

$T_v, T_0$ = vaporization (boiling) temperature and initial temperature, respectively (K)

$L_v$ = latent heat of vaporization ($\mathrm{J\,g^{-1}}$)

$K$ = thermal conductivity ($\mathrm{W\,cm^{-1}\,K^{-1}}$)

Ready [1] concluded that the construction of a good theory is probably of little worth, because of the complications in considering the relaxation oscillations and other details of the process. Since the experimental data vary so widely, depending on the circumstances, it is advisable to consider each case individually.

In summary, for normal pulsed lasers the total energy per pulse is usually much greater than for $Q$-switched lasers. The shape of the normal pulse is that of the long-duration ($\approx 1$ ms) spiked pulse which contains relaxation oscillation pulses of $\approx 50$ ns spaced regularly throughout. Normal pulsed lasers has low power densities (see Table 2). Particle emission from dielectric materials can be treated as ordinary vaporization, and conventional thermal emission phenomena occur. Phenomena similar to surface ionization were involved and a strong correlation was found between ion formation probabilities and ionization energies. The reproducibility of the signals was indicated to be very closely correlated to the surface temperature. In the case of metals the mechanism can be described as follows. As more energy is delivered to the

**Figure 10.** Vaporization by normal-mode laser pulses with ejection of particles. Material, steel; output energy, 0.5 J. From Ch. 2, Ref. [1], reprinted with permission of Akademische Verlagsgesellschaft Geest & Portig K.G., Leipzig.

surface, the initial vaporization process changes to the melting–spraying mechanism, by which a significant amount of material is removed from the crater (Figs. 10 and 11). This process may occur several times during the same laser shot. The stream of vapor leaves the surface with a velocity on the order of $10^4$ cm s$^{-1}$.

### 3.3.2. Vaporization by Giant Pulse or Semi-*Q*-Switched Laser Pulse

*Q*-switched lasers are typified by lower energies, much shorter pulse length, and therefore higher power densities, in the range $10^9$ to $10^{12}$ W cm$^{-2}$ (see Table 2). The temporal distribution of a giant pulse is described by a smooth curve rising to a maximum and falling back to the baseline, and may be approximated by a triangle [1]. The spatial distribution of the pulse may be rather complex [1, 4, 90–93].

The temperature of the vapor leaving the surface is higher than the boiling temperature and therefore partly ionized. A stream of vapor leaves the surface with velocities of $10^{-6}$ cm s$^{-1}$. The microplasma formed above the surface absorbs the later part of the incoming laser spike. Depending on the type of

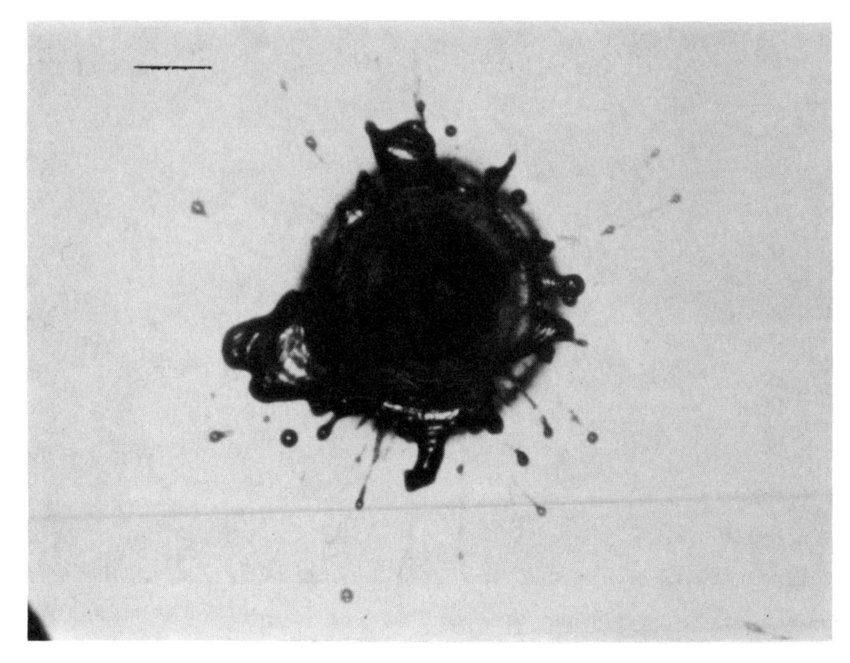

**Figure 11.** Effect of laser impact on a steel sample crater of 100 $\mu$m depth and 110 $\mu$m diameter produced by a nonswitched ruby resonator. From Ch. 2, Ref. [1], reprinted with permission of Akademische Verlagsgesellschaft Geest & Portig K.G., Leipzig.

material, ejection in molten form is also observed (Fig. 12). Full descriptions of these processes have been made previously, and may be found in work by several authors [1, 50, 94–100].

With the action of one single giant pulse as obtainable, for instance, with the help of a Kerr or Pockels cell, very high power and power densities on the surface can be obtained. The result of the interaction with the surface is a very shallow crater with little or no molten residue. The diameter may be rather large. From the appearance of the crater, it can be concluded that most of the material removed is atomized with only a very low fraction of molten or solid particles. The vapor cloud may radiate strongly. The atomic radiation can be used directly for spectrochemical analysis at not too low concentrations.

With the mode of operation of saturable absorbers it is possible to make a choice of the number of spikes through a change of path length or concentration of the dye (Table 4). Figure 13 shows the oscillograms of laser spikes after changing the transmission $T$ of the dye. One giant pulse could be obtained with $T = 55\%$, two spikes with $T = 65\%$, four spikes with $T = 78\%$, 12 spikes with $T = 91\%$, and a normal-mode pulse with $T = 100\%$. Figures 14

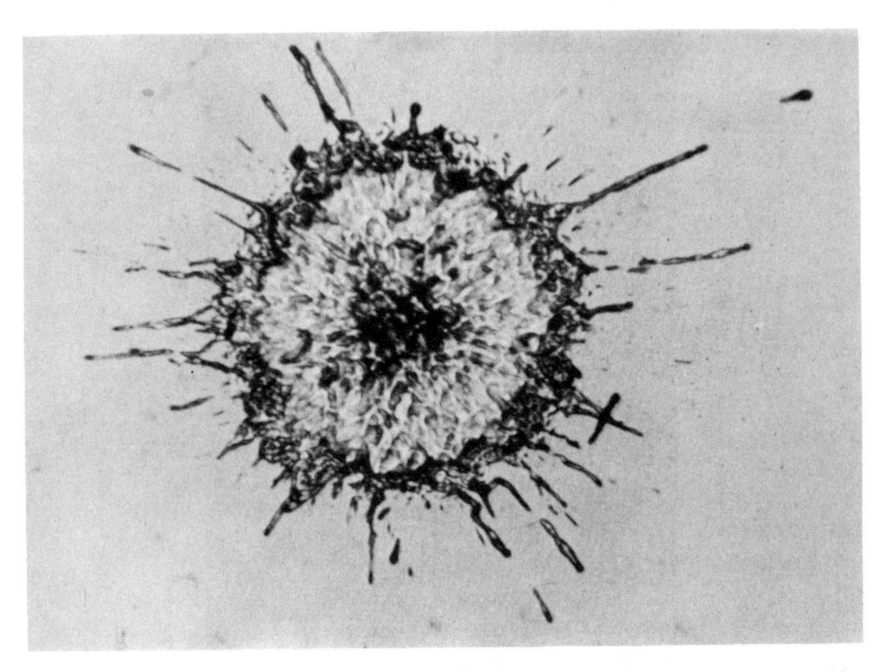

**Figure 12.** Effect of laser impact on a steel sample; shallow crater of 175 $\mu$m diameter produced by $Q$-switched ruby resonator. From Ch. 2, Ref. [1], reprinted with permission of Akademische Verlagsgesellschaft Geest & Portig K. G., Leipzig.

**TABLE 4. Variation of the Concentration of a Saturable Absorber and the Resulting Parameters of a Ruby Laser**

| Concentration of Cryptocyanine[a] (mg/liter) | Transmission of the Dye (%) | Output Energy (mWs) | Number of Spikes | Energy per Spike (mWs) | Calculated Power of All Spikes (MW) |
|---|---|---|---|---|---|
| 1.12 | 55 | 126 | 1 | 126 | 1.5 |
| 0.79 | 65 | 136 | 2 | 68 | 0.8 |
| 0.54 | 74 | 156 | 3 | 52 | 0.7 |
| 0.45 | 78 | 172 | 4 | 43 | 0.6 |
| 0.36 | 81 | 189 | 5 | 38 | 0.5 |
| 0.17 | 89 | 251 | 10 | 25 | 0.4 |
| 0.06 | 95 | 309 | 20 | 15 | 0.2 |
| 0 | 100 | 364 | Approx. 100 | Approx. 4 | 0.01 |

*Source*: Ref. 101–104; reprinted by permission of VEB Gustav Fischer Verlag, Jena.

[a] Bis(1-ethylquinolin-4-yl)trimethinecyanineiodide in methanol.

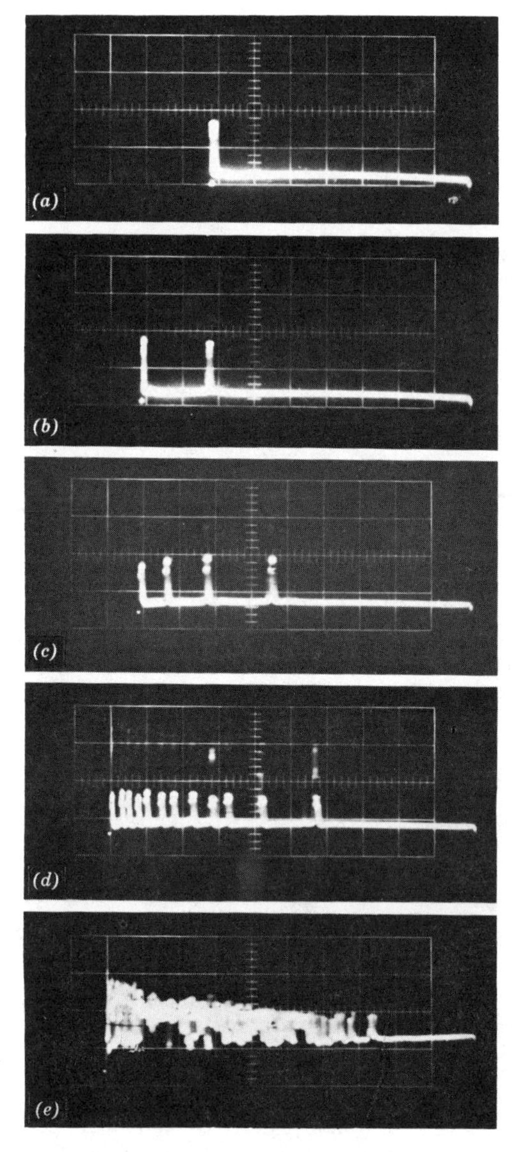


**Figure 13.** Oscillograms showing the dependence between the transmission $T$ of bleachable dye and the number of spikes: (*a*) 1 spike at $T = 55\%$; (*b*) 2 spikes at $T = 65\%$; (*c*) 4 spikes at $T = 78\%$; (*d*) 12 spikes at $T = 91\%$; (*e*) normal-mode pulsed laser emission at $T = 100\%$. From Ref. [102], reprinted with permission by VEB Gustav Fischer Verlag, Jena.

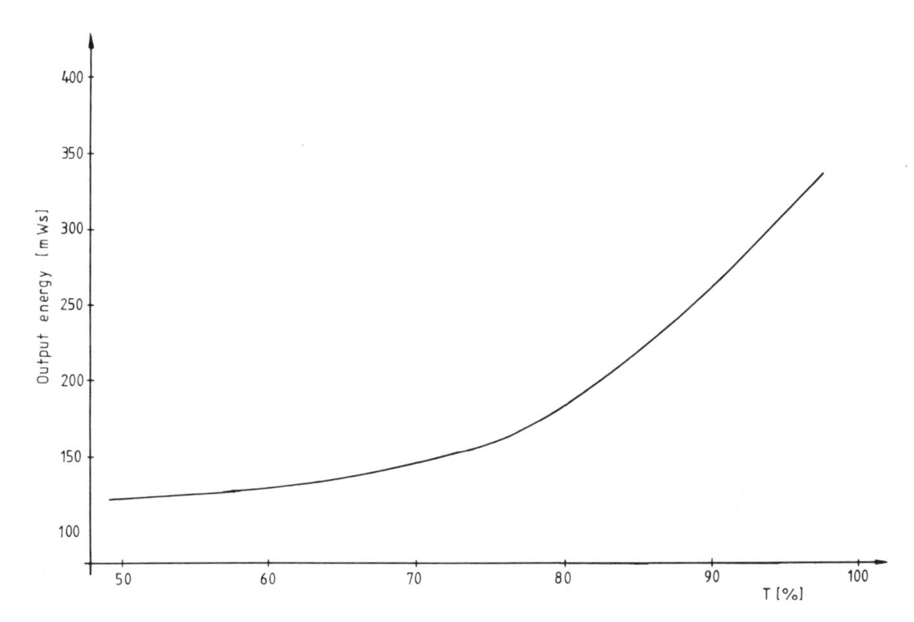


**Figure 14.** Output energy of a passive switched laser versus the transmission $T$ of bleachable dye. From Ref. [102], reprinted with permission of VEB Gustav Fischer Verlag, Jena.

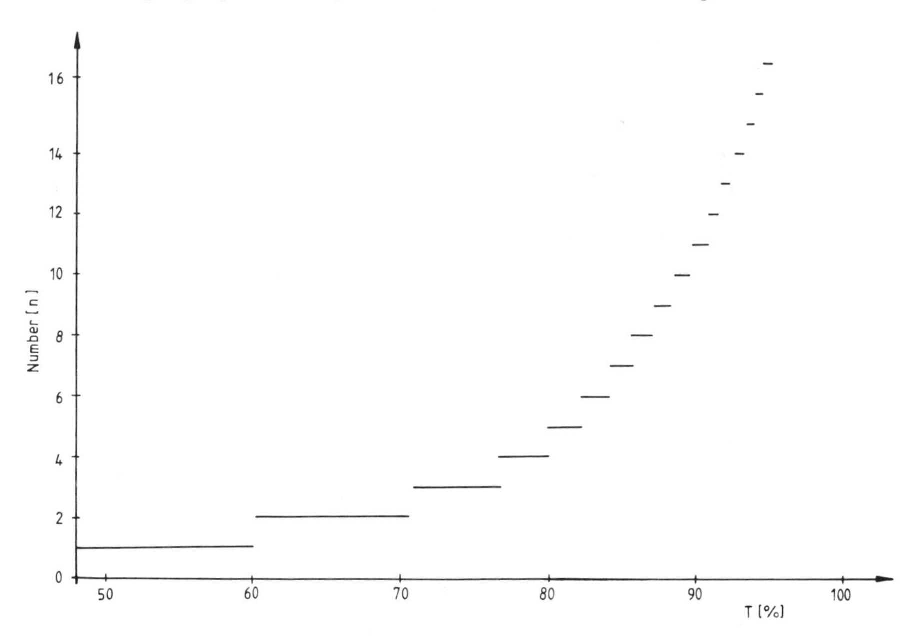


**Figure 15.** Number of spikes versus the transmission of bleachable dye. From Ref. [102], reprinted with permission of VEB Gustav Fischer Verlag, Jena.

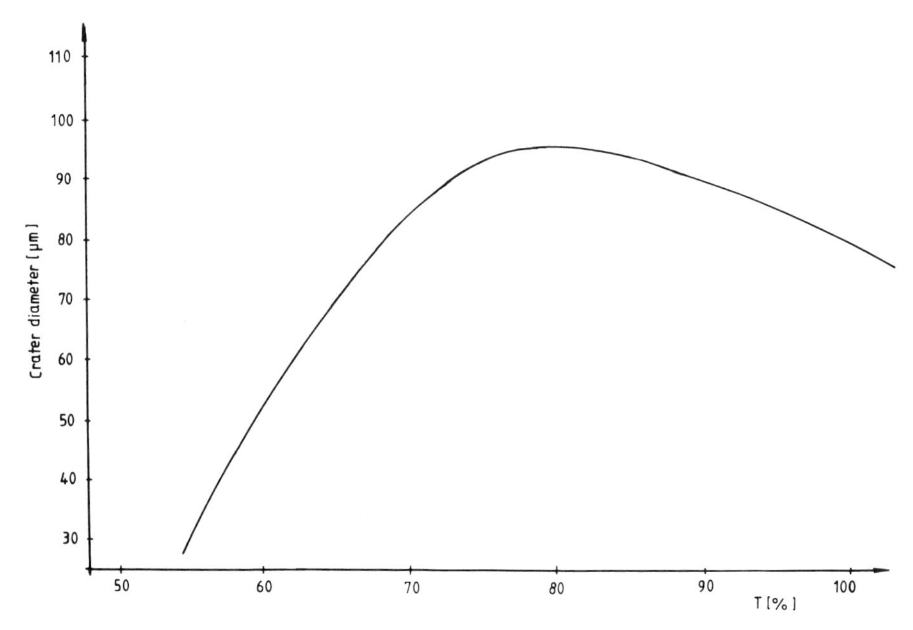


**Figure 16.** Crater diameter in glass SSK 6 versus the transmission $T$ of bleachable dye. From Ref. [102], reprinted with permission of VEB Gustav Fischer Verlag, Jena.

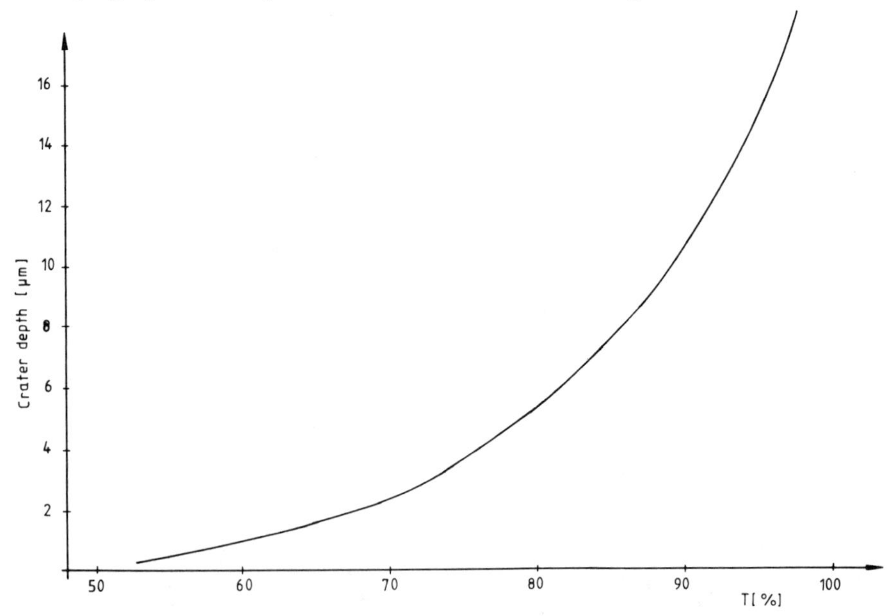


**Figure 17.** Crater depth in glass SSK 6 versus the transmission $T$ of bleachable dye. From Ref. [102], reprinted with permission of VEB Gustav Fischer Verlag, Jena.

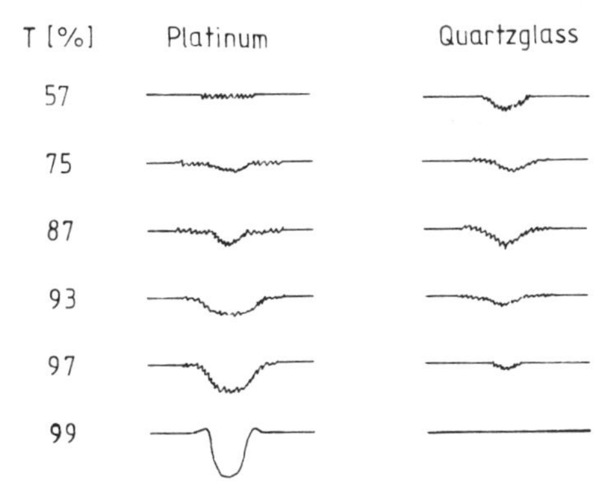


**Figure 18.** Scheme for changing of the crater cross sections of platinum and quartz glass versus the transmission $T$ of bleachable dye. From Ref. [102], reprinted with permission of VEB Gustav Fischer Verlag, Jena.

and 15 show the dependence of output energy and number of spikes on transmission of the dye. If the relation between dye transmission and crater dimensions for a material is ascertained, the crater diameter and the ablation depth can be chosen (see Figs. 16 and 17).

A comparison of semi-$Q$-switched ablation of a metal (Pt) and a nonmetal ($SiO_2$ glass) is given in Fig. 18. The optimal analysis of metals, such as platinum, is performed at $T = 99\%$ and of quartz glass at $T = 87\%$. The evaporation crater obtained in such a case shows no melting ridge; evaporation is more reproducible and spectra could be received without auxiliary excitation; however, spectra show broadened and self-absorbing lines [105].

With $Q$-switched operation employing acousto-optical shutters, the dimensions of the crater and its shape are similar to those obtainable with a free-running laser and semi-$Q$-switched pulse. Each of the 50 to 100 spikes interacts with a solid and rather cool surface. On the other hand, compared to the normal-mode laser, the power of the spikes is higher, and thus also the temperature of the vapor cloud and the degree of excitation and ionization.

### 3.3.3. Vapor Pressure and Temperature

For power densities within the range specified above, the vapor pressure can be so high that the effects of laser absorption must be considered. This absorption is assumed to occur in the thin, dense, partially ionized layer that exists between the solid and vapor phases (Krokhin [10]). The electron density ($n_e$) within this layer approaches the critical electron density ($n_c$) that will prevent laser light reaching the target. Light is reflected from a plasma if its wavelength is greater than the plasma wavelength $\lambda_p$, defined (e.g., Hughes [15]) by

$$\lambda_p \simeq 10^{-15}(n_e)^{1/2} \tag{19}$$

For light of 1-$\mu$m wavelength, $n_e$ is around $10^{27}$ m$^{-3}$; since the density of the bulk solid is around $10^{29}$ cm$^{-3}$, the layer volume must increase by about two orders of magnitude before irradiation can continue. The emitted light at this time would be primarily blackbody radiation from the hot, dense gas. As the density falls below $n_0$, the vapor becomes highly absorbing and enters the so-called "self-regulating" regime (Afanasyev et al. [41], Caruso and Gratton [106]). In this regime, light is absorbed in the vapor by inverse bremsstrahlung, resulting in heating and expansion which, in turn, reduce the vapor density. Laser light can now reach the target and evaporate more material, thereby increasing the vapor density once more. The process repeats itself, with the critical density boundary moving toward the laser at velocities up to 10 km s$^{-1}$ (Alcock et al. [107]), until removal of the laser irradiation allows the plasma to cool and decay. As the plasma cools, line emission tends to dominate the radiation processes, with temperatures as high as 25,000 K being recorded (Adrain et al. [32, 37]) in plasma suitable for microanalysis.

Such high expansion velocities, when coupled with power densities of 10 MW cm$^{-2}$ or more, produce a considerable ablation pressure on the target, which can significantly modify the target interaction (Askaryan and Moroz [108]). Ablation pressure is defined as

$$p_A \simeq \frac{Ev_a}{L_v + v_a^2/2} \tag{20}$$

where $v_a$ is the vapor expansion velocity.

For normal-mode lasers, the continual heating and cooling caused by successive laser spikes allows a pool of molten material to form behind the evaporation front. Irradiances of 1 MW cm$^{-2}$ to 1 GW cm$^{-2}$ coupled with expansion velocities of, for example, 1000 m s$^{-1}$ produce ablation pressures up to 20 MPa, which forces the molten material away from the target with high kinetic energy. Craters about 100 $\mu$m wide by a similar depth are formed in the material, with profiles ranging from hemispherical to conical. The ejection of molten material, however, does not occur immediately but is delayed until about 50 to 100 $\mu$s after the start of irradiation (Chun and Rose [109]), corresponding to about 100 or so laser spikes at the target (Watson [110]).

By contrast, in the case of $Q$-switched-mode plasmas, power densities up to 1 TW cm$^{-2}$ can produce ablation pressures as high as 10 GPa; consequently, evaporation occurs at temperatures significantly above the normal boiling point of the material. Accordingly, the bulk of material removal occurs in the vapor phase, resulting in shallow craters a few micrometers deep.

Spectroscopic temperature measurements of laser plasmas present difficult problems as a result of the inhomogeneous spatial characteristics and highly transient nature of these plasmas. A number of methods have been used by various investigators to determine laser plasma temperatures [111–115]. Bögershausen and Hönle [113], for instance, used the Ornstein two-line method and obtained a maximum temperature of 29,000 K for a 1-MW $Q$-switched laser plasma. Because of the high temperature gradients in laser plasmas, temperature values obtained by this method could depend on the excitation potentials of the spectral lines chosen. This is because the lines do not necessarily radiate from the same region in the plasma. Ideally, a time- and space-resolved readout of the spectral line intensities is required, followed by the necessary Abel transformations. However, the dynamic nature of laser plasmas, the extreme temperature gradients, and the time period in which they exist are factors that would certainly limit the accuracy of such a difficult study.

David and Weichel (112) have shown by interferometric measurements that the maximum electron temperature attained for a laser-heated carbon plasma, using an energy density on the order of 100 J cm$^{-2}$, was approximately 116,000 K. At this relatively low energy density, the plasma was not in local thermodynamic equilibrium, with the electron temperature nearly an order of magnitude higher than the plasma temperature. A decrease in the electron temperature to that of the plasma (Saha) temperature was observed for increasing laser energy density. This decrease implied that the electrons transferred their energy to the more massive particles in the plasma. At 1000 J cm$^{-2}$ the plasma was found to be in thermodynamic equilibrium, with a maximum temperature of approximately 24,000 K.

### 3.3.4. Influences of Sample Parameters on Vaporization

Physical properties of the sample to be analyzed, such as reflectivity, thermal conductivity, vaporization temperature, heat capacity, and latent heats of fusion and vaporization influence the process of absorption of laser radiation at surfaces.

#### *3.3.4.1. Thermal Properties of Solids*

In calculations of the temperature rise, it is to be assumed that the thermal properties of the absorbing material are independent of temperature in many cases of metals. Table 5 gives some values for the thermal conductivity and specific heat capacity of several metals in solid form over a range of temperatures.

**TABLE 5. Thermal Properties of Solid Metals**

| Metal | Thermal Conductivity ($W\,m^{-1}\,K^{-1}$) at Temperature (K): | | | | | | |
|---|---|---|---|---|---|---|---|
| | 273 | 373 | 673 | 973 | | | |
| Ni | 91 | 83 | 64 | 66 | | | |
| W | 190 | 165 | 145 | 120 | | | |
| Al | 238 | 230 | 226 | | | | |
| Cu | 385 | 382 | 376 | 350 | | | |
| Ag | 418 | 417 | 362 | | | | |
| | Specific Heat Capacity ($J\,mol^{-1}\,K^{-1}$) at Temperature (K): | | | | | | |
| | 298 | 400 | 600 | 800 | 1000 | 1500 | 2000 |
| Al | 24.3 | 25.5 | 28.1 | 30.5 | | | |
| Cu | 24.5 | 24.5 | 26.4 | 27.6 | 28.8 | | |
| W | 24.5 | 24.5 | 26.0 | 26.6 | 27.7 | 28.7 | 30.3 |
| Ag | 25.5 | 25.5 | 26.7 | 28.2 | 29.9 | | |
| Ni | 26.0 | 28.7 | 34.6 | | | 36.3 | |

*Source:* Refs. 1, 116, 117.

For a more complete treatment, the variation in heat capacity with temperature over the range from room temperature to the melting point must be taken into account. Ready [1] assumes that the energy lost from the surface by thermal radiation amounts to about $10^3\,W\,cm^{-2}$ or less for solid metals, and that for a laser power density of about $10^6\,W\,cm^{-2}$ and greater, the error could be neglected. But if the pulse is on for a long time, heat can be conducted over a larger area. Then these area contributions to reradiation and the total reradiated power may approach the power absorbed. In addition, heat is conducted in the interior, and the greater the thermal conductivity, the smaller is the vaporized amount. As the laser power density increases, it reaches a value at which the heat is supplied too fast to be conducted away. Then the factor of latent heat of vaporization becomes important. The considerable difference between the behavior of surfaces struck by normal laser pulses of millisecond duration and those of $Q$-switched laser pulses in the nanosecond-to-microsecond range was dealt with earlier.

#### *3.3.4.2. Reflectivity*

Measurement of the reflectivity of metallic surfaces irradiated by a normal pulse laser indicated that the reflectivity dropped to a low value in the first 200 $\mu$s of the pulse, so that most of the pulse energy was absorbed despite the initial high reflectivity of the target [118, 119].

Results were obtained for a $Q$-switched laser irradiating various targets (Teflon, Al, Sn, Cu, C) in similar fashion [120]. Below $10^8\ \mathrm{W\,cm^{-2}}$, the reflectivities are near the values for undisturbed surfaces. Above $10^8\ \mathrm{W\,cm^{-2}}$, the reflectivities decrease, reaching a value of about 0.1 of the normal value. These results indicate that under appropriate conditions laser energy can be coupled effectively into a target even if the original reflectivity is high.

## 3.4. MICROPLASMA FEATURES FOR ANALYTICAL USE

The vapor could removed from the solid surface may consist of molecules, molecular fractions, neutral atoms, excited atoms, ions, and electrons. The composition depends on the initial parameters of the laser radiation, as shown above. An optimization for each analytical method of laser microanalysis is necessary.

### 3.4.1. Foundations

For power densities just above threshold (1 to $10\ \mathrm{MW\,cm^{-2}}$) evaporation from the liquid boundary layer proceeds at the normal boiling temperature of the material [1]. Such low-density vapors are virtually transparent to the incident laser beam; consequently, little heating of the vapor occurs and the expansion velocity and temperature of the plasma are dependent primarily on the thermal properties of the material. Afanasyev et al. [41] predict a temperature of 3200 K for an iron target. Such low temperatures are of use in atomic absorption spectrometry, where free atoms in the ground state are needed. An atom is said to be in the ground state when its electrons are at their lowest energy levels. On the basis of the well-known Boltzmann relation,

$$\frac{N_m}{N_n} = \frac{g_m}{g_n} \exp\left( - \frac{\Delta E}{kT} \right) \tag{21}$$

where $N$ = number of atoms in a state $n$ or $m$

$g$ = statistical weight for a particular state

$k$ = Boltzmann constant

Walsh [134] calculated the ratio $N_m/N_n$ for a number of common atoms over a range of temperature (see Table 6) for the case where $m$ refers to the first excited state and $n$ is the ground state.

The very low proportion of atoms in the first excited state, even at temperatures of 3000 K, indicates that absorption of radiation, other than that originating from a transition involving the ground state, would be very small. Atoms excited by absorption of resonance radiation also reemit the absorbed

**TABLE 6. Values of $N_m/N_n$ for Typical Elemental Resonance Lines**

| | Line (nm) | $\frac{g_m}{g_n}$ | $\frac{N_m}{N_n}$ at 3000 K | $\frac{N_m}{N_n}$ at 5000 K |
|---|---|---|---|---|
| Cs | 852 | 2 | $7 \times 10^{-3}$ | $7 \times 10^{-2}$ |
| Na | 589 | 2 | $6 \times 10^{-4}$ | $1 \times 10^{-2}$ |
| Ca | 423 | 3 | $4 \times 10^{-5}$ | $3 \times 10^{-3}$ |
| Zn | 214 | 3 | $6 \times 10^{-10}$ | $4 \times 10^{-6}$ |

energy. This process is called atomic fluorescence. When excitation is by absorption of resonance radiation, the fluoresence spectrum of an element, occurs simultaneously with the absorption spectrum, although they can be observed and measured separately by appropriate choice of experimental conditions. The very high proportion of ground-state to excited-state atoms when these are in equilibrium, as shown in Table 6, suggests two particular advantages of atomic absorption and fluorescence measurements over emission. Because absorption and fluorescence measurements over emission. Because absorption and fluorescence are direct measurements of the number of ground-state atoms, these would be expected to yield a better sensitivity than that provided by emission.

On the other hand, at power densities of laser radiation well above the threshold (i.e., greater than about 0.1 GW cm$^{-2}$), plasmas of higher temperatures are formed, which are generally useful in emission spectrometry. When an optical transition occurs between two levels of an atom or ion, the observed intensity $I_x$ integrated over the line profile is given by

$$I_x = h\nu_{mn}A_{mn}N_m \tag{22}$$

where $h$ = Planck's constant
$\nu$ = frequency of the optical transition
$A_{mn}$ = Einstein coefficient for spontaneous emission from the upper energy level

In the case of local thermodynamic equilibrium, the Boltzmann equation holds and we have

$$I_x = h\nu_{mn}A_{mn}N_n\frac{g_m}{g_n}\exp\left(-\frac{\Delta E}{kT}\right) \tag{23}$$

The scheme of spectrographic analysis (OES) was given by Mandelstam

[121] as follows:

$$c \rightarrow N \rightarrow N^* \rightarrow J \rightarrow Q \rightarrow I_x \rightarrow S_x \tag{24}$$

where $c$ = concentration of analyte in the sample
$N$ = number of atoms of the analyte in the plasma
$N^*$ = number of analyte atoms in the excitation state
$J$ = total radiation from these atoms
$Q$ = fraction of this radiation entering the spectrograph
$I_x$ = intensity of the radiation of analyte at a given wavelength
$S_x$ = line density produced on the photoemulsion or the signal in spectrometry

Quantitative analysis has as its basis the knowledge of the function:

$$S_x = f(c_x) \tag{25}$$

which expresses the relationship between the density $S_x$ of analytical line and concentration $c_x$ of element $x$ to be analyzed.

Laser micro mass spectrometry needs complete ionization. In the case of local thermodynamic equilibrium, ion concentrations are described by a chemical equilibrium yielding the well-known Saha–Eggert equation [122–124]:

$$\frac{n_+ n_e}{n_0} = \frac{2Z_+}{Z_0} \frac{(2\pi m_e kT)^{3/2}}{h^3} \exp\left(-\frac{I - \Delta I}{kT}\right) \tag{26}$$

where $n_0, n_+, n_e$ = concentration of atoms, ions, electrons
$Z_0, Z_+$ = partition functions of atoms and ions
$m_e$ = electron mass
$T$ = absolute temperature
$k, h$ = Boltzmann and Planck constants
$I$ = ionization energy
$\Delta I$ = lowering of $I$ depending on $n_e$

Fürstenu [22, 26, 27] observed ion yields $\alpha_\pm = n_\pm / n_0$ from single element targets from $10^{-5}$ to $10^{-3}$, depending on target material and power density. Values of free electron density $n_e$ have been measured for metals and semiconductors as $n_e \approx 10^{23}\,\mathrm{cm}^{-3}$ and for insulators as $n_e \approx 10^{19}\,\mathrm{cm}^{-3}$, with power densities of $10^{10}\,\mathrm{W\,cm}^{-2}$ and $\lambda = 532\,\mathrm{nm}$ by Floren [25].

### 3.4.2. Local Thermodynamic Equilibrium

When local thermodynamic equilibrium (LTE) holds, the plasma can be described by a single temperature function. Local thermodynamic equilibrium will hold when collisional excitation and energy transfer processes are sufficiently rapid. High electron densities are desirable to produce equilibrium quickly. In order to have local thermodynamic equilibrium, various criteria on the collision times must be satisfied [2], the most stringent generally being the time for equalization of the electron and ion temperatures. For laser-produced plasmas, the electron densities typically are high, and the electron-ion collision time may be less than $10^{-9}$ s. Thus on the time scale of conventional *Q*-switched laser pulses, it is possible that the plasma may be in LTE (see [1, 33, 37, 42–45, 48, 114–132]).

Some of the authors consider the laser action exclusively as a heating process and consequently ion production as a thermionic emission, Therefore, they propose the use of the Langmuir–Saha equation. Ready [1] stated: "The Langmuir–Saha equation must be applied with caution in such circumstances."

Dietze and Zahn [138] tried to obtain an empirical formula. Kovalev et al. [100] applied a correction factor. Others reported an attempt to apply the local thermodynamic equilibrium model in using the law of mass action of Guldberg–Waage to calculate the dissociation yields of compounds into the plasma. This approach supposes a previous knowledge of all possible compounds in the sample, but that is not generally the case. Deloule and Eloy [47] have considered a new approach to apply a theoretical model of laser ionization for a quantitative determination. It is now accepted that the LTE model is not based on physical arguments but on empirical ones.

Deloule and Eloy [47] explained a two-phase type of laser–target interaction. In the first phase, the energy absorbed by the target is due to the collective excitation of the electrons induced by the electrical field of the laser photons (see also [5] and [52]). There are transitions between different free states of the electrons which absorb the energy and transmit them to the atoms by collision. The creation of a cloud of free electrons brought about by these transitions lasts about $2 \times 10^{-10}$ s, the time duration required to begin vaporization of the target. In the second phase, the light energy is absorbed by the free electrons in the plasma, transmitted either to the ions by collision or to the solid by thermal diffusion. For example, in the case of an aluminum target and for a laser time duration of approximately 30 ns, the yield of laser energy transmitted to the plasma is on the order of 10% [5]. Afterward the vaporization process is governed by the laws of heat propagation. The absorption of energy is localized in two zones (see Fig. 19): the boundary of the opaque vapor, where there is heating of the sample, and the plasma, where ionization of the species occurs.

From the observations of Eloy [33], the use of a laser time duration shorter than 10 ns is recommended, because the loss of thermal energy by conductivity is highly decreasing in the dense medium (Fig. 19). The conditions required for

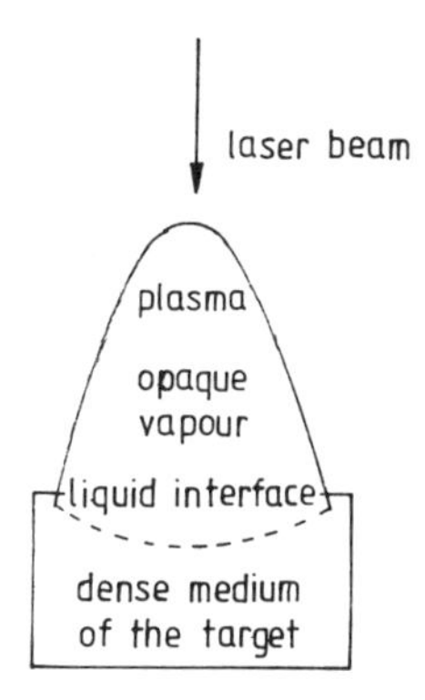


**Figure 19.** Assumption of laser target interaction. From Ref. [47], reprinted with permission of J. F. Eloy.

a LTE plasma by Venugopalan [133] should be fulfilled in the ion source design and the procedure of Eloy [33].

### 3.4.3. Energy Balance; Energy Losses in and above the Target

The interaction of focused laser radiation with solids is obviously a highly complex phenomenon (Fig. 20). Energy balance measurements [135] have shown that apart from the losses of laser output energy in the plume, the energy losses in and above the target are high. The absolute height of energy losses depends on the target material and the mode of laser action. Therefore, knowledge of the energy situation is important for optimization of analytical

**TABLE 7. Energy Losses by Reflection and Scattering of Laser Light from the Surface of Metal Samples**

| Mode of Laser Action | Energy of Incident Laser radiation (J) | Percentage of Reflected and Scattered Energy: Steel (Cr 18%) | Al | Cu |
|---|---|---|---|---|
| Normal mode | 0.26 | 2.0 | 5.7 | 2.4 |
| Semi-$Q$-switched | 0.25 | 2.3 | 5.5 | 1.6 |
| with decreasing | 0.22 | 2.7 | 4.9 | 1.1 |
| number of spikes | 0.18 | 3.0 | 4.2 | 0.8 |
| | 0.10 | 2.8 | 5.9 | 0.1 |
| Single pulse | 0.06 | 2.3 | 6.5 | — |

*Source*: Ref. 135.

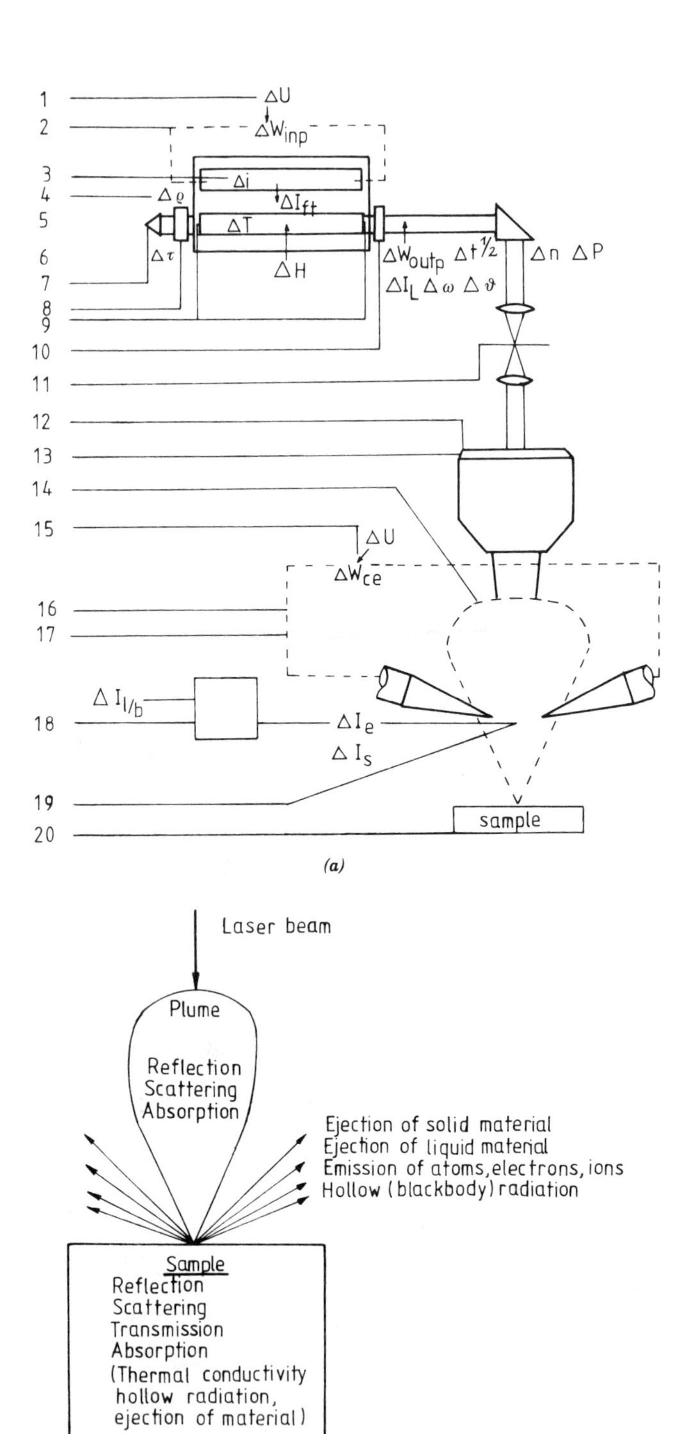
1
2
3
4
5
6
7
8
9
10
11
12
13
14
15
16
17
18
19
20
ΔU
ΔW inp
Δi
ΔI ft
Δϱ
ΔT
Δτ
ΔH
ΔW outp
Δt ½
Δn ΔP
ΔI L Δω Δϑ
ΔU
ΔW ce
ΔI l/b
ΔI e
ΔI s
sample
Laser beam
Plume
Reflection
Scattering
Absorption
Ejection of solid material
Ejection of liquid material
Emission of atoms, electrons, ions
Hollow (blackbody) radiation
Sample
Reflection
Scattering
Transmission
Absorption
(Thermal conductivity
hollow radiation,
ejection of material)

*(a)*

*(b)*

conditions. In [135], energy losses were estimated with the help of physical measurements (Table 7). Metal samples, especially steel, were used and the laser action was varified from normal-mode pulses to semi-$Q$-switched and single-pulse $Q$-switched pulses using saturable absorbers in six steps of transmission (see Section 2.4.3). A simplified calculation of the remainder energy $E_{\text{rem}}$ is possible:

$$E_{\text{rem}} = E_{\text{laser(output)}} - E_{\text{losses(plume)}} - E_{\text{losses(sample)}}$$

and

$$E_{\text{losses(sample)}} = E_{\text{refl + scat}} + E_{\text{therm cond}} + E_{\text{melt}} + E_{\text{kin}} + E_{\text{holl rad}} + E_{\text{vap}}$$

Figure 21 is a schematic of the equipment for measurement of reflected and scattered laser energy from the surface of a sample. Only in the case of copper could a dependence be obtained. The reflection losses are normally less significant for $Q$-switched laser interactions than for free-running laser interactions (see also [136]). Beyond this, the reflection and scattering depend on the degree of surface roughness [137].

For the following calculations the mean value of 2 to 3% reflection and scattering for steel will be considered as $E_{\text{refl + scat}}$ [135]. The energy lost by thermal conductivity $E_{\text{therm cond}}$ in steel was independent of the laster mode and run to 45%. The thermal conductivity was measured by a thermistor, including a small steel sample in a bore such that good contact was assured. After calibration, changes in temperature could be measured down to 0.01°C with a reproducibility of $\leqslant 5\%$ [135].

As described in Section 3.3, some models of interaction between a laser beam and a solid sample surface neglect the fusion process in order to simplify the calculation of total energy requirements for complete vaporization of the

**Figure 20.** Parameters that influence the energy balance of interaction between laser radiation and solid sample. (*a*) 1, $U$ = voltage; 2, $W_{\text{inp}}$ = input energy; 3, $i$ = intensity of current (flashtube), $I_{\text{ft}}$ = radiation intensity of the flashtube; 4, $\rho$ = reflection power of the mirror foil; 5, $T$ = temperature of the resonator rod; 6, $H$ = homogeneity of the resonator material; 7, $\tau$ = stability of the prism; 8, $T$ = transmission of saturable dye; 9, $W_{\text{outp}}$ = output energy; 10, $t_{1/2}$ = half-width of spikes, $n$ = number of spikes, $P$ = power of spikes, $\omega$ = space angle, $\theta$ = radiation density; 11, stability of the screen; 12, focal length of the objective; 13, aperture of the objective; 14, geometry, temperature, density, mass, ($T_e$ = electron temperature, $n_e$ = electron density, $n_i$ = ion density), of the laser micro plume; 15, $U$ = voltage of the auxiliary spark, $W_{\text{ce}}$ = energy of the current electrodes; 16, atmosphere; 17, material of electrodes, form and distance; 18, $I_{l/b}$ = intensity ratio line/background; 19, spatial and temporal synchronization of plasma plume and spark discharge; 20, physical parameters of the sample as shown in (*b*). Fig. 20a is from Ref. [101] reprinted with permission of Springer-Verlag, Wien.

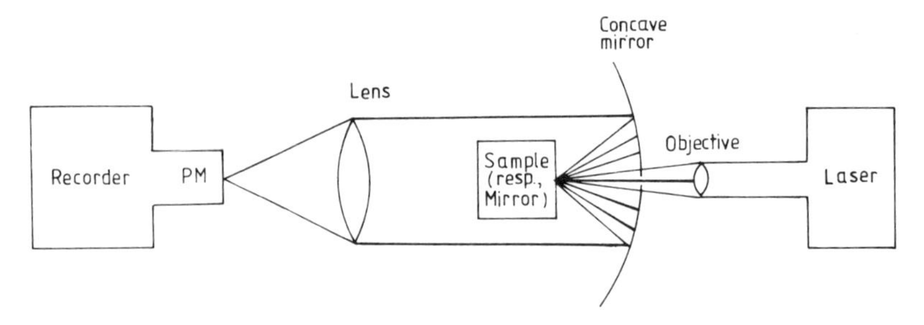


**Figure 21.** Unit for measuring reflection and scattering of laser radiation on sample surface. Lens: $f = 77.5$ mm, $\phi = 43$ mm; concave mirror: $f' = 16.8$ mm, bore: $\phi = 1$ mm.

displaced material. However, in [135], corresponding to [69], it was shown empirically that the process responsible for removal of most of the crater material is ablation of molten metal and that the fusion process occurs as a physical aspect that cannot be ignored.

In [135] the mass of molten and ejected material (steel) was picked up in a small cylindrical glass chamber and determined with the help of a microscope. The results are shown in Table 8, including the consumption of energy for melting $E_{melt}$.

The kinetic energy $E_{kin}$ of the splashes is calculated in Table 9. The measured speeds of the ejected splashes correspond with the results of Chun and Rose [109].

The energy losses by hollow black body radiation $E_{holl\,rad}$ of the laser crater

**TABLE 8. Energy Losses by Ejection of Liquid Material after Interaction of Laser Radiation with the Surface of a Steel Sample**

| Mode of Laser Action | Energy of Incident Laser Radiation (J) | Crater Volume ($\mu m^3$) | Crater Bulb ($\mu m^3$) | Splash Volume ($\mu m^3$) | Melting Energy (J) | Percentage of Laser Energy |
|---|---|---|---|---|---|---|
| Normal mode | 0.26 | $9.3 \times 10^6$ | $7.6 \times 10^6$ | $16.3 \times 10^4$ | 0.11 | 42 |
| Semi-*Q*-switched | 0.25 | $7.0 \times 10^6$ | $6.8 \times 10^6$ | $11.8 \times 10^4$ | 0.081 | 33 |
| with decreasing | 0.22 | $3.0 \times 10^6$ | $2.9 \times 10^6$ | $9.4 \times 10^4$ | 0.034 | 16 |
| number of | 0.18 | $1.5 \times 10^6$ | — | $4.4 \times 10^4$ | 0.003 | 1.6 |
| spikes | 0.10 | $1.0 \times 10^6$ | — | $2.4 \times 10^4$ | 0.003 | 0.3 |
| Single pulse | 0.06 | n.d.[a] | — | — | — | — |

*Source*: Ref. 135.

[a]n.d. = not determined.

**TABLE 9. Energy Losses by Kinetic Energy of the Ejected Splashes after Interaction of Laser Radiation with the Surface of a Steel Sample**

| Mode of Laser Action | Energy of Incident Laser Radiation (J) | Speed of Ejected Splashes ($cm\,s^{-1}$) | Kinetic Energy of the Splashes (J) | Percentage of Laser Energy |
|---|---|---|---|---|
| Normal mode | 0.26 | $0.9 \times 10^4$ | $1.8 \times 10^{-4}$ | 0.07 |
| Semi-$Q$-switch | 0.25 | $1.2 \times 10^4$ | $2.0 \times 10^{-4}$ | 0.08 |
| with decreasing | 0.22 | $1.6 \times 10^4$ | $2.6 \times 10^{-4}$ | 0.12 |
| number of spikes | 0.18 | $1.5 \times 10^4$ | $1.2 \times 10^{-4}$ | 0.07 |
| | 0.10 | $0.9 \times 10^4$ | $0.3 \times 10^{-4}$ | 0.03 |
| Single pulse | 0.06 | — | — | — |

*Source*: Ref. 135.

**TABLE 10. Energy Losses by Hollow Radiation of the Laser Crater**

| Mode of Laser Action | Energy of Incident Laser Radiation (J) | Area of Crater Surface ($cm^2$) | Duration of Laser Radiation ($\mu s$) | Emitted Energy by Hollow Radiation (J) | Percentage of Laser Radiation |
|---|---|---|---|---|---|
| Normal mode | 0.26 | $7 \times 10^{-4}$ | 300 | $2.4 \times 10^{-3}$ | 0.9 |
| Semi-$Q$-switched | 0.25 | $6 \times 10^{-4}$ | 250 | $2.0 \times 10^{-3}$ | 0.8 |
| with decreasing | 0.22 | $5 \times 10^{-4}$ | 100 | $0.6 \times 10^{-3}$ | 0.3 |
| number of | 0.18 | $4 \times 10^{-4}$ | 50 | $0.2 \times 10^{-3}$ | 0.1 |
| spikes | 0.10 | $3 \times 10^{-4}$ | 20 | | |
| Single pulse | 0.06 | $3 \times 10^{-4}$ | 0.1 | | |

*Source*: Ref. 135.

**TABLE 11. Energy Losses of Vapor outside the Plume**

| Mode of Laser Action | Energy of Incident Laser Radiation (J) | Volume of Vapor Condensation ($\mu m^3$) | Vaporization Energy (J) | Percentage of Laser Radiation |
|---|---|---|---|---|
| Normal mode | 0.26 | $1289 \times 10^2$ | $7.83 \times 10^{-3}$ | 3.0 |
| Semi-$Q$-switched | 0.25 | $662 \times 10^2$ | $4.02 \times 10^{-3}$ | 1.6 |
| with decreasing | 0.22 | $17 \times 10^2$ | $0.87 \times 10^{-3}$ | 0.4 |
| number of | 0.18 | $16 \times 10^2$ | $0.79 \times 10^{-3}$ | 0.4 |
| spikes | 0.10 | $14 \times 10^2$ | $0.63 \times 10^{-3}$ | 0.6 |
| Single pulse | 0.06 | | | |

*Source*: Ref. 135.

**TABLE 12. Remainder Energy as a Difference of Incident Laser Energy and Sum of the Energy Losses in and above the Target**

| Mode of Laser Action | Energy of Incident Laser Radiation (J) | Sum of Percent of Energy Losses | Remainder Percentage of Energy | Remainder Energy (J) |
|---|---|---|---|---|
| Normal mode | 0.26 | 94.0 | 6.0 | 0.016 |
| Semi-$Q$-switched | 0.25 | 82.6 | 17.4 | 0.044 |
| with decreasing | 0.22 | 65.0 | 35.0 | 0.078 |
| number of spikes | 0.18 | 50.3 | 49.7 | 0.089 |
| | 0.10 | 49.0 | 51.0 | 0.051 |
| Single pulse | 0.06 | 48.2 | 51.8 | 0.031 |

*Source*: Ref. 135.

are calculated by the Stefan–Boltzmann law, assuming a black radiator:

$$W = \sigma T^4 A e \tag{27}$$

where $\sigma$ = Stefan–Boltzmann constant $5.67 \times 10^{-12}\,\mathrm{W\,cm^{-2}\,K^{-4}}$
$T$ = absolute temperature
$A$ = total radiating area
$e$ = emission factor of radiation: 0.5

Finally, energy losses of vapor outside the plume $E_{vap}$ should be mentioned. Condensation of vapor could be observed with the aid of a glass cylinder ring, 2 mm in diameter and 1 mm in height, covered with a glass ring.

By summing up all the energy losses in and above the target, neglecting the energy losses caused by the plume (because the incident flux density reaches only low values in this case), we obtain the remainder energy $E_{rem}$ (Table 12). The energy losses are lowest in the single-pulse mode. The remainder energy has its maximum value in the semi-$Q$-switched mode.

## 3.5. ADDITIONAL EXCITATION AND IONIZATION OF LASER-BEAM-PRODUCED VAPOR BY SPARK DISCHARGES AND OTHER SOURCES

As already mentioned, generation of excited atoms and ions by exciting laser-beam-produced vapor is necessary for laser microanalysis based on optical emission spectroscopy. Focused radiation of normal-mode and semi-$Q$-switched laser pulses produces more or less excited vapor. Beyond this, resonance lines are often self-reserved, indicating the presence of a central region of excited atoms surrounded by a region of relatively unexcited atoms that absorbs the photons. On the other hand, emission lines from the hottest region of laser plumes are usually relatively broad because of pressure and Doppler broadening.

Cross excitation of a laser plume with another excitation source, such as a high-voltage spark, improves the degrees of excitation and ionization, tends to eliminate self-reversal, and commonly produces narrower lines. The setup for additional spark excitation consists of two pointed carbon electrodes, arranged at a variable distance of 0.5 to 2 mm, and aligned about 1 to 2 mm above the sample surface. A 1- to 25-$\mu$F capicator charged to a voltage ranging from 1 to 6 kV is connected to the electrodes. For a normal-mode laser with long duration, an inductance of 30 to 1000 $\mu$H extends the spark oscillation for several hundreds of microseconds [138–148]. The electrode spark is triggered either by ionized material arriving between the electrodes or by an external cricuit timed to the laser pulse (see Fig. 22). In the first case a capacitive circuit is charged to a potential difference just below that of the air breakdown potential (about 3 kV for a 1-mm gap). The rising vapor plume intercepts the electrode gap and initiates a capacitative discharge releasing about 25 J (for a 2-$\mu$F capacitor) into the vapor. In the second case spark initiation can be externally triggered such that vapor enhancement can be timed to coincide with the highest vapor densities [13, 149]. The change of density of platinum spectrum lines with a change of time interval between a laser pulse and an auxiliary electrode discharge is shown in Fig. 23. This could be obtained by a delay stage, which is connected with the external ignition, thus allowing the moment of ignition to be varied [13, 149].

For a study of the effect of the excitation energy of the auxiliary spark gap on line intensity, the electrode separation, capacity, and inductances were kept constant and the voltage varied between 2 and 7 kV (Fig. 24). External triggered cross excitation allows the electrode separation to be set independently from the driving voltage. The change in density of different spectrum

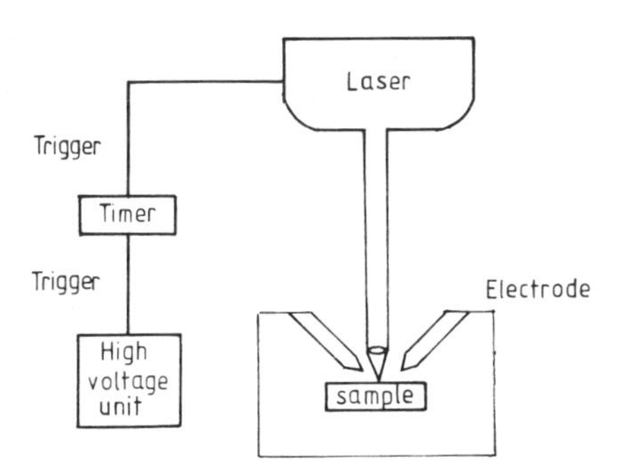


**Figure 22.** Schematic diagram of a laser microprobe system with cross excitation.

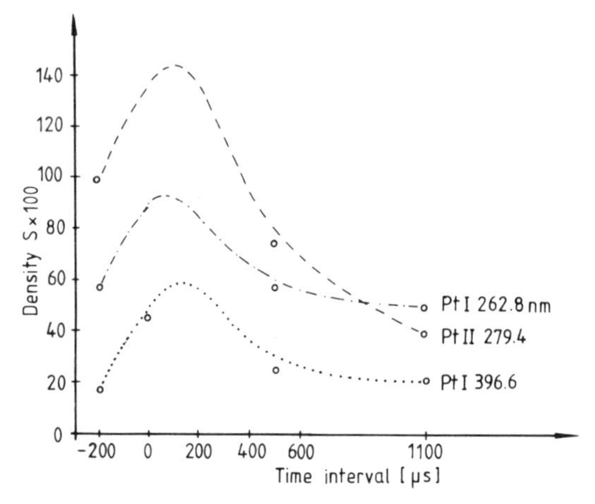


**Figure 23.** Change of density of platinum spectrum lines with change of time interval between laser pulse and auxiliary electrode discharge. From Refs. [13] and [149], reprinted with permission of VEB Deutscher Verlag der Wissenschaften, Berlin.

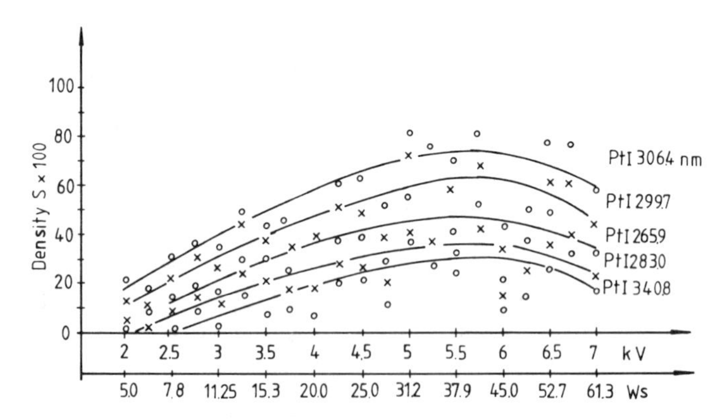


**Figure 24.** Change of density of platinum spectrum lines with change of electrode voltage. Capacity, 2.5 $\mu$F; inductance, 125 $\mu$H. From Refs. [13] and [149], reprinted with permission of VEB Deutscher Verlag der Wissenschaften, Berlin.

lines with electrode separation is illustrated in Fig. 25. In general, the best electrode separation is about 1 mm.

The influence of capacity and inductance on the discharge duration of the spark, the emission duration of the cross-excited microplasma, and the appearance of atom and ion lines are shown in Table 13. Compared with the one-step method of vaporization, excitation, and ionization, where the microplasma is produced by a $Q$-switched laser, the analytical detectability is

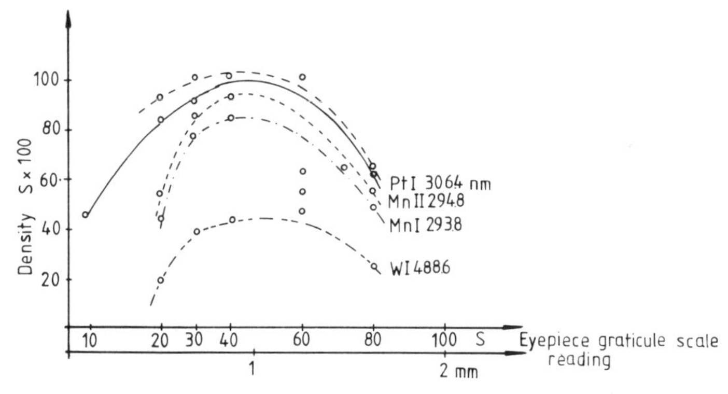


**Figure 25.** Change of spectrum line density with change of electrode gap separation. From Refs. [13] and [149], reprinted with permission of VEB Deutscher Verlag der Wissenschaften, Berlin.

**TABLE 13. Discharge Duration of the Spark and Emission Duration of Cross-Excited Microplasma in Relation to Capacity and Inductance[a]**

| Capacity (μF) | Inductance (μH) | Discharge Duration of Spark (μs) | Emission Duration of Cross-Excited Microplasma (μs): Cd I, 283.6 nm | Cd I, 346.6 nm | Cd II, 274.8 nm |
|---|---|---|---|---|---|
| 1 | 30 | 250 | 200 | | |
| | 125 | 750 | 600 | | |
| | 500 | 1750 | 1000 | | |
| | 1000 | 2500 | 1500 | | |
| 2.5 | 30 | 450 | 300 | 350 | 400 |
| | 125 | 1150 | 950 | 900 | 900 |
| | 500 | 2500 | 2000 | 2500 | 1050 |
| | 1000 | 4000 | 2500 | 3000 | 1400 |

*Source*: Ref. 144.

[a]Normal-mode laser of 1 J, electrode distance 1 mm, electrode height above sample surface 1 mm, sample CdS.

increased by a factor of 10 to 100 using a two-step method [13, 16, 19, 32, 34, 35, 149–161].

In [144], results of an investigation of the expansion velocity of a normal-mode (1 J)-laser-induced plume with cross excitation (5 kV) are given. The velocity of the element Cd decreases from $2 \times 10^3\ \mathrm{m\,s^{-1}}$ to $1.4 \times 10^3\ \mathrm{m\,s^{-1}}$ to $0.8 \times 10^3\ \mathrm{m\,s^{-1}}$, corresponding to the distance between the surface and the electrodes of 1, 2 and 3 mm. The device used was similar to that illustrated in Fig. 26. Comparable results are obtained (see Table 14).

The sequence of events in the two-step process is illustrated in Fig. 27. The xenon flash tube emits incoherent pumping radiation (chain-line curve), causing laser activity which can, in the case of an unswitched resonator

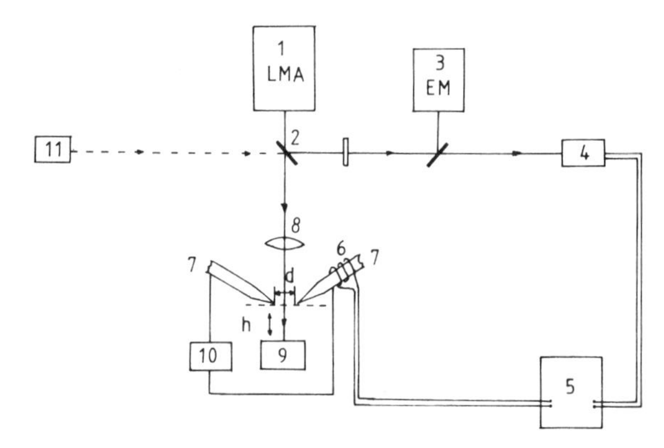


**Figure 26.** Setup for determining the vapor velocity. 1, Laser microanalyzer; 2, deflection plate; 3, energy-measuring device; 4, photocell; 5, oscillograph; 6, coil; 7, electrode; 8, lens; 9, sample; 10, supply unit; 11, tuning laser. From Refs. [160, 161] reprinted with permission of G. Dimitrov.

**TABLE 14. Entry Velocity of a 1-J Normal-Mode-Laser-Induced Microplume in the Pulse Discharge Zone of Two Electrodes (5 kV) in Relation to Height above the Sample Surface**

| | | Velocity ($m\,s^{-1}$) at Height (mm): | | | |
|---|---|---|---|---|---|
| Element | Height | 0.5 | 1.0 | 2.0 | 3.0 |
| C | | 500 | 500 | 230 | 190 |
| Pb | | 500 | 500 | 140 | 80 |
| Sn | | 500 | 360 | 130 | 50 |
| Fe | | 500 | 190 | 110 | 90 |
| Cu | | 110 | 80 | 80 | 50 |
| Zn | | 160 | 70 | 60 | 50 |
| Al | | 210 | 60 | 40 | 30 |
| W | | 80 | 50 | 40 | 30 |
| Ge | | 100 | 40 | 30 | 20 |

*Source*: After Refs. 160–162; reprinted by permission.

discussed, be demonstrated on an oscillograph as a series of spikes given off in an interval of a few hundred microseconds.

The first particles in the ascending vapor cloud arrive in the space between the charged electrodes in about 6 $\mu$s. The peak intensity of the microplasma (dotted curve) is reached in about 100 to 150 $\mu$s. The spark discharge between the microscope objective and the sample surface is released between 5 and 100 $\mu$s after the start of the laser emission.

The electrical parameters of the auxiliary excitation illustrated in Fig. 27 were 2.5 $\mu$F and 125 $\mu$H [145]. The duration of the spark discharge can be

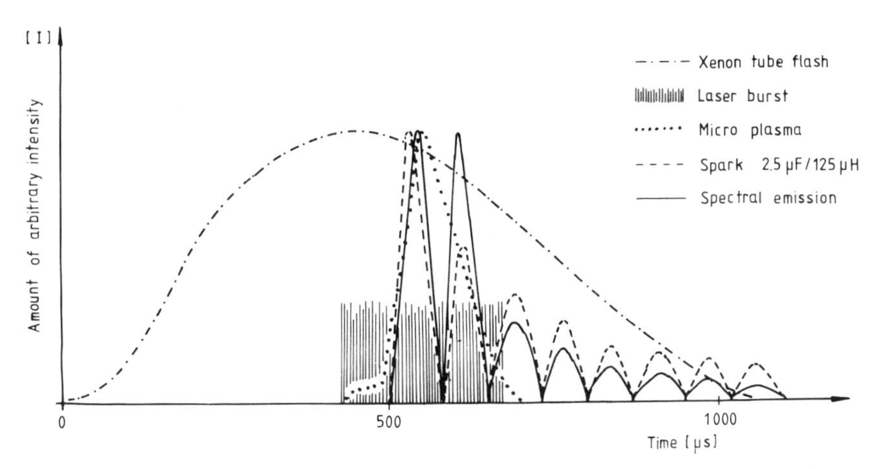


**Figure 27.** Time sequence of events in a two-step working procedure: pumping flash, laser burst, microplasma without cross excitation, spark discharge, and spectral emission of microplasma excited by this discharge. From Ref. [13], reprinted with permission of Akademische Verlagsgesellschaft Geest & Portig K.G., Leipzig.

altered through appropriate choice of the parameters. Spectrum lines can be determined immediately after start of the discharge.

An alternative to the cross discharge discussed above is axial discharge, in which a single electrode is positioned 3 to 4 mm above the target, with the target itself connected to earth [156–158]. For voltages of around 4 kV across a 0.5-$\mu$F capacitor, about 4 $J$ can be discharged into the vapor plume. The discharge so formed occurs almost axially with the vapor plume, with the peak intensity of the plume. Both unipolar and bipolar discharges have been employed, although there is a tendency for the discharge to wander in the bipolar case. Also, a unipolar discharge can remove the possibility of cathode jet effects and hence considerably reduces the variance in recorded intensity ratio. Just as in the two-electrode system, a triggered spark gap allows better control over the plasma, thus giving rise to more reproducible results [149]. A variable external ignition of the auxiliary spark gap permits the maximum electrode voltage (6 kV, for instance) to be utilized at any electrode distance and therefore enables a high-discharge energy, together with the preselectable synchronization between laser pulse and the spark function, to be achieved. The discharge energy $W$ is proportional to the square of the voltage $V$, according to the formula

$$W = \tfrac{1}{2} C V^2 \tag{28}$$

where $C$ is the capacity.

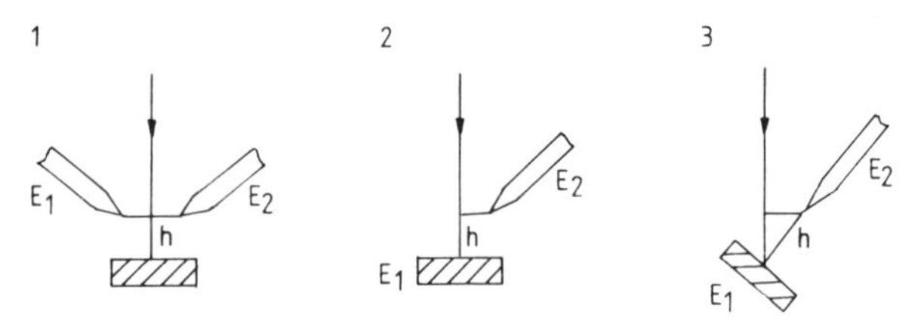


**Figure 28.** Three positions of electrodes and samples corresponding to Table 15. From Refs. [160–162], reprinted by permission of G. Dimitrov.

**TABLE 15. Amplification Factors for Some Spectral Lines for the Positions of Electrodes and Sample Shown in Fig. 28**

| $L = 1000\,\mu$H: $h = 2$ mm: | Cu I 324.75 nm | Pb I 368.35 nm | Si I 288.16 nm | Mg I 277.98 nm | Zn I 334.50 nm | Mn II 259.37 nm |
|---|---|---|---|---|---|---|
| $K_{21}$ | 1.02 | 1.65 | 1.42 | 1.32 | 1.26 | 1.16 |
| $K_{32}$ | 1.23 | 1.17 | 1.40 | 1.28 | 1.35 | 1.15 |
| $K_{31}$ | 1.25 | 1.93 | 1.98 | 1.69 | 1.71 | 1.34 |

*Source*: After Refs, 160–162; reprinted by permission.

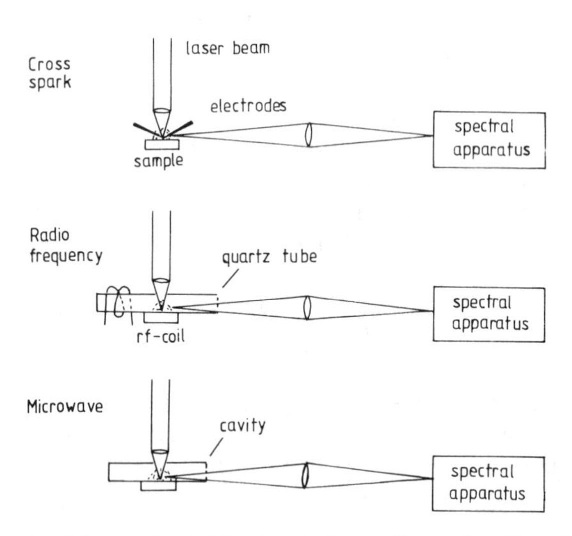


**Figure 29.** Principles of devices for additional excitation and ionization of laser-beam-produced vapor by spark discharge between two electrodes, by high-frequency discharge with radio frequencies, and by microwaves. From Ref. [83], reprinted by permission of Pergamon Journals Ltd.

From the point of view that energy losses by absorption in the laser-induced plume are to be expected [5, 40]. Dimitrov [160–162] tried to shorten the path by inclining the sample and by using only one electrode compared to the one and two electrodes used in a perpendicular orientation (Fig. 28 and Table 15). The removal of one electrode and the use of the sample as an electrode (for electroconductive samples only) leads to greater stability in the pulse discharge channel. For a copper sample the ratios of relative standard deviation ratio of 19.4% (position 1):9.5% (position 2):10.2% (position 3) was obtained. The amplification factors for some spectral lines are noteworthy.

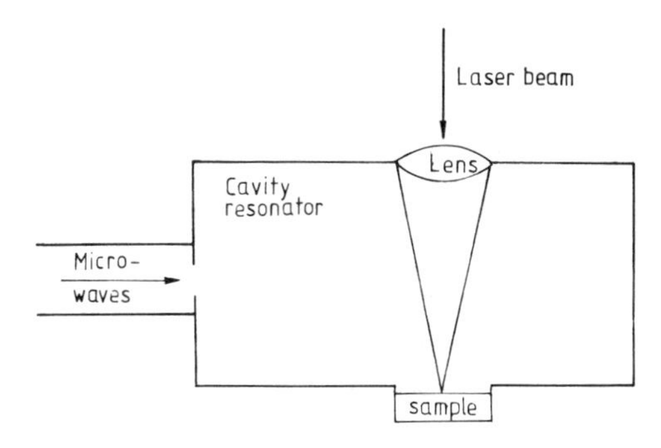


**Figure 30.** Device for additional excitation of laser-induced vapor in a cavity resonator by microwaves. From Ref. [163].

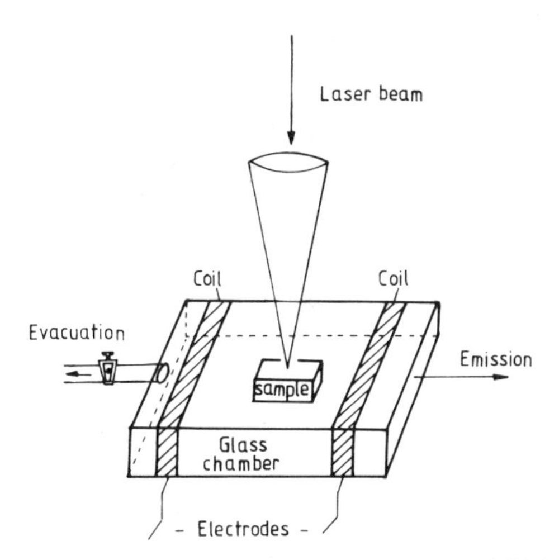


**Figure 31.** Device for additional excitation of laser-induced vapor with high-frequency alternating voltage. From Ref. [166].

A disadvantage of cross excitation with carbon or graphite electrodes is that of contamination by evaporation of electrode material. This can be avoid using electrodeless techniques, as shown in Fig. 29.

Excitation of laser-induced plasma by microwaves was patented in 1966 by Karthe et al. [163] (see Fig. 30). A rectangular resonant cavity ($TE_{102}$) was used by Leis [164, 165] for the excitation of laser-produced vapor in a microwave discharge (2450 MHz) in an air and an argon atmosphere.

Excitation of laser-induced plasma by high-frequency alternating voltage at 10 to 100 MHz was proposed by Wengler and patented in 1970 [166] (see Fig. 31]. Möde [167] used a similar method with a radio-frequency discharge of 80 MHz.

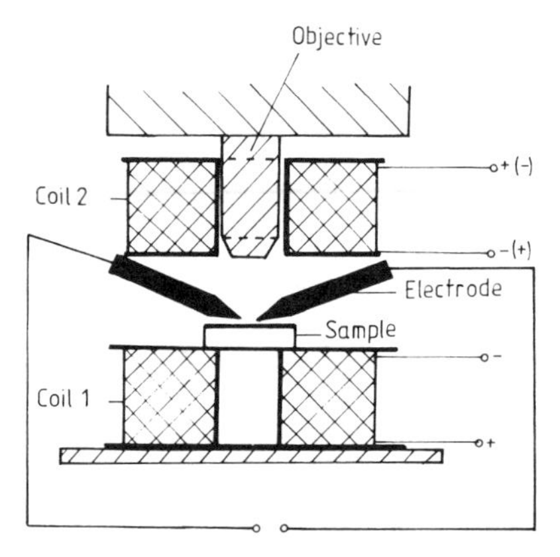


**Figure 32.** Equipment for studying the influences of magnetic field of the discharge between carbon electrodes and the spectral line intensity. From Ref. [171].

**TABLE 16. Amplification Factors of Spectral Line Intensities Using Laser-Induced and Cross-Excited Microplasmas with and without a Magnetic Field in Four Variants**

| Switch of the Coils | Intensity Factors, $I_m/I_0$ Fe I at 345.0 nm | Cu I at 296.1 nm | Ni I at 346.1 nm | Magnetic Field Strength ($A\,m^{-1}$) |
|---|---|---|---|---|
| *Two Coils* | | | | |
| Coordinate | 1.46 | 1.7 | 1.28 | 40,100–66,800 |
| Opposite | 1.57 | 1.57 | 1.35 | 40,100 |
| *One Coil* | | | | |
| Coil 1 | 1.0 | 1.59 | 1.0 | 13,370 |
| Coil 2 | 1.0 | 1.0 | 1.0 | 0 |

*Source*: Ref. 171.

Petrakiev et al. [168–170] demonstrated the influence of an inhomogeneous magnetic field on the intensity of spectral lines and the line/background ratio.

Moenke-Blankenburg et al. [171] built equipment (Fig. 32) for studying the influences of magnetic fields on the discharge between carbon electrodes in four variants. Table 16 represents the intensity factors $I_m/I_0$ ($I_m$ is the intensity of spectral line obtained with a magnetic field, $I_0$ is the intensity obtained without a magnetic field).

## 3.6. VAPORIZATION AND EXCITATION UNDER VARIABLE PRESSURE AND IN VARIOUS GAS MEDIA

The first studies of laser evaporation—variation of crater diameter and amount of vaporized material in relation to gas pressure—were carried out by Piepmeier and Osten [172, 68]. Some years later Manaba and Piepmeier [173] continued the work with another laser system. Also in 1971, Treytl et al. [174] began their investigation of various gas media, including helium, nitrogen, oxygen, and argon. The highest intensities were obtained in an argon atmosphere at normal pressure.

Petrakiev, Dimitrov, and co-workers [175–180, 160] worked in this field from 1972 until 1983. Dimitrov published an extensive investigation [160 concerning the influence of various gas media, such as Ar, $O_2$, $N_2$, and air, in laser microanalysis. Following is a summary of this work.

1. Investigations to determine the influence of the protective gas medium on the intensities of the spectral lines, the spatial and time distribution of the laser microplasma glow, as well as on the quantity of eroded products, the plasma flame volume, and the discharge plasma cloud during laser microspectral analysis were carried out for the first time.

2. Considerable amplification of the intensities was obtained on ion as well as an atomic spectral lines for evaporation and substance excitation in a protective gas medium with a higher ionization potential than the effective air ionization potential. It was determined that the gas medium influences both the emission characteristics and the physical and chemical processes taking place in the plasma flame and the sector affected by the laser radiation.

3. The influence of the protective gas medium on the spatial and temporal distribution of the spectral line glow and temperature changes for different operating conditions was investigated.

4. It was determined that the laser-induced plasma flame and discharge plasma cloud increase in volume for gases with lower heat conductivity,

depending on the heat conductivity of the discharge gas used. Greater capacitances and smaller inductances in the discharge circuit also lead to an increase in volume. Depending on the degree of expanded volume of the discharge cloud and the height of the electrodes above the sample, it is possible to apply additional heating to the sample and to increase the quantity of substance evaporated.

5. For the first time experiments were carried out for decreasing absorption in the plasma flame by inclining the sample, which leads to improvements in reproducibility, concentration sensitivity, and detection limits for laser microspectral analysis.

6. It was determined that with the increase in the difference between the breakdown and the operating voltage there appears for a fixed value of the latter, desynchronization between the pulse discharge and the laser generation (i.e., the discharge is delayed with respect to the laser generation). This leads to an increase in the intensities of the spectral lines for delays that is no larger than the time required to reach maximum values of laser generation.

7. A more complete orientation and direction of the erosion products toward the discharge zone was achieved by means of a blended evaporation of the substance, whereby the small praticles and drops of molten metal are evaporated further and excited, and as a result more intensive spectra are obtained.

8. There was developed and applied an original method for the determination of the average rate of diffusion of laser-induced microplasma, where the "ignition" delay of the pulse discharge with respect to the start of the laser generation is measured depending on the height of the electrodes above the sample.

Systematic investigations were undertaken after construction of vacuum chambers by Moenke-Blankenburg and Moenke [105, 181, 182], Mehl et al. [183], and Mohr [184]. The greatest advantage of laser microanalysis is that the bulk of the chemical elements in solid specimens of all types can be determined without a vacuum (i.e., in air). Nevertheless, the first considerations concerning the construction of vacuum chambers are the following:

- To extend the spectral range into the far-UV field, particularly to perform C, P, and S analysis in metallic and mineral specimens, since, as is well known, their analytical lines (and also those of other elements, such as Cl, Br, I, O, H, and N) lie in the spectral region $< 200$ nm
- To lower the limits of detection in the entire spectral range by using gas purgings during the evaporation and excitation process.

The vacuum chambers developed by the above-mentioned authors were developed for the study of the mode of action of subatmospheric, normal, and superatmospheric pressure on the propagation and radiation conditions of microplasma in a range of pressure greater than six orders of magnitude. The technical details are given in Table 17 and the results in Table 18.

Using a pressure range of $10^3$ to $10^6$ Pa in an argon atmosphere has the following advantages:

- No cyanogen bands and a decrease in the intensity of the interfering spectrum
- Higher intensity of the spectral lines with and without auxiliary spark discharge (Fig. 33)
- Choice of the degree of excitation and ionization (Fig. 34)
- Possibility of optimizing the analysis of trace components or the main components of the specimen by varying the pressure in the sample chamber
- Elimination of self-reversal and self-absorption, and steeper slope of the calibration curves in quantitative analysis (Fig. 35)
- Laser emission microanalysis in the far-UV spectral region in combination with a vacuum spectrograph (Fig. 36)

**TABLE 17. Technical Details of a Sample Chamber**

*Gas pressure* $2.5 \times 10^3$ to $2 \times 10^5$ Pa

*Maximum size of the sample (door opening)* 64 mm (height); 160 mm (width)

*Depth of the sample space* 120 mm

*Maximum mass of the sample* 1 kg

*Movement of the sample* Horizontal: $\pm$ 10 mm in the $x$ and $y$ directions. Vertical: 10 mm continuously variable. By means of supports of different stage thicknesses, it is possible to compensate for another 56 mm of difference in sample heights.

*Electrodes* Spectral carbon, diameter 5 mm, length 40 mm, unilaterally tapered. Electrode distance can be decreased smoothly from 6 mm down.

*Ignition voltage of the auxiliary spark* 0.8 to 6 kV, selectable as a function of the distance between the electrode tips, and of the type and pressure of the gas in the sample chamber.

*Gases capable of use* Noble gases, such as argon, and molecular gases, such as carbon dioxide, and mixtures of the two types of gases.

*Objective* Catoptric objective $\times$ 40/0.5. Range of movement, $\pm$ 5 mm.

*Adaptation to spectrograph* Quartz condenser, $f = 80$ mm, diameter 40 mm

*Adaptation to a vacuum spectrograph* LiF condenser

*Source*: Ref. 184; reprinted with permission of VEB Deutscher Verlag der Wissenschaften, Berlin.

**TABLE 18. Laser Microemission Spectral Analyses under Variable Pressure**

| Pressure Range | Effects in Laser Microplume | Effects in Laser Microplume with Auxiliary Spark Discharge | Analytical Results |
|---|---|---|---|
| *Subatmospheric* | | | |
| $p < 4\text{Pa}$ | Too high propagation velocity of the cloud of specimen vapour | Flashover voltage too high; oscillating discharge becomes detached | Spectra too weak |
| $4\text{Pa} < p < 3 \times 10^3$ Pa | | Glow discharge region: too low flashover voltage | Unsuitable for analytical purposes |
| *Normal* | | | |
| $3 \times 10^3$ Pa $< p < 10^5$ Pa | Propagation and radiation conditions capable of analytical evaluation | Flashover voltage and course of the discharge suitable for analytical purposes | Avoidance of self-absorption and self-reversal; 45° steeper calibration curves; advantageous analysis of main components |
| *Superatmospheric* | | | |
| $10^5$ Pa $< p < 10^6$ Pa | | | Advantageous analysis of trace components |
| $p > 10^6$ Pa | Absorption of the laser radiation in the microplasma; less removal of material | | Spectra too weak; high spectral background |

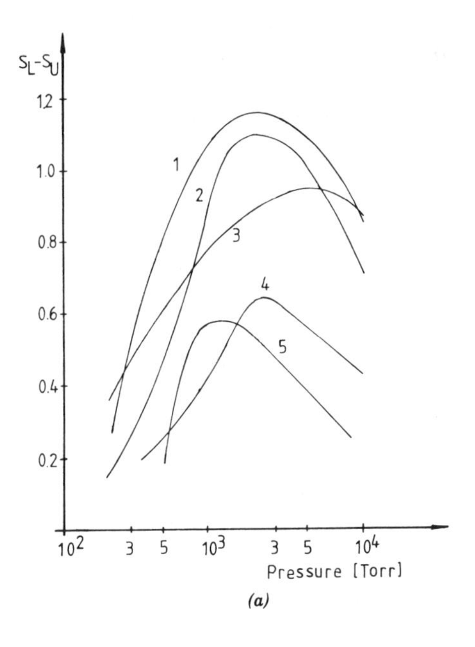


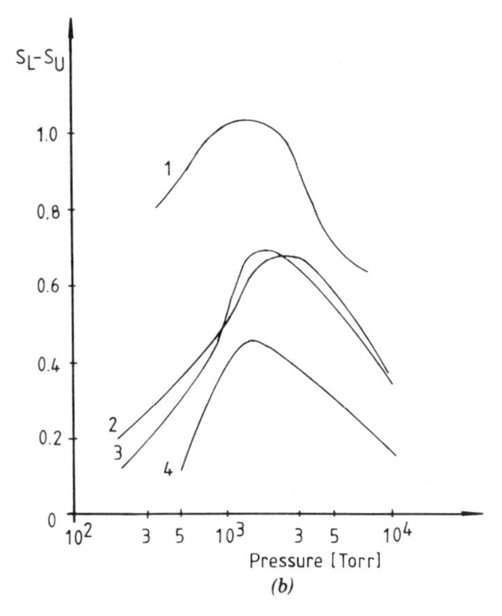


**Figure 33.** Dependence of spectral line intensity on pressure of argon atmosphere. (*a*) Laser-induced plume with cross excitation: 1, Mg 279.6 nm; 2, Ca 393.4 nm; 3, Pt 306.5 nm; 4, Si 250.6 nm; 5, Mn 257.6 nm. (*b*) Laser-induced plume without cross excitation: 1, Si 288.2 nm; 2, Pt 306.5 nm; 3, Ca 393.4 nm; 4, Mg 279.6 nm. From Ref. [184], reprinted with permission of VEB Deutscher Verlag der Wissenschaften, Berlin.

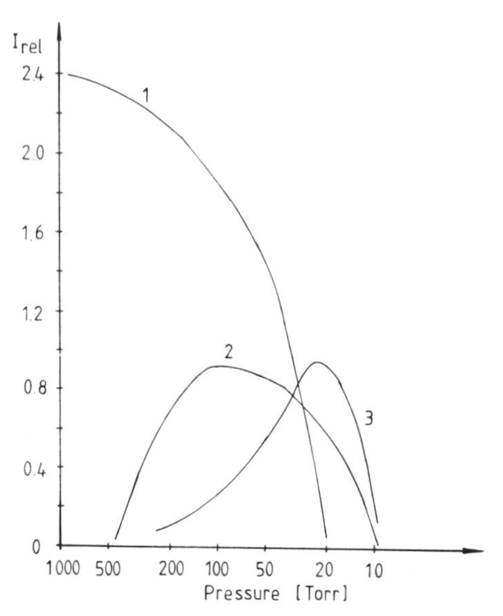


**Figure 34.** Intensity of Pb(I) = 283.3 nm (1), Pb(III) = 304.4 nm (2), and Pb(IV) = 286.5 nm (3) versus pressure of argon atmosphere. From Ref. [184], reprinted with permission of VEB Deutscher Verlag der Wissenschaften, Berlin.

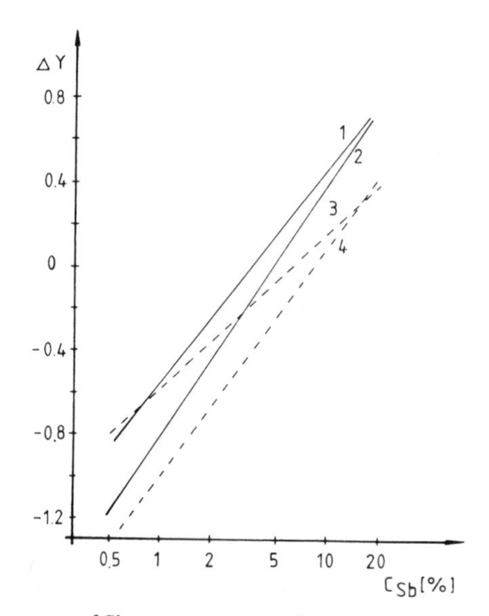


**Figure 35.** Calibration curve of Sb versus pressure of argon atmosphere. 1, 760 torr; 2, 200 torr (with cross excitation); 3, 760 torr; 4, 200 torr (without cross excitation). From Ref. [184], reprinted with permission of VEB Deutscher Verlag der Wissenschaften, Berlin.

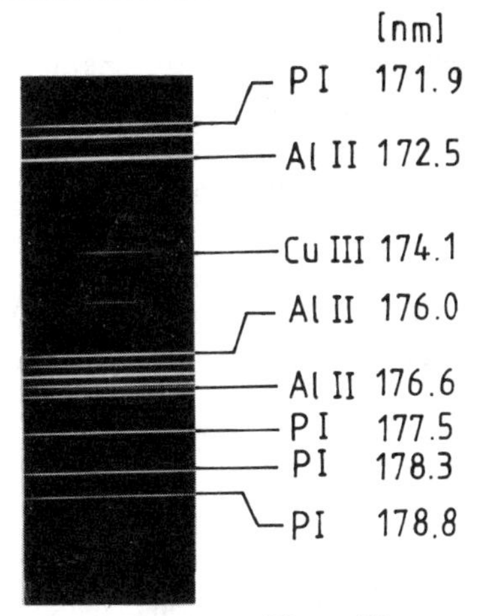


**Figure 36.** Spectrum of turquoise in the range 170 to 180 nm recorded with a vacuum spectrograph after laser-induced vaporization and excitation. From Ref. [184], reprinted with permission of VEB Deutscher Verlag der Wissenschaften, Berlin.

Laser microanalysis under reactor conditions (40 atm He) have been investigated by Stupp [185, 186]. More than 50% of the normal-mode laser energy is absorbed in the plasma. Here the inverse absorption of bremsstrahlung pressure, the dominating excitation mechanism, pressure effects, and plasma inhomogeneities are expressed by self-absorption, inversion, asymmetric and symmetric line broadening, and shifting.

Karjakin et al. [187] showed that the absorption values of manganese (in a target of pelleted graphite powder) varied strongly with the pressure of hydrogen and oxygen in the atmosphere. At different heights of observation, the absorbance increased with increased hydrogen pressure. In the presence of oxygen, the Mn absorbance fell off with increased $O_2$ pressure, but only near the target surface. At a distance of 15 mm in front of the target surface, they observed an inverted behavior of oxygen. They deduced that this was due to the reaction of the carbon matrix and oxygen, forming CO and $CO_2$ in this cooler sphere of the plume.

Kagawa and Yokoi [188] investigated the effect of air pressure on the intensity of the plasma plume when a $N_2$ laser was focused onto metallic targets. They found that both the diameter of the plasma plume and the characteristics of the emission spectra vary with gas pressure. At 1 atm they noted an intense continuous spectrum and small intensities in atomic emission. At 1 torr the background signal decreased, whereas the atomic emission increased.

Dörge et al. [40] concluded his results in the following way: The self-absorption appearing in the laser plasma can be avoided using an $O_2$

atmosphere, which leads to an increase in line intensities compared to the number in an air atmosphere. This is caused by an exothermal $C/O_2$ reaction which prevents a high temperature gradient. In $N_2$ and Ar atmospheres the self-absorption is amplified. The temperatures range from 8000 K (in air) to 11,000 K (in $O_2$). In the laser-spark plasma the flow of the laser plasma is preserved. Compared to air there are amplifications, particularly in an $Ar/O_2$ atmosphere of volume ratio 4:1; in an He atmosphere the intensities decrease. The amplification factors depend on the excitation and ionization energies of the lines. The excitation temperatures range from 20,000 to 30,000 K.

Atomic fluorescence observations of Li and Cu in a laser microprobe plume were made by Lewis et al. [189, 190] in order to study the chemical and physical influences of a reactive atmosphere of oxygen relative to an atmosphere of argon.

The vapor plume produced by the action of a focused laser beam on a solid sample of an aluminum alloy or zirconium was studied in a controlled atmosphere. The formation of binary compounds by chemical reaction between the vapor plume and the atmosphere was followed by the use of emission, absorption, and fluorescence detection methods. Most experiments were performed at 150 torr in 100% $O_2$, 50% $O_2$/50% Ar, and 100% Ar [191].

## 3.7. MATRIX EFFECTS

Before discussing matrix effects, a distinction must be made between physical and chemical matrix effects. The term "physical matrix effects" refers to the influence exerted on the removal process by the nature of the sample in its widest physical sense. Chemical matrix effects originate in the dependence on the chemical composition of the matrix of the excitation conditions in the plasma.

Physical matrix effects are dealt with in Sections 3.3 and 3.4. Marich et al. [192] found that the matrix effects are principally physical rather than chemical in nature. Cerrai and Trucco [193] established a relationship between spectral line intensities of macrocomponents of different materials and their recrystallization conditions, grain size, and mechanical strain present in the sample. Kirchheim et al. [58] used high-purity single-crystal metals and alloys with different orientations; the latter were found to be an important factor in density and reproducibility of characteristic lines.

Referring to chemical matrix effects, differing statements are given in the literature. Panteleev et al. [194] observed selective volatilization but less marked than those obtained in arc analysis. Margoshes et al. [195] and Whitehead and Heady [142] reported that laser microanalyses are relatively free of matrix effects. However, Scott and Strasheim [111], Ishizuka [57],

Karyakin and Kaigorodov [196], and Buravlev et al. [197] do mention the occurrence of signal suppression by matrix components. Dittrich and Wennrich [198] and Smith et al. [199] described plasma reactions inside the laser plume (e.g., dissociation of clusters, molecules, and radicals), as well as the formation of stable molecules or radicals. Formation of carbides was discussed by Karyakin et al. [200]. Wennrich et al. [201] found a decrease in absorption values of some elements in LM-AAS in the presence of small amounts of halides. The degree of atomization of Co, Mn, Mo, Ti, U, V, and W, which form stable carbides, and additional elements which form stable oxides was calculated and tested by Wulfson et al. [202]. In particular the effect of the stable oxides of Mg and Fe were tested by Quentmeier et al. [203] with laser micro atomic absorption spectroscopy. Marich et al. [192] described the matrix effect of sodium acetate, sodium sulfate, and some organic compounds.

With LM-MS the presence of molecules was shown in studies of Dietze et al. [204] and Heinen [205], where mass spectra of selected samples are presented and discussed which demonstrate the characteristic influence of the chemical nature of the compounds on the types of ions occurring in the spectra. In papers using laser micro mass spectrometry, the mode of action of sulfates, oxides silicates, and sulfides has been reported [206–213].

## REFERENCES

1. J. F. Ready, *Appl. Phys. Lett.*, **3**, 11 (1963); *J. Appl. Phys.*, **36**, 462 (1965); *Effects of High-Power Laser Radiation*, Academic Press, Inc., New York, 1971; *Industrial Applications of Lasers*, Academic Press, Inc., New York, 1978.
2. H. R. Griem, *Plasma Spectroscopy*, McGraw-Hill Book Company, New York, 1964.
3. Y. V. Afanasyev and O. N. Krokhin, *Sov. Phys.-JETP*, **25**, 639 (1967).
4. S. Anisimov, *Sov. Phys.-JETP*, **27**, 182 (1968).
5. W. Bögershausen and R. Vesper, *Spectrochim. Acta*, **24B**, 103 (1969).
6. T. J. Bastov, *Nature* (*London*), **222**, 5192 (1969).
7. A. Maitland and M. H. Dunn, *Laser Physics*, North-Holland Publishing Company, Amsterdam, 1969.
8. C. D. Michelis, *IEEE J. Quantum Electron., QE-6630-41* (1970).
9. J. Cooper, *Rep. Prog. Phys.*, **29**, 35 (1971).
10. O. N. Krokhin, in *Laser Handbook*, Vol. 2, F. T. Arecchi and E. O. Schulz-duBois, Eds., North-Holland Publishing Company, Amsterdam, 1972, p. 1371.
11. V. A. Batanov, F. V. Bunkin, A. M. Prochorov, and V. B. Fedorov, *Zh. Eksp. Teor. Fiz.*, **63**, 586 (1972).
12. E. C. Beahm, An investigation of the laser source mass spectrometer, Thesis, Pennsylvania State University, University Park, Pa., 1973.

13. H. Moenke and L. Moenke-Blankenburg, *Laser Microspectrochemical Analysis*, Adam Hilger Ltd., London, England, 1973, and Crane, Russak & Co., Inc., New York, 1973.
14. C. Bar Isaac and M. Korn, *Appl. Phys.*, **3**, 45 (1974).
15. T. P. Hughes, *Plasmas and Laser Light*, Adam Hilger Ltd., Bristol, England, and John Wiley & Sons, Inc., New York, 1975.
16. R. H. Scott and A. Strasheim, in *Applied Atomic Spectroscopy*, Vol. 1, E. L. Grove, Ed., Plenum Press, New York, 1978, pp. 73–118.
17. M. L. Petukh and A. A. Yankovskii, *Zh. Prikl. Spektrosk.*, **29**, 1109, (1978); *J. Appl. Spectrosc.* (*USSR*), **29**, 1527 (1978).
18. C. A. Sacchi and O. Svelto, in *Analytical Laser Spectroscopy*, N. Omenetto, Ed., Vol. 50 in *Chemical Analysis*, P. J. Elving, J. D. Winefordner, and I. M. Kolthoff, Eds., John Wiley & Sons, Inc., New York, 1979, pp. 1–46.
19. K. Laqua, in *Analytical Laser Spectroscopy*, N. Omenetto, Ed., Vol. 50 in *Chemical Analysis*, P. J. Elving, J. D. Winefordner, and I. M. Kolthoff, Eds., John Wiley & Sons, Inc., New York, 1979, pp. 47–118.
20. N. Bloembergen, in *Laser–Solid Interactions*, Cambridge University Press, Cambridge, 1979, pp. 1–10.
21. D. H. Auston, J. A. Golovchenko, A. L. Simons, R. E. Slusher, P. R. Smith, C. M. Surko, and T. N. C. Venkatesan, in *Laser–Solid Interactions*, Cambridge University Press, Cambridge, 1979, pp. 11–26.
22. N. Fürstenau, F. Hillenkamp, and P. Nitsche, *Int. J. Mass Spectrom. Ion Phys.*, **31**, 85 (1979).
23. Y. P. Raizer, *Usp. Fiz. Nauk*, **132**, 549 (1980); *Sov. Phys. Usp.*, **23**, 789 (1980).
24. E. J. Yoffa, *Phys. Rev.*, **B21**, 2115 (1980).
25. T. Floren, Diploma thesis, Frankfurt, FRG, 1980.
26. N. Fürstenau, Doctoral thesis, Frankfurt, 1981.
27. N. Fürstenau and F. Hillenkamp, *Int. J. Mass Spectrom. Ion Phys.*, **37**, 135 (1981).
28. H. Hora, *Physics of Laser Driven Plasmas*, John Wiley & Sons, Inc., New York, 1981.
29. R. F. Wood and G. E. Giles, *Phys. Rev.*, **23B**, 2923 (1981).
30. W. Demtröder, *Laser Spectroscopy*, Springer-Verlag, Berlin, 1982.
31. J. F. Young, J. E. Sipe, J. S. Preston, and H. M. Van Driel, *Appl. Phys. Lett.*, **41**, 261 (1982).
32. R. S. Adrain and J. Watson, *J. Phys. D Appl. Phys.*, **17**, 1915 (1984).
33. J. F. Eloy, *Scanning Electron Microsc.*, **11**, 563 (1985); and *Les Lasers de Puissance*, Masson Editeur, Paris, 1985.
34. E. H. Piepmeier, Eds., *Analytical Applications of Lasers*, Vol. 87 in *Chemical Analysis*, P. J. Elving, J. D. Winefordner, and I. M. Kolthoff, Eds., John Wiley & Sons, Inc., New York, 1986.
35. D. A. Cremers and L. J. Radziemski, in *Laser Spectroscopy and Its Applications*,

L. I. Radziemski, R. W. Solarz, and J. A. Paisner, Eds., Marcel Dekker Inc., New York, 1987.

36. E. H. Piepmeier and H. V. Malmstadt, *Anal. Chem.*, **41**, 700 (1969).
37. R. S. Adrain, J. Watson, P. H. Richards, and A. Maitland, *Opt. Laser Technol.*, **12**, 137 (1980).
38. G. Palitzsch and A. Visser, *VDI Z.*, **111** (25), 1111 (1968).
39. F. A. Peuser, M. Mazurkiewicz, and H. Nickel, *Bericht der Kernforschungsanlage*, Jülich, FRG, Jül-1439, 1977.
40. W. Dörge, M. Mazurkiewicz, and H. Nickel, *Bericht der Kernforschungsanlage*, Jülich, FRG, Jül-1881, 1983.
41. Y. V. Afanasyev, O. N. Krokhin, and G. V. Sklizkov, *IEEE J. Quantum Electron.*, *QE-2*, 483 (1966).
42. W. Leising, Laser prepulse vaporization of aluminium wire targets, Ph.D. thesis, University of Rochester, Rochester, N.Y., 1973; University microfilm, Ann Arbor, Michigen, No. 73–14, pp. 1–168.
43. C. A. Andersen and J. R. Hinthorne, *Anal. Chem.*, **45**, 1421 (1973).
44. D. S. Simons, J. E. Baker, and C. A. Evans, Jr., *Anal. Chem.*, **48**, 1341 (1976).
45. D. E. Newbury, *Proc. 2nd International Conf. Secondary Ion Mass Spectrometry*, Standford, Calif., A. Benninghoven et al., Eds., Springer-Verlag, Berlin, 1979, p. 53.
46. J. F. Eloy, *Proc. 5th International Symp. High Purity Material in Science and Technology*, Dresden, GDR, Vol. II, 1980, p. 96.
47. E. Deloule and J. F. Eloy, *SEAPC Technical Report 81*, SGN 001 MGA, Commissariat à l'Energie Atomique, Grenoble, France, 1981, pp. 1–52.
48. J. Suba and A. Stopka, *Proc. 3rd International Conf. SIMS*, A. Benninghoven et al., Eds., Springer-Verlag, Berlin, 1982, p. 164.
49. L. I. Grechikin and L. Y. Minko, *Zh. Teckh. Fiz.*, **37**, 1169 (1967); English transl.: *Sov. Phys.-Tech. Phys.*, **12**, 846 (1967).
50. J. F. Eloy, *C.E.A. Rep 4777*, Commissariat à l'Energie Atomique, Grenoble, France, 1976.
51. H. Nickel, F. A. Peuser, and M. Mazurkiewicz, *Spectrochim. Acta*, **33B**, 675 (1978); H. Nickel, F. A. Peuser, M. Mazurkiewicz, and W. Dörge, *Jena Rev.*, **24**, 199 (1979).
52. V. A. Ageev, S. S. Pryakhin, and Y. V. Khlopkov, *Zh. Prikl. Spektrosk.*, **37**, 727 (1982); see also V. A. Ageev, A. V. Kolesnik, and A. A. Jankovsky, *Zh. Prikl. Spektrosk.*, **26**, 360, 417 (1977).
53. R. Ishida and M. Kubota, *J. Spectrosc. Soc. Jpn.*, **21**, 16 (1972).
54. H. Kawaguchi, J. Xu, and T. Tanaka, *Bunseki Kagaku*, **31**, E185 (1982).
55. C. D. Allemand, *Spectrochim. Acta*, **27B**, 185 (1972).
56. M. Margoshes, in *Microprobe Analysis*, C. A. Anderson, Ed., John Wiley & Sons, Inc., New York, 1972.
57. T. Ishizuka, *Anal. Chem.*, **45**, 538 (1973).

58. R. Kirchheim, U. Nagorny, K. Maier, and G. Tölg, *Anal. Chem.*, **48**, 1505 (1976).
59. A. Y. Vorobyev and V. M. Kuzmichov, *Kvantovaya Elektron.* (*Moscow*), **7**, 183 (1980).
60. C. Y. Ho, R. W. Powell, and P. W. Liley, *J. Phys. Chem.*, Ref. Data I, 279 (1972).
61. A. M. Prochorov, V. A. Batanov, F. V. Bunkin, and V. B. Fedorov, *IEEE J. Quantum Electron.*, **QE-9**, 503 (1973).
62. V. P. Veiko, E. A. Krutenkova, and B. M. Yurkewich, *Fiz. Khim. Obrab. Mater.*, **21**, 4 (1982).
63. E. Kocher, L. Tschudi, J. Steffen, and G. Herziger, *IEEE J. Quantum Electron.*, **QE-8**, 120 (1972).
64. U. C. Paek and F. P. Gagliano, *IEEE J. Quantum Electron.*, **QE-8**, 112 (1972).
65. P. I. Uljanov, *Fiz. Khim. Obrab. Mater.*, **20**, 19 (1981).
66. A. K. Jain, V. N. Kulkarni, and D. K. Sood, *Thin Solid Films*, **86**, 1 (1981).
67. A. K. Jain, V. N. Kulkarni, D. K. Sood, M. Sundararaman, and R. D. S. Yadar, *Nucl. Instrum. Methods*, **168**, 275 (1980).
68. D. E. Osten and E. H. Piepmeier, *Appl. Spectrosc.*, **27**, 165 (1973).
69. J. M. Baldwin, *Appl. Spectrosc.*, **24**, 429 (1970); **25**, 642 (1971); *J. Appl. Phys.*, **44**, 3362 (1973).
70. S. D. Rasberry, B. F. Scribner, and M. Margoshes, *Appl. Opt.*, **6**, 81 (1967).
71. J. Desserre and J. F. Eloy, *C.E.A. Rep. 204*, Commissariat à l'Energie Atomique, Grenoble, France, 1975, p. 71.
72. M. Moldovan, D. Barbulescu, M. Dinescu, I. Apostol, V. Draganescu, I. N. Michailescu, and I. Morjan, *Rev. Roum. Chim. Phys.*, **26**, 1075 (1981).
73. A. K. Jain, V. N. Kulkarni, and D. K. Sood, *Appl. Phys.*, **25**, 127 (1981).
74. J. Magill and R. W. Ohse, *J. Nucl. Mater.*, **71**, 191 (1977).
75. V. G. Braginskii, I. I. Minakova, and V. N. Rudenko, *Zh. Tekh. Fiz.*, **37**, 753 (1967).
76. P. D. Zavidsanos, *GE Rep. R 67SD11* (1967).
77. S. I. Anisimov et al., *Zh. Tekh. Fiz.*, **36**, 1273 (1966); English transl.: *Sov. Phys.-Tech. Phys.*, **11**, 945 (1967).
78. V. P. Veiko et al., *Zh. Tekh. Fiz.*, **37**, 1920 (1967); English transl.: *Sov. Phys.-Tech. Phys.*, **12**, 1410 (1968).
79. F. P. Gagliano, R. M. Lumley, and L. S. Watkins, *Proc. IEEE*, **57**, 114 (1969).
80. A. I. Akimov and L. I. Mirkin, *Dokl. Akad. Nauk SSSR*, **183**, 562 (1968); English transl.: *Sov. Phys.-Dokl.*, **13**, 1162 (1969).
81. H. G. Landau, *Q. J. Appl. Math.*, **8**, 81 (1950).
82. L. Moenke-Blankenburg, in *Advances in Optical and Electron Microscopy*, Vol. 9, R. Barer and V. K. Cosslett, Eds., Academic Press, Inc., London, 1984, pp. 243–322.
83. L. Moenke-Blankenburg, *Prog. Anal. Spectrosc.*, **9**, 335 (1986).
84. H. Klocke, *Spectrochim. Acta*, **24B**, 263 (1969).

85. M. Newstein and N. Solimene, *IEEE J. Quantum Electron.*, **QE-17**, 2085 (1981).
86. M. Bertolotti and C. Sibilia, *IEEE J. Quantum Electron.*, **QE-17**, 1980 (1981).
87. F. W. Dabby and U. C. Peak, *IEEE J. Quantum Electron.*, **QE-8**, 106 (1972).
88. F. Leis and K. Laqua, *Spectrochim. Acta*, **27B**, 27 (1972).
89. E. K. Wulfson, W. I. Dvorkin, and A. V. Karjakin, *Zh. Prikl. Spektrosk.*, **29**, 781 (1978); **32**, 414 (1980); see also *Spectrochim. Acta*, **35B**, 11 (1980).
90. Y. V. Afanasjev and O. N. Krokhin, *J. Exp. Theor. Phys.*, **52**, 966 (1967); English transl.: *Sov. Phys.-JETP*, **25**, 639 (1967).
91. S. I. Anisimov, *J. Exp. Theor. Phys.*, **54**, 339 (1968); English transl.: *Sov. Phys.-JETP*, **27**, 182 (1968).
92. S. I. Anisimov, *Teplofiz. Vys. Temp.*, **6**, 116 (1968); English transl.: *High Temp.*, **6**, 110 (1968).
93. P. I. Uljakov, *J. Exp. Theor. Phys.*, **52**, 820 (1967); English transl.: *Sov. Phys.-JETP*, **25**, 537 (1967).
94. J. L. Dumas, Étude de la Photoionisation des Cibles Métalliques en Vue d'Application à la Spectrométrie de Masse, Thesis, University of Grenoble, Grenoble, France, 1970.
95. Y. A. Bykovskii, N. N. Degtyarenko, V. F. Elesin, Y. P. Kozyrev, and S. M. Silnov, *Zh. Eksp. Teor. Fiz.*, **60**, 1306 (1971); English transl.: *Sov. Phys.-JETP*, **33**, 706 (1971).
96. Y. A. Gykovskii, G. I. Zhurarlev, V. I. Belousev, V. M. Gladskoi, V. G. Degtyarev, Y. N. Kolosov, and V. N. Nevolin, *Fiz. Plazmy*, **4**, 323 (1978); *Sov. J. Plasma Phys.*, **4**, 180 (1978).
97. J. F. Eloy, *Les Lasers de Puissance*, Masson Éditeur, Paris, 1985.
98. A. I. Busygin, *Pis'ma Zh. Tekh. Fiz.*, **3**, 1116 (1977); English transl.: *Sov. Tech. Phys. Lett.*, **3**, 459 (1977).
99. A. I. Busygin, *Pis'ma Zh. Tekh. Fiz.*, **3**, 1137 (1977); English transl.: *Sov. Tech. Phys. Lett.*, **3**, 468 (1977).
100. I. D. Kovalev, G. A. Maksimov, A. I. Suchkov, and N. V. Larin, *Int. J. Mass Spectrom. Ion Phys.*, **27**, 101 (1978).
101. H. Moenke, L. Moenke-Blankenburg, J. Mohr, and W. Quillfeldt, *Mikrochim. Acta*, **6**, 1154 (1970).
102. L. Moenke-Blankenburg and J. Mohr, *Jenaer Jahrb.*, 1970, p. 195.
103. J. Mohr, *Jena Rev.*, 1971; DD patent WP 90401, 1972.
104. J. Mohr and H. Schmidt, *Jena Rev.*, 93 (1973).
105. L. Moenke-Blankenburg, *Nouv. Rev. Opt. Appl.*, **3** (5), 243 (1972).
106. A. Caruso and R. Gratton, *Plasma Phys.*, **10**, 867 (1968).
107. A. J. Alcock, C. De Michelis, K. Hamel, and B. A. Tozer, *IEEE J. Quantum Electron.*, **QE-4**, 593 (1968).
108. G. A. Askaryan and E. M. Moroz, *Sov. Phys.-JETP*, **16**, 1638 (1963).
109. M. K. Chun and K. Rose, *J. Appl. Phys.*, **41**, 614 (1970).

110. J. Watson, *UKAEA Rep. ND-R-332(D)*.
111. R. H. Scott and A. Strasheim, *Spectrochim. Acta*, **25B**, 311 (1970), and **26B**, 707 (1971).
112. C. D. David, Jr., and H. Weichel, *J. Appl. Phys.*, **40**, 3674 (1969).
113. W. Bögershausen and K. Hönle, *Spectrochim. Acta*, **24B**, 71 (1969).
114. R. E. Honig and J. R. Watson, Appl. Phys. Lett., **2**, 138 (1963).
115. C. M. Verber and A. H. Adelmann, *J. Appl. Phys.*, **36**, 1522 (1965).
116. J. F. Ready, *Phys. Rev.*, **137A**, 620 (1965).
117. R. Brdička, *Grundlagen der physikalischen Chemie*, VEB Deutscher Verlag der Wissenschaften, Berlin, 1982.
118. M. K. Chun, *IEEE J. Quantum Electron.*, **QE-5**, 316 (1969); *J. Appl. Phys.*, **41**, 2 (1970).
119. A. M. Bronch-Bruevich, S. E. Potanov, and Ya. A. Imas, *Zh. Tekh. Fiz.*, **38**, 1116 (1968); *Sov. Phys.-Tech. Phys.*, **13**, 925 (1969).
120. N. G. Basov, V. A. Boiko, O. N. Krokhin, O. G. Semeno, and G. V. Sklizkov, *Zh. Tekh. Fiz.*, **38**, 1973 (1968), and *Sov. Phys.-Tech. Phys.*, **13**, 1581 (1969).
121. S. L. Mandelstam, *Izv. Akad. Nauk SSSR Ser. Fiz.*, **26**, 848 (1962).
122. J. Eggert, *Phys. Z.*, **21**, 570 (1919).
123. M. N. Saha, *Proc. R. Soc. London*, **A99**, 135 (1921).
124. M. N. Saha, *Philos. Mag.*, **44**, 1128 (1922).
125. J. K. Cobb and J. J. Muray, *Br. J. Appl. Phys.*, **16**, 271 (1965).
126. J. Franzen and K. D. Schuy, *Proc. 7th International Conference on Phenomena in Ionized Gases*, Vol. III, Gradjevinska Knjiga Belgrade, 1966, p. 242.
127. J. F. Tonon, Étude de l'écoulement de plasma créé par interaction du rayonnement laser sur une cible solide, Thesis of Cl. Bernard, Lyon University, Lyon, France, 1973, pp. 1–202.
128. J. M. Schroeer, T. M. Rhodin, and R. C. Bradley, *Surf. Sci.*, **34**(3), 571 (1973).
129. G. Bekefi, *Principles of Laser Plasmas*, John Wiley & Sons, Inc., New York 1976.
130. J. Watson, Ph.D. thesis, University of St. Andrews, St. Andrews, Scotland, 1978.
131. R. J. Conzemius and J. M. Capellen, *Int. J. Mass Spectrom. Ion Phys.*, **34**, 197 (1980).
132. M. Landmann, Studium der Plasmaeigenschaften und Erarbeitung optimaler Bedingungen für eine gezielte Anwendung der Laser-Mikro-Spektralanalyse in der Denkmalpflege und Restaurierung, Doctoral thesis, Martin Luther University Halle-Wittenberg, Halle, GDR, 1988.
133. M. Venugopalan, *Reaction under Plasma Conditions*, Wiley-Interscience, New York, 1970, pp. 1–545.
134. A. Walsh, *Spectrochim. Acta*, **7**, 108 (1955).
135. L. Moenke-Blankenburg and M. Hüfner, laboratory report.
136. N. G. Vasov, V. A. Boiko, O. N. Krokhin, O. G. Semenov, and G. V. Skilizkov, *Sov. Phys.-Tech. Phys.* (Engl. Transl.), **13**, 1581 (1969).

137. L. Moenke-Blankenburg, W. Quillfeldt, and E. Berlinghoff, laboratory report.

138. H.-J. Dietze and H. Zahn, *Exp. Tech. Phys.*, **20**(5), 389 (1972).

139. F. Brech, *Appl. Spectrosc.*, **16**, 59 (1962).

140. H. Moenke and L. Moenke-Blankenburg, *Einführung in die Laser-Mikro-Emissionsspektralanalyse*, Akademische Verlagsgesellschaft Geest und Portig KG, Leipzig, GDR, 1966.

141. S. D. Rasbery, B. F. Scribner, and M. Margoshes, *Appl. Opt.*, **6**, 81 (1967).

142. A. B. Whitehead and H. H. Heady, *Appl. Spectrosc.*, **22**, 7 (1968).

143. W. W. Panteleev and A. A. Yankovskij, *J. Appl. Spectrosc.*, **8**, 905 (1968).

144. H. Moenke, L. Moenke-Blankenburg, and W. Quillfeldt, *Mikrochim. Acta, Suppl. III*, 221 (1968).

145. L. Moenke-Blankenburg, *Feingeraetetechnik*, **17**, 193 (1968).

146. R. H. Scott and A. Strasheim, *Proc. 15th Colloquium Spectroscopicum Internationale*, Madrid, 1969, Abstract 318.

147. E. S. Beatrice, I. Harding-Barlow, and D. Glick, *Appl. Spectrosc.*, **23**, 257 (1969).

148. V. V. Panteleev, M. L. Petuch, O. I. Putrenko, T. A. Jankovskaja, and A. A. Jankovsky, *Zh. Prikl. Spektrosk.*, **12**, 1106 (1970).

149. W. Quillfeldt, *Exp. Tech. Phys.*, **17**, 415 (1969); WP 78290 (GDR-patent), 1970.

150. W. J. Treytl, J. B. Orenberg, A. J. Saffir, and D. Glick, *Anal. Chem.*, **44**, 1903 (1972).

151. T. Kubona, *J. Appl. Phys.*, **49**, 5790 (1978).

152. R. Klockenkämper and K. Laqua, *Spectrochim. Acta*, **32B**, 207 (1977).

153. N. V. Korolev, V. V. Ryukhin, and G. B. Lodin, *J. Appl. Spectrosc.*, **19**, 21 (1973).

154. T. Yamane and S. Matsushita, *Spectrochim. Acta*, **27B**, 27 (1972).

155. Zh. Guosheng, P. Donkai, Bao Liang, F. Jingui, G. Weichun, and D. Aimai, *Laser J. (Shanghai)*, **7**, 43 (1980).

156. N. V. Korolev, V. V. Ryukhin, and G. B. Lodin, *J. Appl. Spectrosc.*, **19**, 400 (1973).

157. R. S. Adrain, D. R. Airey, and E. J. Ormerod, *Proc. 5th IEE Conf. Gas Discharges*, **165**, 70 (1978); see also R. S. Adrain, R. C. Klewe, and E. J. Ormerod, *Proc. Conf. Electro-Optics Laser International '80*, ICP Science and Technology Press, Ltd., p. 375.

158. W. Quillfeldt, DD patent WP 78392, 1970.

159. A. Petrakiev, L. Georgieva, and G. Dimitrov, *C. R. Acad. Bulg. Sci.*, **22**(9), 983 (1969); *Spectrosc. Lett.*, **2**, 97 (1969).

160. G. Dimitrov, An investigation of the processes of evaporation, entry and excitation of the substance and the influence of the gas medium in laser microspectral analysis, Doctoral thesis, University of Sofia, Sofia, Bulgaria, 1983.

161. G. Dimitrov and L. Moenke-Blankenburg, Lecture given at the *3rd Conf. Methods and Application of Laser Micro Analysis*, Halle, German Democratic Republic, 1987.

162. G. Dimitrov and Ts. Zhelev, *Spectrochim. Acta*, **39B**, 1209 (1984).

163. W. Karthe, R. Neubert, and G. Wiederhold, DD patent WP 48059, 1966.

164. F. Leis, Ph.D. thesis, University of Düsseldorf, Düsseldorf, Federal Republic of Germany, 1976.
165. F. Leis and K. Laqua, *Spectrochim. Acta*, **33B**, 727 (1978); **34B**, 307 (1979).
166. P. Wengler, DD patent WP 74362, 1970.
167. U. Möde, Extension and enhancement of the spectral emission of radiation from samples vaporized by laser radiation, Ph.D. thesis, University of Münster, Münster, Federal Republic of Germany, 1970.
168. A. Petrakiev, R. Milanova, and L. Georgieva, *C. R. Acad. Bulg. Sci.*, **21**(10) (1968); **22**(9) (1969).
169. L. Georgieva and A. Petrakiev, *Bol. Geol. Min.*, **85**, 491 (1969).
170. A. Petrakiev, G. Dimitrov, and L. Georgieva, *Spectrosc. Lett.*, **2**, 97 (1969).
171. L. Moenke-Blankenburg, W. Quillfeldt, and C. Lindner, laboratory report.
172. E. H. Piepmeier and D. E. Osten, *Appl. Spectrosc.*, **25**, 642 (1971).
173. R. M. Manaba and E. H. Piepmeier, *Anal. Chem.*, **51**, 2066 (1979).
174. W. J. Treytl, K. W. Marich, J. B. Orenberg, P. W. Carr, D. C. Miller, and D. Glick, *Anal. Chem.*, **43**, 1452 (1971).
175. A. Petrakiev and G. Dimitrov, *Proc. 16th Colloquium Spectroscopicum Internationale*, Heidelberg, Federal Republic of Germany, 1971, Vol. 1, p. 186.
176. G. Dimitrov and A. Petrakiev, *Proc. 11th International Conf. Phenomena in Ionized Gases*, Prague, Czechoslovakia, 1973, p. 255.
177. G. Dimitrov and V. Gagov, *Spectrosc. Lett.*, **10**, 337 (1977).
178. G. Dimitrov, V. Gagov, and S. Aslam, *Proc. 13th International Conf. Phenomenon in Ionized Gases*, Berlin, 1977, Part I, p. 179.
179. G. Dimitrov, L. Nikolova, and Y. Vasilev, *Mikrochim. Acta*, **15**, 503 (1979).
180. G. Dimitrov and T. Maximova, *Spectrosc. Lett.*, **14**, 734 (1981).
181. L. Moenke-Blankenburg and H. Moenke, *CZ-Chem.-Tech.*, **2**, 297 (1973).
182. H. Moenke and L. Moenke-Blankenburg, *Mikrochim. Acta, Suppl. V*, 377 (1974).
183. L. Mehl, W. Mikkeleit, H. Moenke, L. Moenke-Blankenburg, and J. Mohr, *Exp. Tech. Phys.*, **22**, 161 (1974).
184. J. Mohr, *Exp. Tech. Phys.*, **22**, 327 (1974); see also *Jena Rev.*, 245 (1979).
185. H. J. Stupp, *Berichte der Kernforschungsanlage*, Jülich, FRG, Jül-933-RG, 1973.
186. H. J. Stupp and T. Overhoff, *Spectrochim. Acta*, **29B**, 77 (1974); **30B**, 77 (1975).
187. A. V. Karjakin, E. K. Wulfson, and A. F. Januschkewicz, *Proc. Festkörperanalytik II*, Karl-Marx-Stadt, GDR, 1977, p. 7.
188. K. Kagawa and S. Yokoi, *Spectrochim. Acta*, **37B**, 789 (1982).
189. A. L. Lewis II, G. J. Beenen, J. W. Hosch, and E. H. Piepmeier, *Appl. Spectrosc.*, **37**(3), 263 (1983).
190. A. L. Lewis II and E. H. Piepmeier, *Appl. Spectrosc.*, **37**(6), 523 (1983).
191. G. J. Beenen and E. H. Piepmeier, *Appl. Spectrosc.*, **38**(6), 851 (1984).

192. K. W. Marich, P. W. Carr, W. J. Freytl, and D. Glick, *Anal. Chem.*, **42**, 1775 (1970).

193. E. Cerrai and R. Trucco, *Energ. Nucl.* (*Milan*), **15**, 581 (1968).

194. V. V. Panteleev and A. A. Yankovskii, *J. Appl. Spectrosc.*, **3**, 70 (1965).

195. M. Margoshes, B. F. Scribner, and S. D. Rasberry, *Appl. Opt.*, **6**, 87 (1967).

196. A. V. Karyakin and V. A. Kaigorodov, *J. Anal. Chem. USSR*, **22**, 504 (1967).

197. Y. M. Buravlev, B. P. Nadezna, and L. N. Babanskaya, *Zavod. Lab.*, **40**, 165 (1974).

198. K. Dittrich and R. Wennrich, *Prog. Anal. At. Spectrosc.*, **7**, 139 (1984).

199. B. Smith, N. Omenetto, and J. D. Winefordner, *Proc. Laser Applications to Chemical Analysis*, Lake Tahoe, Nev., 1987.

200. A. V. Karyakin, E. K. Wulfson, and A. F. Januschkewicz, *Proc. Festkörperanalytik II*, Karl-Marx-Stadt, GDR, 1977, p. 7.

201. R. Wennrich, K. Dittrich, and U. Bonitz, *Spectrochim. Acta*, **38B**, 657 (1984).

202. E. K. Wulfson, V. I. Dvorkin, A. V. Karyakin, and P. Y. Misakov, *Zh. Prikl. Spektrosk.*, **35**, 415 (1981).

203. A. Quentmeier, K. Laqua, and W.-D. Hagenah, *Spectrochim. Acta*, **34B**, 117 (1979); **35B**, 139 (1980).

204. H.-J. Dietze, S. Becker, L. Opanszky, L. Matus, I. Nyary, and J. Frecska, *ZFI-Mitt.*, **48** (1981).

205. H. J. Heinen, *Int. J. Mass Spectrom. Ion Phys.*, **38**, 309 (1981).

206. B. Schneler, F. R. Krueger, and P. Feigl, *Int. J. Mass Spectrom. Ion Phys.*, **47**, 3 (1983).

207. H. J. Heinen, S. Meier, H. Vogt, and R. Wechsung, *Adv. Mass Spectrom.*, **8A**, 942 (1980).

208. E. Denoyer, R. van Grieken, F. Adams, and D. F. S. Natusch, *Anal. Chem.*, **54**, 26A (1982).

209. D. M. Hercules, R. J. Day, K. Balasamugam, T. A. Dang, and C. P. Li, *Anal. Chem.*, **54**, 280A (1982).

210. R. Kaufmann, P. Wieser, and R. Wurster, *Scanning Electron Microscopy*, Part II, AMF O'Hare, Chicago, 1980, p. 606.

211. E. Michiels, A. Celis, and R. Gijbels, in *Microbeam Analysis*, K. F. J Heinrich, Ed., San Francisco Press, Inc., San Francisco, 1982, p. 383; *Int. J. Mass Spectrom. Ion Phys.*, **47**, 23 (1983).

212. J. M. Beusen, P. Surkyn, R. Gijbels, and F. Adams, *Spectrochim. Acta*, **38B**, 843 (1983).

213. F. J. Bruynseels and R. E. van Grieken, *Spectrochim. Acta*, **38B**, 853 (1983).

CHAPTER

4

# PRESUMPTIONS FOR QUANTITATIVE LASER MICROANALYSIS OF SOLIDS:

## STATISTICS IN ANALYTICAL CHEMISTRY

Generally, in quantitative analysis, information is acquired on the composition of a sample by experimental methods. The process consists of two consecutive steps: the analysis itself and the evaluation of results, that is, decoding the information represented by the results [1–14]. The performance of an analytical chemical method is characterized especially by the precision of the analytical procedure and the accuracy of the results.

Accuracy characterizes the agreement between measured values (or their mean $\bar{x}$) and the true value $x$. There is a systematic error when a permanent difference can be found between the measured values (or their mean value) and the true value. In this case the absolute error is the difference between $\bar{x}$ and $x$.

Precision measures the agreement between the results of repeated measurements. In this case, random (statistical) errors occur and for a given analytical procedure are represented by the standard deviation. As an estimation of the true standard deviation, $\sigma$, the value of $s$ can be used, which is given by the following expression:

$$s = \sqrt{\frac{1}{n-1}\sum_{i=1}^{n}(x_i - \bar{x})^2} \tag{29}$$

where $n$ is the number of replicate measurements and $\bar{x}$ is the mean of measured $x_i$ values.

The result is characterized by the confidence interval [15], $\bar{x} \pm \Delta\bar{x}$, where

$$\Delta\bar{x} = \frac{st(P, f)}{\sqrt{n}} \tag{30}$$

where $t(P, f)$ is the value of Student's $t$-distribution for a $P = 1 - \alpha$ statistical confidence level ($\alpha$ is the error probability) and $f = n - 1$ degrees of freedom. The confidence interval defined by eq. (30) can be interpreted as meaning that the "true value" will be in the region ($\bar{x} - \Delta\bar{x}$ and $\bar{x} + \Delta\bar{x}$) with a probability of $P$.

## 4.1. ACCURACY AND PRECISION OF LASER MICROANALYSIS

Laser microanalysis (LMA) was originally developed to enable qualitative local analysis with high spatial resolution of material of varying composition, including conducting and nonconducting materials. Since the method was shown to be suitable to this type of analysis, efforts have been made to use LMA in quantitative analysis as well (see Chapter 1, Refs 1 to 12; Chapter 2, Refs 1 and 15 to 22; and the references in Chapters 5 to 8), inevitably requiring a systematic study of the problems associated with the following aspects:

- Optimization and stability of the laser output, and therefore reproducibility of the sample removal processes
- If used, optimization and stability of auxiliary excitation
- Homogeneity of reference material
- Reliability of signal measurements and calculations

The necessity for determination of the optimum laser parameters for quantitative analysis as related to the nature of the analyte and possibilities for corrections is discussed in Sections 2.4 and 2.5. The important role of sample parameters and plasma relationship is indicated in Section 3.3 and 3.4. The use of additional excitation, and working under special conditions with the aim of improving accuracy and precision, are reported in Sections 3.5 and 3.6.

From the studies noted above, it is seen that quantitative analysis involves relating the signal intensity of an element in the plasma to the concentration of that element in the target. The signal intensity emitted is actually a measure of the population of a given energy level in the plasma, which is assumed to be a reliable indicator of the sample composition. Whether or not this is true depends on the combined effects of the physical and chemical properties of the target on the plasma composition. These effects, known as "matrix effects," have been the subject of much discussion in all branches of spectrochemical analysis (see Section 3.7).

Since laser microanalysis is not an absolute method, the problem of preparation, evaluation, and testing of standards (i.e., reference materials) arises [16]. The main problem is to get homogeneous reference samples. To achieve homogeneity or to destroy the crystalline structure of, for example, a nonmetallic sample, it may be heated with a fusing reagent or fusing mixture to form a fused sample [17]. The molten, fused sample may be poured onto a cold, flat surface to produce a glass that can be analyzed directly, or it may be cooled, ground, and then analyzed. To facilitate grinding, the hot, brittle solid can be shattered by dropping it into cold water. A metallic test sample, commercially available, may be used directly if the sample is sufficiently homogeneous [18–21]. Methods to prove the homogeneity are given in detail in Section 4.2. Another way to get test samples with lower requirements for accuracy and precision of results is to use powder standards [22–24] or to mix synthetic samples and briquette these without or with a binder to pressed disks or pellets. Other proposals have been made by Rosan [25] and by Mohaupt and Pätzmann [26]. The reliability of signal measurements and calculations is demonstrated in the following sections, which deal with the various instruments and their applications.

Laser micro mass analysis also has limitations. Since the accuracy of every analytical device depends on reproducibility of the processes involved, and since the interaction of laser light with a specimen is a highly nonlinear process, LMA is rather sensitive to statistical fluctuations of the laser parameters and to inhomogeneity of the specimen. Therefore, absolute quantitative analysis is possible only if the laser output is stabilized as far as possible and the specimen is relatively homogeneous with respect to mass density thickness, spatial orientation, and optical absorption. In a standard specimen (0.3 to 1-$\mu$m sections of epoxy resin), fairly linear calibration plots have been obtained between element concentrations and the intensity of the recorded mass signals. Standard deviation in these specimens was $< 5\%$ at higher concentrations ($> 10$ ppm) and $< 15\%$ at low concentration levels. In biological specimens the required homogeneity cannot always be obtained. This usually prevents absolute quantitative analysis but does not exclude a relative quantification (i.e., the determination of elemental concentration ratios), which is often the information required.

## 4.2. HOMOGENEITY TESTS OF REFERENCE MATERIALS

Chemical homogeneity is a relative property of a solid which depends on the spatial resolution and precision of the analytical procedure being used for investigation [27]. Solids are defined as being analytical homogeneous if fluctuations in chemical composition over the entire sample volume determined in various areas of the sample are not significantly larger than the error of the analytical procedure. These relationships correspond to the sampling condition for representative samples in bulk analysis.

Some authors [20, 23] have used or adapted the formula of Baule and Benedetti-Pichler [28]:

$$s_{s(\mathrm{rel})} = \frac{d_1 d_2}{100\bar{d}^2}\sqrt{\frac{\bar{a}^3}{e\bar{x}}(100d_1 - \bar{x}d)} \tag{31}$$

where $s_{s(\mathrm{rel})}$ = relative standard deviation of sampling
$d_1, d_2$ = densities of the two components of a mixture AB

$a$ = mean edge length of the sampling particle
$e$ = mass of the sampling particle
$\bar{x}$ = mean content of A in the mixture

The formula for estimating the sampling error is also valid for investigations of analytical homogeneity of solids. In this case the sampling error is replaced by the inhomogeneity error, that is, by the relative standard deviation of the result caused by sample inhomogeneity. Wilson [29] described the relation between the properties of a binary mixture sample AB and the sampling error. Thus the equation can be written as

$$\left(\frac{\sigma_i}{\bar{x}}\right)^2 = \frac{(c_A - \bar{x})(\bar{x} - c_B)}{\bar{x}^2} \frac{P_A P_B}{\bar{p}^2} \frac{\bar{V}}{V_p} \tag{32}$$

where $\sigma_i$ = standard deviation caused by the sample inhomogeneity
$\bar{x}$ = mean concentration of the element
$c_A, c_B$ = concentration of this element in the constituents A and B
$P_A, P_B$ = densities of constituents A and B
$\bar{p}$ = mean density of the sample
$\bar{V}$ = mean volume of homogeneous regions
$V_p$ = volume of the subsample (e.g., an excitation volume)

For $P_A = P_B = \bar{p}$, eq. (32) becomes

$$\left(\frac{\sigma_i}{\bar{x}}\right)^2 = \frac{(c_A - \bar{x})(\bar{x} - c_B)}{\bar{x}^2} \frac{\bar{V}}{V_p} \tag{33}$$

because

$$\left(\frac{\sigma_t}{\bar{x}}\right)^2 = \left(\frac{\sigma_i}{\bar{x}}\right)^2 + \left(\frac{\sigma_a}{\bar{x}}\right)^2 \tag{34}$$

it follows that

$$\left(\frac{\sigma_t}{\bar{x}}\right)^2 = \frac{(c_A - \bar{x})(\bar{x} - c_B)}{\bar{x}^2} \frac{\bar{V}}{V_p} + \left(\frac{\sigma_a}{\bar{x}}\right)^2 \tag{35}$$

The material is considered to be chemically homogeneous if

$$\left(\frac{\sigma_i}{\bar{x}}\right)^2 < \left(\frac{\sigma_a}{\bar{x}}\right)^2 \approx \left(\frac{\sigma_t}{\bar{x}}\right)^2 \tag{36}$$

and inhomogeneous if

$$\left(\frac{\sigma_t}{\bar{x}}\right)^2 > \left(\frac{\sigma_a}{\bar{x}}\right)^2 \tag{37}$$

where $\sigma_t$ is the total standard deviation and $\sigma_a$ is the standard deviation caused by the analytical procedure.

The relative standard deviation of the analytical results caused by sample inhomogeneity increases with decreasing concentration and subsample volume. For a given concentration and a given subsample volume, the proof of homogeneity will be the stronger the samller the statistical error of the analytical procedure [30, 31].

From eqs. (36) and (37); the investigation of chemical homogeneity has to be based on testing the null hypothesis $H_0: \sigma_t^2 \approx \sigma_a^2$ (homogeneity) against the alternative hypothesis $H_1: \sigma_t^2 > \sigma_a^2$ (inhomogeneity). In practice, $s$, as an estimate of $\sigma$, is used, because only a finite number of measurements are available. The null hypothesis must be checked by the Fisher test. Inhomogeneity must be accepted if

$$F = \left(\frac{s_t}{s_a}\right)^2 > \hat{F}(\bar{P}; f_t; f_a) \tag{38}$$

where $f_t$ and $f_a$ are the degrees of freedom of $s$ and $s_a$, respectively, and $\bar{P}$ is the one-sided probability [8].

The testing of the significance of homogeneity is based on different mathematical models for inhomogeneity, which can be distributed randomly (stochastically), systematically (concentration gradients), or periodically [30]. In analytical practice the testing of homogeneity is carried out by measuring $n$ replicates on each of $m$ different sample areas. The error of the analytical procedure will be calculated according to

$$s_a^2 = [m(n-1)]^{-1} \sum_{i=1}^{m} \sum_{j=1}^{n} (x_{ij} - \bar{x}_i)^2 \tag{39}$$

The total error $s_t$ will be obtained from the dispersion between the $m$ different sample points:

$$s_t^2 = (mn-1)^{-1} \sum_{i=1}^{m} \sum_{j=1}^{n} (x_{ij} - \bar{\bar{x}})^2 \tag{40}$$

where $\bar{\bar{x}}$ is the total mean. For the $F$-test the numbers of the degrees of freedom are $f_t(=f_1) = mn - 1$ and $f_a(=f_2) = m(n-1)$.

In analytical practice the testing of homogeneity is without problems if a nondestructive analytical procedure is used, because independent replicate measurements on one and the same sample area are possible for estimation of the procedural error. If a destructive method is employed, and that is the case of LMA, such independent replicate measurements are not possible. In this case the error of the analytical procedure can only be approximated [32].

The following mathematical models for testing chemical homogeneity have been recommended: the one-way analysis of variance, based on [8, 31, 32]

$$x_i = \bar{x} + \alpha_i + \varepsilon_i \tag{41}$$

where $x_i$ = measured value
$\bar{x}$ = mean value
$\alpha_i$ = error contribution resulting from the local fluctuations in the composition
$\varepsilon_i$ = measuring error

and the two-way analysis of variance, based on [8, 31]

$$x_{ij} = \bar{\bar{x}} + \alpha_i + \beta_i + \gamma_{ij} + \varepsilon_{ij} \tag{42}$$

where $x_{ij}$ = measured values in rows and columns
$\bar{\bar{x}}$ = total mean
$\alpha_i$ = row deviation
$\beta_i$ = column deviation,
$\gamma_{ij}$ = interaction contribution
$\varepsilon_{ij}$ = analytical procedural error

Further mathematical models for this purpose are the regression analysis, based on

$$x_i = \alpha_0 + \alpha_1 Z_{1i} + \alpha_2 Z_{2i} + \alpha_3 Z_{1i}^2 + \alpha_4 Z_{2i}^2 + \alpha_5 Z_{1i} Z_{2i} \tag{43}$$

(see [8, 30–33]); the gradients method, based on

$$x_i = \beta_0 + \beta_1 Z_{1i} + \beta_2 Z_{2i} + \beta_{12} Z_{1i} Z_{2i} \tag{44}$$

created by Parczewski [34, 35]; and the pattern cognition and pattern recognition methods [30, 36–42].

In laser microanalysis an approximate value of $s_a^2$ is to be estimated in an indirect manner. Figure 37 shows the schematic representations of laser spot arrangements for different sampling methods [32].

1. Testing of several adjacent sampling points can be done by assuming that the composition of the sample does not significantly change in adjacent areas; the laser spots will be arranged in groups.

2. By multiple laser spots at the same sample area it is presumed that the sample composition does not change or changes only insignificantly in a small range of depth.

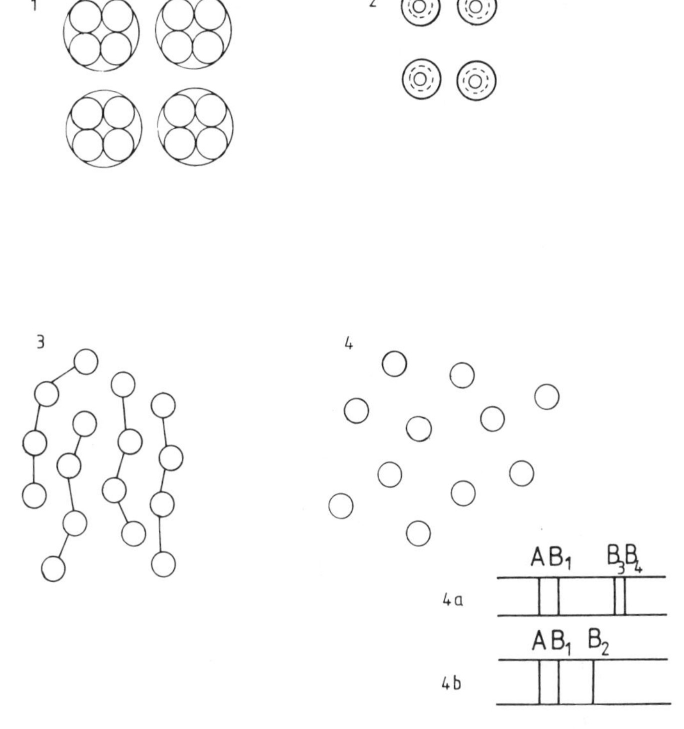


**Figure 37.** Schematic representations of laser spot arrangements for different sampling methods for investigation of homogeneity by simple analysis of variance and also for methods using reference lines. See the text for an explanation of 1 to 4. From Ref. [32], reprinted with permission of Pergamon Journals Ltd.

3. Local integration of measured values is used in one-way analysis of variance [see eq. (41)]. The basic idea of an integration method is the elimination of the error contribution $\alpha_i$ by moving the sample during the excitation. Because

$$\sum_{i=1}^{m} \alpha_i = 0 \tag{45}$$

the local averaging of values measured at the largest possible number $m$ of different sampling points yields the best estimate of the analytical procedural error. If the measuring points will be recorded photographically one after another on the same spot on the photographic plate, each with an exposure time $t/m$, then analogously, $s_a$ is computed from $n$ repeated measurements.

**TABLE 19. Arrangement of Measuring Points on the Sample Surface for Two-Way Analysis of Variance with Quasi-repetition Measurements**

| | | | | | | | | |
|---|---|---|---|---|---|---|---|---|
| $x_{111}$ | $x_{112}$ | $x_{121}$ | $x_{122}$ | $x_{131}$ | $x_{132}$ | $\cdots$ | $x_{1a1}$ | $x_{1q2}$ |
| $x_{113}$ | $x_{114}$ | $x_{123}$ | $x_{124}$ | $x_{133}$ | $x_{134}$ | $\cdots$ | $x_{1q3}$ | $x_{1q4}$ |
| $x_{211}$ | $x_{212}$ | $x_{221}$ | $x_{222}$ | $\cdots$ | | $\cdots$ | $x_{2q1}$ | $x_{2q2}$ |
| $x_{213}$ | $x_{214}$ | $x_{223}$ | $x_{224}$ | $\cdots$ | | $\cdots$ | $x_{2q3}$ | $x_{2q4}$ |
| $\vdots$ | $\vdots$ | $\vdots$ | $\vdots$ | | | | $\vdots$ | $\vdots$ |
| $x_{p11}$ | $x_{p12}$ | $x_{p21}$ | $x_{p22}$ | $\cdots$ | | $\cdots$ | $x_{pq1}$ | $x_{pq2}$ |
| $x_{p13}$ | $x_{p14}$ | $x_{p23}$ | $x_{p24}$ | $\cdots$ | | $\cdots$ | $x_{pq3}$ | $x_{pq4}$ |

*Source*: Ref. 1.

4. Comparison using reference lines is the basic idea of two methods, which consists of simultaneous estimation of the total and procedural errors from the same spectrum by evaluation of different pairs of spectral lines. According to the method of double pairs of lines by Hemschick and Schuffenhauer [43], the total error will be estimated from deviations of the intensities of line pairs A and $B_1$ and the procedural error by $B_3$ and $B_4$. With Skogerboe's [44] three-line method, $s$ is determined from A and $B_1$, but $s_A$ is determined from $B_1$ and $B_2$. For applications of these methods, see Refs. 20, 21, 45, and 46.

A prerequisite for the application of two-way analysis of variance is the regular arrangement of the measuring points on the sample surface in rows and columns. For two-way classification, $rc$ subsamples are arranged on the sample surface regularly, as shown in Table 19 and Fig. 37.

Comparisons of the above-mentioned models of mathematical homogeneity tests and practical application in laser microanalysis are given by Pacher [47] and Moenke-Blankenburg [48, 49]. They prepared reference material on the basis of borax melting.

$$Na_2CO_3 + 4H_3BO_3 \rightarrow Na_2B_4O_7 + CO_2 + 6H_2O$$

(15 min by 1000°C). The composition of the test sample was 61.06% $B_2O_3$, 7.63% $SiO_2$, 30.55% $Na_2O$, and 0.76% $Co_2O_3$. The homogeneity of Co distribution in the borosilicate matrix were studied with a laser microanalyzer for using such samples as reference materials for laser local analysis in glass systems. The laser spot diameters were 40 and 80 $\mu$m. The laser spots were distributed over the entire surface. Four spots were always grouped closely together for quasi-repetition measurements. These groups were arranged in rows and columns. The results are given in Table 20 and show homogeneity in both cases, pointed out by one-way and two-way analysis of variance. It is seen that homogeneity index $H$ increases (i.e., homogeneity decreases) with decreasing analyzed area and increasing spatial resolving power, respectively.

**TABLE 20. Homogeneity-Test Results of Cobalt Distribution in Borosilicate Glass by Laser Microanalysis**

| | Crater Diameter ($\mu$m)[a] | | | |
|---|---|---|---|---|
| | 80 | | 40 | |
| *One-Way Analysis of Variance* | | | | |
| Total variance | 0.0013 | | 0.0030 | |
| Analytical variance | 0.0011 | | 0.0022 | |
| Calculated $F$ | 1.179 | | 1.357 | |
| Tabulated $\hat{F}$ | 1.46 | | 1.44 | |
| $H = F/\hat{F}$ | | 0.807 | | 0.943 |
| *Two-Way Analysis of Variance* | | | | |
| Variance within rows | 0.0109 | | 0.0129 | |
| *Calculated F* | 3.238 | | 5.814 | |
| Tabulated $\hat{F}$ | 2.74 | | 2.50 | |
| $H = F/\hat{F}$ | | 1.182 | | 2.326 |
| Variance within columns | 0.0039 | | 0.0031 | |
| Calculated $F$ | 0.688 | | 1.409 | |
| Tabulated $F$ | 2.35 | | 2.50 | |
| $H$ | | 0.293 | | 0.563 |
| Variance of interactions | 0.030 | | 0.0042 | |
| Calculated $F$ | 1.788 | | 1.905 | |
| Tabulated $\hat{F}$ | 1.80 | | 1.79 | |
| $H$ | | 0.993 | | 1.064 |
| Total $H$ | | 0.813 | | 0.936 |

[a]Using the line-blackening difference: $\Delta S = S_{Si} - S_{Co}$; distance between close adjoining craters, 100 $\mu$m; distance between groups, 30 $\mu$m; number of rows, 10; number of columns, 10.

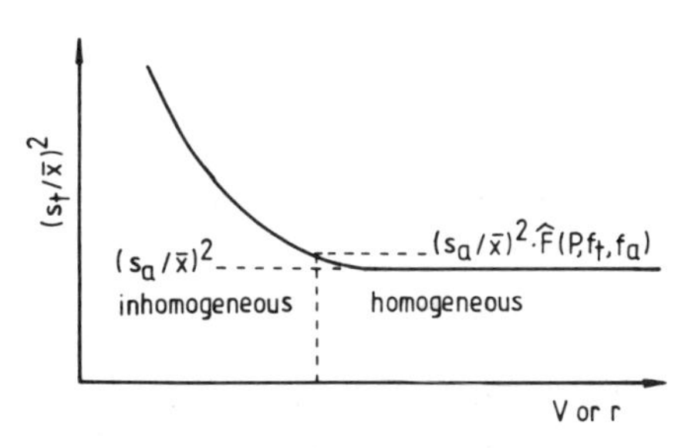


**Figure 38.** Homogeneity versus spatial resolution and precision of the analytical procedure. $(s_t/\bar{x})^2$ = variance of the total error, $(s_a/\bar{x})^2$ = variance of analytical procedure error, $\hat{F}$ = Fisher test, tabulated values, $V$ = volume of the analyte, $r$ = radius of laser crater. From Ref. [50], reprinted with permission of Pergamon Journals Ltd.

The principle of the dependence of homogeneity on the size of analyzed area or volume by a given precision of the analytical procedure is shown in Fig. 38. Further literature about homogeneity tests are given in Refs. 51 to 63.

## 4.3. SUMMARY OF USEFUL STATISTICS FOR SPECTROANALYTICAL CHEMISTRY

### 4.3.1. Linear Regression

Linear regression is used by analytical chemists to obtain calibration lines, to compare two analytical procedures, or to relate analytical results to an outside variable. This technique, together with the $t$-test for comparing means of series of observations, is the most used statistical method for analytical purposes. It is usually carried out using the least-squares technique, a basic idea proposed by Gauss of estimating quantities by minimizing a sum of squares of deviations (or "errors").

In spectrochemical analysis the relationship between the measured intensity $y$ and the concentration $x$ is established by calibration. As the first step the calibration function is determined from measurements on known samples. In the following step of the analysis, the inverse function—the function of analysis—is used to transform the measured intensity $y_a$ into the concentration of the analyzed sample $x_a$. For both purposes, analysis of regression is used. The main problem is to find the best fit for the calibration function.

It has to be assumed that the particular values $x_i$ and $y_i$ of two variables, $X$ and $Y$, are related by the equations

$$\begin{aligned} y_1 &= a + bx_1 \\ y_2 &= a + bx_2 \\ &\vdots \\ y_m &= a + bx_m \end{aligned} \tag{46}$$

where $a$ is the intercept and $b$ is the slope of the underlying straight-line relationship. $X$ is fixed equal to $x_i$ and has no error attached to it: It is a mathematical variable, not a random variable. In addition to depending on $X$, there is a random error in determining $Y$, so that the value $y_i$ corresponding to a given fixed $x_i$ consists of two components, $a$ and $b$.

The method of least squares is applied to the model (46) in the following way:

$$\sum(y_i - Y_i)^2 = \sum(y_i - a - bx_i)^2 \tag{47}$$

The minimum can be found using partial differentiation [6]. It follows from the results

$$b = \frac{m\sum x_i y_i - \sum x_i \sum y_i}{m\sum x_i^2 - (\sum x_i)^2} \tag{48}$$

$$a = \frac{\sum y_i - b\sum x_i}{m} = \bar{y} - b\bar{x} \tag{49}$$

The estimated line $y = a + b\bar{x}$ may therefore be written

$$y - \bar{y} = b(x - \bar{x}) \tag{50}$$

Because $a$ and $b$ are estimated values with random errors, the standard deviations and the confidence intervals have to be calculated:

$$s_0^2 = \frac{\sum_1^m (y_i - Y_i)^2}{m - 2} \tag{51}$$

The number of degrees of freedom is $f = m - 2$, because a line needs at least two points.

$$s_b^2 = \frac{s_0^2}{\sum(x_i - \bar{x})^2} = \frac{ms_0^2}{m\sum x_1^2 - (\sum x_i)^2} \tag{52}$$

$$s_a^2 = \frac{s_0^2 \sum x_i^2}{m\sum(x_i - \bar{x})^2} = \frac{s_0^2 \sum x_i^2}{m\sum x_i^2 - (\sum x_i)^2} = \frac{s_b^2}{m}\sum x_i^2 \tag{53}$$

The confidence intervals are

$$\Delta b = t(P, f)s_b \qquad \text{and} \qquad \Delta a = t(P, f)s_a \tag{54}$$

Taking the standard deviations $s_a$ and $s_b$ into consideration, the confidence interval of the calculated value $Y_k$ for a given $x_k$ will be

$$\Delta Y_k = t(P, f)\sqrt{s_0^2\left[\frac{1}{m} + \frac{(x_k - \bar{x})^2}{\sum(x_i - \bar{x})^2}\right]}$$

$$= t(P, f)\sqrt{s_0^2\left[\frac{1}{m} + \frac{m(x_k - \bar{x})^2}{m\sum x_i^2 - (\sum x_i)^2}\right]} \tag{55}$$

### 4.3.2. Linear Correlation

Some of the foregoing statistical methods deal with one variate only. However, often several characteristics are measured on each part of a sample, and it may be of interest to determine whether the variates the interrelated.

The degree of correlation is measured by a correlation coefficient. The correlation coefficient between $n$ pairs of observations whose values are $(x_i, y_i)$ is

$$r = \frac{\sum_{i=1}^{n} (x_i - \bar{x})(Y_i - \bar{Y})}{\sqrt{\left[\sum_{i=1}^{i} (x_i - \bar{x})^2\right]\left[\sum_{i=1}^{n} (y_i - \bar{y})^2\right]}} = \frac{\sum d_x d_y}{\sqrt{(\sum d_x^2)(\sum d_y^2)}} \tag{56}$$

It can be shown that $r$ takes values only in the interval $-1$ to $+1$. It takes the values $\pm 1$ when there is an exact straight-line relation $Y = mX + c$ connecting the two variates: $r = +1$ when $m$ is positive and $r = -1$ when $m$ is negative. When there is no correlation, $r$ will take a value close to zero; when there is a strong correlation, $r$ will take a value near to $\pm 1$. Intermediate values need a test for significance. The distribution of $r$ depends on the amount of data available for calculating $r$. The density function for $r$ contains a parameter $(f = n - 2)$, where $n$ is the number of pairs of observations. The amount of $|r|$ is to be compared with tabulated limit values of $r(P, f)$. The significance of the relationship is evident if $|r| > r(P, f)$.

### 4.3.3. One-Way Analysis of Variance

As described in Section 4.1, the measure for precision is the estimate of the standard deviation for a finite number of measurements under the following conditions:

- The analytical method has been given in full.
- The material to be analyzed is clearly defined.
- The conditions for analysis are standardized.

Consequently, there can be only one source of random errors—the error of the analytical procedure. If there is a second reason for errors, such as inhomogeneity of the sample (see Section 4.2), it leads to a total error as a sum of the single variances. The total error can be calculated by one-way analysis of variance. The results are divided into $m$ groups with the same random error and calculated in the following way, shown in Table 21. Under the assumption

**TABLE. 21. One-Way Analysis of Variance**

| Source of Variation | Sum of Squares | Degrees of Freedom | Variances | Components of Variances |
|---|---|---|---|---|
| Deviation between $m$ groups | $SQ_1 = \sum_{j=1}^{m} n_j(\bar{x}_j - \bar{\bar{x}})^2$ | $f_1 = m - 1$ | $s_1^2 = \frac{SQ_1}{f_1}$ | $s_1^2 = s_2^2 + s_3^2$ |
| Deviation within $m$ groups (analytical error) | $SQ_2 = \sum_{j=1}^{m} \sum_{i=1}^{n} (x_{ji} - \bar{x}_j)^2$ | $f_2 = n - \mathrm{m}$ | $s_2^2 = \frac{SQ_2}{f_2}$ | |
| Total deviation | $SQ_T = SQ_1 + SQ_2 = \sum (x_{ji} - \bar{x})^2$ | $f = f_1 + f_2 = n - 1$ | | |

of fulfillment of the null hypothesis $s_1^2 = s_2^2 (s_3^2 = 0)$, it is proved by the Fisher test, $F = s_1^2/s^2$ when $F < \hat{F}(\bar{P}; f_1; f_2)$. When the null hypothesis has to be rejected, $F > \hat{F}(\bar{P}; f_1; f_2)$, $s_1^2 \neq s_2^2$ and $s_3^2 \neq 0$ (i.e., the sample is not homogeneous).

### 4.3.4. Two-Way Analysis of Variance

Experiments to compare different varieties are carried out by dividing a field into a number of small units ("plots") and larger units ("blocks"). Each block consists of sufficient plots. Plots within a block are as alike as possible, so that the variation between the blocks effectively removes the systematic variation present in the field. This scheme is called a randomized block design. The method used in its analysis is called two-way analysis of variance since it must recognize variation between blocks as well as between varieties.

In Section 4.2 two-way analysis of variance was used in LMA for homogeneity tests of reference materials. A trial plan with quasi-repetition measurements (adjacent plots in blocks) arranged in rows and columns is described in Table 22. In the table the sums and means are

$$S_{ij.} = \sum_{l=1}^{r} x_{ijl} \qquad \bar{x}_{ij.} = \frac{S_{ij}}{r}$$

$$S_{i..} = \sum_{j=1}^{s} \sum_{l=1}^{r} x_{ijl} \qquad \bar{x}_{i..} = \frac{S_{i..}}{s \cdot r}$$

$$S_{.j.} = \sum_{i=1}^{c} \sum_{l=1}^{r} x_{ijl} \qquad \bar{x}_{.j.} = \frac{S_{.j.}}{c \cdot r}$$

$$S_{...} = \sum_{i=1}^{c} \sum_{j=1}^{s} \sum_{l=1}^{r} x_{ijl} \qquad \bar{x}_{...} = \frac{S_{...}}{n}$$

where $S_{...}$ is the grand total sum. Now the total sum of squares ($SQ_T$) is to be calculated:

$$SQ_T = SQ_A + SQ_B + SQ_{AB} + SQ_R$$

where

$$SQ_T = \sum_{i=1}^{c} \sum_{j=1}^{s} \sum_{l=1}^{r} (x_{ijl} - \bar{x}_{...})^2$$

and the residual sum of squares is

$$SQ_R = \sum_{i=1}^{c} \sum_{j=1}^{s} \sum_{l=1}^{r} (x_{ijl} - \bar{x}_{ij.})^2 = SQ_T - SQ_A - SQ_B - SQ_{AB}$$

**TABLE 22. Two-Way Analysis of Variance with Multivariate Filling**

| Groups of Factor B | | Groups of Factor A 1 | 2 | ⋯ | c | $S_j$ | $\bar{x}_j$ |
|---|---|---|---|---|---|---|---|
| 1 | 1 | $x_{111}$ | $x_{211}$ | ⋯ | $x_{c11}$ | | |
| | 2 | . | . | | . | | |
| | . | . | . | | . | | |
| | . | . | . | | . | | |
| | . | . | . | | . | | |
| | $r$ | $x_{11r}$ | $x_{21r}$ | ⋯ | $x_{c1r}$ | | |
| | $S_{i1.}$ | $S_{11.}$ | $S_{21.}$ | | $S_{c1.}$ | $S_{.1.}$ | $\bar{x}_{.1.}$ |
| | . | . | . | ⋯ | . | | |
| | . | . | . | ⋯ | . | | |
| | . | . | . | ⋯ | . | | |
| $s$ | 1 | $x_{1s1}$ | $x_{2s1}$ | ⋯ | $x_{cs1}$ | | |
| | 2 | . | . | | . | | |
| | . | . | . | | . | | |
| | . | . | . | | . | | |
| | . | . | . | | . | | |
| | $r$ | $x_{1sr}$ | $x_{2sr}$ | ⋯ | $x_{csr}$ | | |
| | $S_{is.}$ | $S_{1s}$ | $S_{2s}$ | ⋯ | $S_{cs.}$ | $S_{.s.}$ | $\bar{x}_{.s.}$ |
| | $S_{i..}$ | $S_{1..}$ | $S_{2..}$ | ⋯ | $S_{c..}$ | $S_{...}$ | |
| | $\bar{x}_{i..}$ | $\bar{x}_{1..}$ | $\bar{x}_{2..}$ | ⋯ | $\bar{x}_{c..}$ | | $\bar{x}_{...}$ |

The variances or mean squares are obtained by dividing by the degrees of freedom:

$$\mathrm{MQ}_A = \frac{\mathrm{SQ}_A}{c-1} \qquad \mathrm{MQ}_B = \frac{\mathrm{SQ}_B}{s-1} \qquad \mathrm{MQ}_{AB} = \frac{\mathrm{SQ}_{AB}}{(c-1)(s-1)}$$

$$\mathrm{MQ}_R = \frac{\mathrm{SQ}_R}{n-cs}$$

The $F$-tests are performed as follows [3–6, 30–33, 47]:

$$F_T = \frac{\mathrm{MQ}_T}{\mathrm{MQ}_R} \qquad F_B = \frac{\mathrm{MQ}_B}{\mathrm{MQ}_R}$$

$$F_A = \frac{\mathrm{MQ}_A}{\mathrm{MQ}_R} \qquad F_{AB} = \frac{\mathrm{MQ}_{AB}}{\mathrm{MQ}_R}$$

## REFERENCES

1. IUPAC Commission V. 4. (1969), *Nomenclature, Symbols, Units, and Their Usage in Spectrochemical Analysis*, Part I, Butterworth & Company (Publishers) Ltd., London, 1971; IUPAC Anal. Chem. Div., Commission on Spectrochemical and Other Optical Procedures for Analysis, *Pure Appl. Chem.*, **45**, 99 (1976); IUPAC, *Pure Appl. Chem.*, **18**, 437 (1969); *Mitteilungsblatt der Chem. Ges. der. DDR*, **42** (1981); *IUPAC Compendium of Analytical Nomenclature*, Pergamon Press, Inc., Elmsford, N.Y., 1978; IUPAC Commission 1.4, Contribution of Y. Maschiko, K. Iizuka, and S. Seaki, *Precision and Accuracy of Physico-Chemical Measurements and the Role of Certified Materials* (1982).
2. ISO, *Statistical Interpretation of Data-Techniques of Estimation and Tests Relating the Means and Variances*, ISO 2854-1976 (E), International Standards Organization.
3. D. L. Massart, A. Dijkstra, and L. Kaufman, *Evaluation and Optimization of Laboratory Methods and Analytical Procedures*, Elsevier Science Publishing Co., Inc., New York, 1978.
4. K. Eckschlager and V. Štepánek, *Information Theory as Applied to Chemical Analysis*, John Wiley & Sons, Inc., New York, 1979.
5. K. Doerffel and K. Eckschlager, *Optimale Strategien in der Analytik*, VEB Deutscher Verlag für Grundstoffindustrie, Leipzig, GDR, 1981.
6. K. Doerffel, *Statistik in der Analytischen Chemie*, VEB Deutscher Verlag für Grundstoffindustrie, Leipzig, GDR, 1987.
7. H. Kaiser, *Spectrochim. Acta*, **2**, 1 (1941).
8. H. Kaiser and H. Specker, *Z. Anal. Chem.*, **149**, 46 (1956).
9. W. J. Youden, *Anal. Chem.*, **19**, 946 (1946).
10. A. Bljum, *Zavod. Lab.*, **11**, 1289 (1976).
11. A. Makulov, *Zavod. Lab.*, **42**, 1457 (1976).
12. W. E. Harris, *Int. Lab.*, **53** (1978).
13. L. B. Rogers, *Acta Pharm. Suec.*, **18**, 75 (1982).
14. R. Klockenämper and H. Bubert, *Spectrochim. Acta*, **37B**, 127 (1982); H. Bubert and R. Klockenkämper, *Spectrochim. Acta*, **38B**, 1087 (1983); H. Bubert, R. Klockenkämper, and H. Waechter, *Spectrochim. Acta*, **39B**, 1465 (1984).
15. G. Ehrlich, *Chem. Anal.* (*Warsaw*), **21**, 303 (1976).
16. A. H. Gillieson, in *Applied Atomic Spectroscopy*, Vol. I, E. W. Grove, Ed., Plenum Press, New York, 1978, p. 237.
17. K. Ohls, K.-H. Koch, and G. Becker, *Z. Anal. Chem.*, **264**, 97 (1973).
18. S. D. Rasbery, B. F. Scribner, and M. Margoshes, *Appl. Opt.*, **6**, 81 (1967).
19. H. Schroth, *Z. Anal. Chem.*, **261**, 121 (1972).
20. L. Moenke-Blankenburg, *Proc. 7th Hüttenmännische Materialprüfertagung*, Balatonszeplak, Hungary, 1973.
21 W. Maul and W. Quillfeldt, *Jena Rev.*, 234 (1977).

22. W. Schrön, *Z. Angew. Geol.*, **18**, 350 (1972).
23. K. Berka, Theoretische und experimentelle Untersuchungen zur Eichprobenvorbereitung in der Emissions- und Lasermikrospektralanalyse, Eng. thesis, Köthen, GDR, 1975.
24. W. Quillfeldt and K. Berka, *Jena Rev.*, 302 (1977).
25. C. Rosan, *Appl. Spectrosc.*, **19**, 97 (1965).
26. G. Mohaupt and G. Pätzmann, *Jena Rev.*, 252 (1974).
27. K. Danzer, K. Doerffel, H. Ehrhardt, M. Geisler, G. Ehrlich, and P. Gadow, *Anal. Chim. Acta*, **105**, 1 (1979).
28. B. Baule and A. Benedetti-Pichler, *Z. Anal. Chem.*, **74**, 442 (1928).
29. A. D. Wilson, *Analyst*, **89**, 18 (1964).
30. K. Danzer and G. Ehrlich, *Proc. 4th Conf. Festkörperanalytik*, Karl-Marx-Stadt, GDR, 1984, p. 547.
31. K. Danzer and G. Marx, *Anal. Chim. Acta*, **110**, 145 (1979).
32. K. Danzer, *Spectrochim. Acta*, **39B**, 949 (1984).
33. R. Singer and K. Danzer, *Z. Chem.*, **24**, 339 (1984); K. Danzer, and R. Singer, *Mikrochim. Acta* (*Wien*), **I**, 219 (1985).
34. A. Parczewski and P. Koscielniak, *Fresenius Z. Anal. Chem.*, **297**, 148 (1979).
35. A. Parczewski, *Anal. Chim. Acta*, **130**, 221 (1981).
36. P. C. Jurs and T. L. Isenhour, *Chemical Applications of Pattern Recognition*, Wiley-Interscience, New York, 1975.
37. C. F. Bender and B. R. Kowalski, *Anal. Chem.*, **45**, 590 (1973).
38. B. R. Kowalski and C. F. Bender, *J. Am. Chem. Soc.*, **94**, 5632 (1972); **95**, 686 (1973).
39. P. C. Jurs, B. R. Kowalski, T. L. Isenhour, and C. N. Reilley, *Anal. Chem.*, **41**, 690 (1969).
40. K. Varmuza, *Pattern Recognition in Chemistry*, Lecture Notes in Chemistry 21, Springer-Verlag, Berlin, 1980.
41. J. Borszéki, J. Inczédy, E. Gegus, and F. Ovári, *Fresenius Z. Anal. Chem.*, **314**, 410 (1983).
42. J. Borszéki, J. Inczédy, F. Ovári, and E. Gegus, *Acta Archaeol. Acad. Sci. Hung.*, **35**, 1 (1983).
43. H. Hemschick and W. Schuffenhauer, *Exp. Tech. Phys.*, **11**, 214 (1963).
44. R. K. Skogerboe, *Appal. Spectrosc.*, **25**, 259 (1971).
45. K. Danzer, G. Marx, and L. Küchler, *Talanta*, **26**, 365 (1978).
46. G. Bajnózi and R. Major, *Magy. Kem. Lapja*, **37**, 385 (1982).
47. M. Pacher, Diploma thesis, University Halle–Wittenberg, Halle, German Democratic Republic, 1985.
48. L. Moenke-Blankenburg, Laser local analysis of silicate systems, lecture, Sirava, Czechoslovakia, 1985.
49. L. Moenke-Blankenburg and M. Pacher, *Acta Chim.* (*Hung.*), in press.
50. K. Danzer and L. Küchler, *Talanta*, **24**, 561 (1977).

51. H. Malissa, *Z. Anal. Chem.*, **273**, 449 (1975).
52. K. Danzer and G. Marx, *Proc. 2nd Conf. Festkörperanalytik*, Karl-Marx-Stadt, GDR, 1978, p. 24.
53. G. Ehrlich and H. Mai, *Proc. 3rd Conf. Festkörperanalytik*, Karl-Marx-Stadt, GDR, 1981, p. 97.
54. J. Springer, Doctoral thesis, University of Gliwice, Gliwice, Poland, 1981.
55. J. Inczedy, *Talanta*, **29**, 643 (1982).
56. K. Danzer, F. Mäurer, F. Liese, and R. Singer, *Proc. 4th Conf. Festkörperanalytik*, Karl-Marx-Stadt, GDR, 1984, p. 8; K. Danzer and G. Ehrlich, *Tagungsberichte der TH*, Karl-Marx-Stadt, GDR, 1985, p. 547.
57. R. Sutarno, Procedure for statistical evaluation of analytical data resulting from international test, *ISO/TC 102/SC2*, International Organization for Standardization, Australia, 1984.
58. A. Priemuth, Doctoral thesis, Martin–Luther–University, Halle–Wittenberg, Halle, GDR, 1978.
59. A. Priemuth, *Proc. 5th Conf. Mikrosonde*, Leipzig, GDR, 1981, p. 276.
60. J. W. Mitchell, J. E. Riley, Jr., and B. S. Carpenter, *Mikrochim. Acta* (*Wien*), **3**, 253 (1983).
61. G. Ehrlich, K. Danzer, and W. Kluge, *Proc. 6th International Symposium of High Purity Materials in Science and Technology*, Dresden, GDR, 1985.
62. T. Tsolov, C. Pravtcheva, R. Milenkov, and J. Pirov, *Fresenius Z. Anal. Chem.*, **323**, 228 (1986).
63. D. Wallaca and B. Kratochvil, *Anal. Chem.*, **59**, 226 (1987).

CHAPTER

5

# LASER MICROANALYSIS BASED ON OPTICAL EMISSION SPECTROMETRY

Simultaneous emission spectrochemical analysis has matured over some decades and today is one of the most commonly used methods, especially for the automated analysis of metals [1]. Optical atomic techniques with different excitation sources are available for control and research, such as the spark source, which is the most frequently used analytical tool for production and process control applications in the ferrous and nonferrous metal industries. The dc arc source is used primarily for trace element analysis in nonmetallic and metallic powders. The glow discharge lamp utilizes an electrical discharge in a low-pressure argon atmosphere, with atoms being sputtered from the sample surface by the bombardment of argon ions, and permits even surface analysis to be undertaken. Hollow cathode discharges can be an effective technique for the determination of trace and ultratrace levels of many elements present in metals. The advantages of laser vaporization include the ability of investigate heterogeneities in solids and to investigate surfaces. ICP and dc-coupled plasmas can be applied effectively to analytes in solutions. The ICP and DCP techniques can be used together with ablation techniques to produce aerosols using arc, spark, or laser pulses from solid samples. ICP excitation provides very low detection limits, a wide linear dynamic range, and relative freedom from interelement effects.

## 5.1. INSTRUMENTS

The combination of a solid-state laser and a special microscope takes the place of the usual discharge stand and source unit of conventional spectrochemical analysis. In most cases commercial spectrographs, monochromators with photoelectric devices or spectrometers, are used which as a rule are suitable without modification for conventional analysis of quantities in the gram and milligram range as well as for laser microanalysis of micrograms.

### 5.1.1. Technical Details of Laser Microanalyzers

Table 23 provides information on the technical details and functional values of three commercially manufactured microanalyzers [2]. The laser irradiance at

TABLE 23. Comparison of the Parameters of Laser Microanalyzers

| Details, Functional Values | LMA 10 VEB Carl Zeiss (GDR) | Mark III Jarrell-Ash Div. US | JLM 200 JEOL Ltd. (Japan) |
|---|---|---|---|
| | *Technical Details* | | |
| 1. *Laser* | | | |
| Rod material | Ruby | $Nd^{3+}$ glass | $Nd^{3+}$ glass (replacement by ruby possible) |
| Rod dimensions | 7 × 75 mm | | 6.5 × 100 mm |
| Laser wavelength | 694 nm | 1060 nm | 1060 nm |
| Length of resonator | 190 mm | | |
| Modes of operation of laser | (1) Unswitched operation; (2) Passive $Q$-switching with stepped cell in six modes of operation | Active $Q$-switching with Pockels cell | Variable in five steps |
| Laser output energy | 0.1 to 1.2 J | 0.1 to 1 J | Max. 2 J |
| Laser output power | 2 MW to 10 kW | | |
| Duration of a sequence of spikes | $\geqslant \mu s$ | 1 to 2 $\mu s$ | |
| Duration of a single spike | 50 ns | | |
| Repetition frequency of laser shots | 4 $min^{-1}$ | 4 $min^{-1}$ | 4 $min^{-1}$ |
| Divergence of laser beam | | 20 min | |
| Flashtube | Rod | Rod | Helix |

| | | | |
|---|---|---|---|
| Flashtube voltage | Continuously adjustable to 1 kV | | |
| Cooling | Air and water (thermostat) | | Water, 3 liters/min (connection to water pipe) |
| Reflector | Elliptical-cylindrical | | |
| 2. *Auxiliary spark gap* | | | |
| | Synchronous operation (triggering of discharge of auxiliary spark by microplasma) | Triggering of synchronous operation of auxiliary spark discharge by microplasma | |
| | External ignition (set off by triggering of the flashtube; time delay of discharge of the auxiliary spark in relation to laser pulse is adjustable) | | Triggering by laser pulse |
| Voltage | 0 to 6 kV continuously adjustable | 1.5, 1.75, 2 kV | 0 to 3 kV, variable |
| Capacity | 2.5 $\mu$F | 20 $\mu$F | 1, 2, 3, …, 10 $\mu$F |
| Inductance | 30, 125, 500 $\mu$H | 150 $\mu$H | 25, 50, 100, 125, 150 $\mu$H |
| Resistance | | | 1, 2, 3, 4, 5, $\Omega$ |
| Electrode distance | Continuously adjustable | | |
| Electrode material | Spectral carbon | | |
| Electrode shape | 5 mm × 40 mm, unilaterally tapered | | |

**TABLE 23.** (*Contd.*)

| Details, Functional Values | LMA 10 VEB Carl Zeis (GDR) | Mark III Jarrell-Ash Div. US | JLM 200 JEOL Ltd. (Japan) |
|---|---|---|---|
| 3. *Microscope* | | | |
| Methods of observation | | | |
| | Incident light | Incident light | Incident light |
| | Transmitted light | Transmitted light (accessory) | Transmitted light (accessory) |
| | Bright field | Bright field | |
| | Qualitative polarization | Qualitative polarization (accessory) | Qualitative polarization (accessory) |
| | Photomicrography (accessory) | Photomicrography, projection (accessory) Polaroid (accessory) | Photomicrography |
| | Electrode projection | | |
| | Autocollimation for adjustment purposes | | |
| Objectives | | | |
| | Flat-field achromat $\times 4/0.05$ $\infty/0$ | Achromat $\times 4$ (focal length $f = 32$ mm) | $\times 2.5$ |
| | Flat-field achromat $\times 16/0.20$ $\infty/0$ (focal length $f = 15.6$ mm) | | $\times 5$ |
| | Catoptric objective $\times 40/0.50$ $\infty/0$ (focal length $f = 6.3$ mm) | Achromat $\times 10$ (focal length $f = 18$ mm) | $\times 40$ |

| | | | |
|---|---|---|---|
| Eyepiece | × 12.5 (16) | × 15 | × 7<br>× 10 (wide field)<br>× 15 |
| Total magnification | × 50, × 200, × 500 | | × 25, × 50, × 400 |
| Diameter of the field of view (at the object) | × 4:4 mm<br>× 16:1 mm<br>× 40:0.4 mm | × 4:3 mm | |
| Working distances | × 16:20 mm without front protective plate; 14 mm with front protective plate<br>× 40:18.8 mm without front protective plate; 15.7 mm with front protective plate | × 4:19.1 mm<br>× 10:9.1 mm | |
| Stage | Circular, centerable, rotatable in two coordinates with verniers | Coordinate stage | Coordinate stage, 125 mm × 125 mm (range of movement 70 mm × 25 mm; vertical 33 mm; verniers); rotary stage (diameter 140 mm, rotatable by 360°, vernier) |
| Sample size (max.) | 70 × 180 × 200 mm | | 100 × 30 mm |
| Mass of sample (max.) | 1 kg | | |
| 4. *Other accessories* | | | |
| | Measuring apparatus for energy and number of spikes | | |
| | AAS coupling unit | | |

**TABLE 23.** (*Contd.*)

| Details, Functional Values | LMA 10 VEB Carl Zeiss (GDR) | Mark III Jarrell-Ash Div. US | JLM 200 JEOL Ltd. (Japan) |
|---|---|---|---|
| | *Physical Performance* | | |
| 1. Dimensions of the analytical region in local analyses | | | |
| Diameter | 10 to 250 μm | 25 μm to 1 mm | 10 to 300 μm |
| Depth | 1 to 250 μm | 8 to 10 μm at small diameters<br>1 μm at large diameters | |
| 2. Consumption of material | ng to μg | ng to μg | ng to μg |
| 3. Number of elements determinable simultaneously | > 60 | > 60 | > 60 |
| 4. Limits of detection (average) | 10 to 100 ppm | 75 ppm | 10 to 100 ppm |
| 5. Reproducibility of spectral line intensities | 5 to 25% | 5 to 25% | |

*Source*: Company publications.

the target surface is calculated from an estimate of the diameter to which the output beam can be focused.

Since the use of short-focal-length lenses implies short working distances from the target, some protective cover (e.g., a quartz plate) is required to prevent the lenses from being coated by ablated target material. Multielement optics (such as microscope objectives) should be air-spacd objectives or mirror lenses to prevent damage by absorption of the high-powered pulse. In commercial instruments, the irradiation optics are incorporated into a microscope to allow pre- and postevaporation examination of the target. Features such as viewing by transmitted, reflected, or polarized light may be included.

An essential feature of a laser microscope is that the objectives for both the observation of the specimens and for focusing the laser radiation must be suitable for the vaporization and excitation of the material. Specifically, for the proposed use in laser micro emission spectroscopy, there is an auxiliary spark gap incorporated between the objective and the surface of the specimen and also an optical pathway for imaging the electrodes of this spark gap on the entrance slit of the spectrograph.

The construction of a microscope and the path of the beams are shown in Fig. 39. A microscope consists, for example, of a base containing the optical system for the reflected beam and the electrode-imaging system, a transverse carrier with the second part of the optical pathway of the reflected beam, and

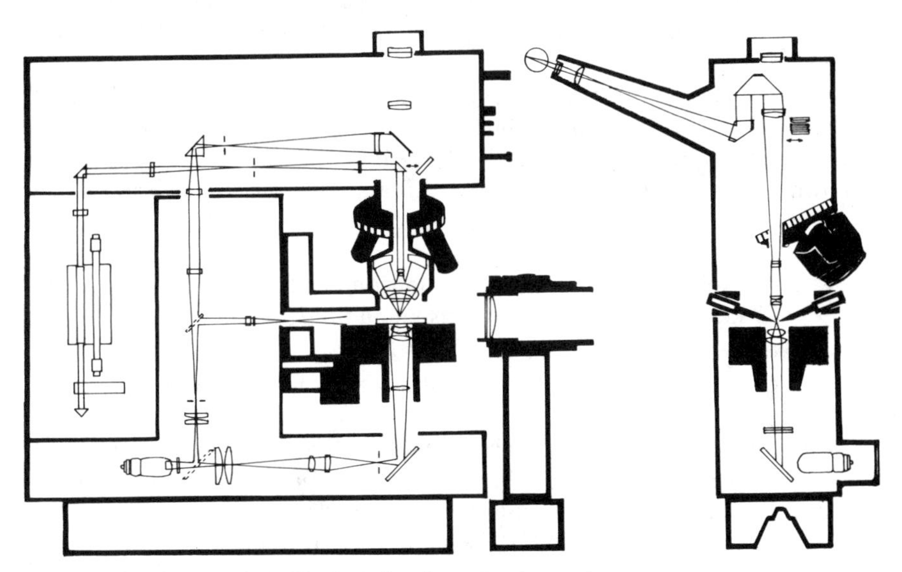

**Figure 39.** Optical pathway in a laser microscope.

the optical system for the pathway of the laser beam and autocollimation, as well as that for the observation beam and photomicrography.

On the left of the vertical carrier is the laser chamber between the base and the transverse carrier. It contains a ruby, a neodymium-doped glass, or a garnet as the active medium and a flashtube as the optical pump arranged at the focus of a reflector, and also the constructional elements of the resonator and the *Q*-switched device. To the right of the vertical carrier is the specimen chamber with the condenser for transmitted light, the specimen stage, the turret head with objectives, and the auxiliary spark gap. In front of the opening on the side turned toward the spectrometer there is a condenser, which images the radiation emitted from the cloud of vapor of the material under investigation onto the spectral apparatus.

A laser microscope may have the following possibilities:

- Optical path for transmitted light
- Optical path for incident illumination
- Observation in polarized light
- Photomicrography
- Projection of the electrodes by an imaging condenser
- Autocollimation beam path for adjusting the laser and the optical adjustment diaphrams

Figure 40 shows the components and principles of operation diagram of a laser microanalyzer. For distribution and bulk analysis it is necessary to use a scanning stage.

Felske et al. [3] reported on the technical requirements and possibilities of application of spectrochemical bulk analysis with laser pulses of relatively high repetition frequency as early as 1966 (Fig. 41a). Starting from the well-known process of spark emission spectral analysis of metallic samples, the following were seen as the advantages of the new method:

- Direct analysis even of electrically nonconducting or poorly conducting solid specimens without time-consuming preparation
- Avoidance of the errors that can arise in analyses with electrical discharges (e.g., glow discharges, diffuse, nonvaporizing discharges, etc.)

For this purpose, a laser-micro device with electronically controlled scanning stage was provided which caused automatic displacement of the specimen in an $x - y$ raster motion. A ruby laser emitted an energy of 0.1 J at a pulse repetition frequency of 1 Hz. Each shot produced craters with a diameter and depth of 20 to 50 $\mu$m. By averaging over 100 shots, it was proposed to reduce the relative scatter of the spectral line intensities of individual shots by a factor of 10, in accordance with the $\sqrt{n}$ law. A factor of 3 was achieved.

A different solution was realized by Moenke and Moenke-Blankenburg [4]. A fast-repeating ruby laser with a maximum pulse repetition frequency of 5 Hz at a maximum output energy of 0.15 J was used for excitation. Laser

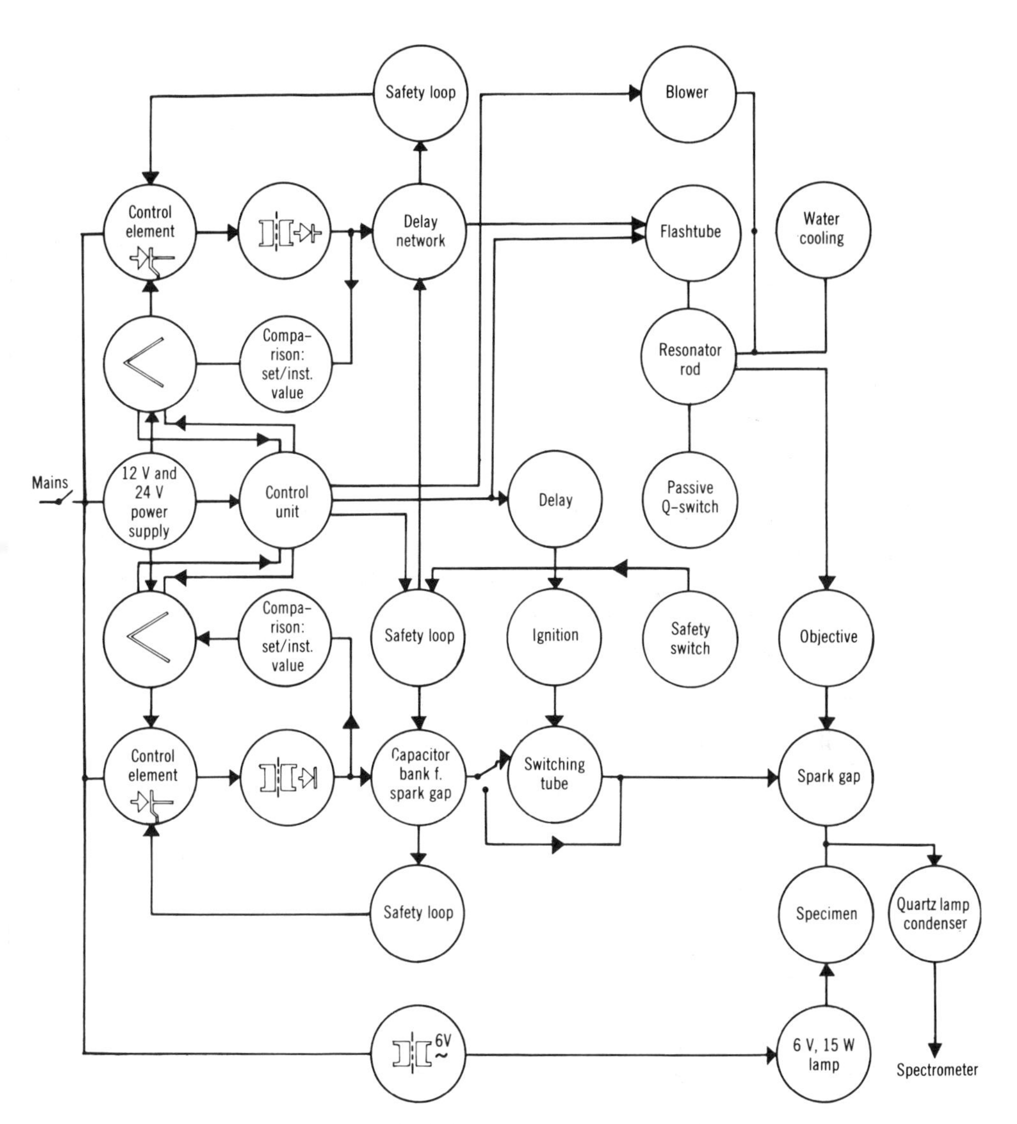


**Figure 40.** Block diagram of a laser microanalyzer.

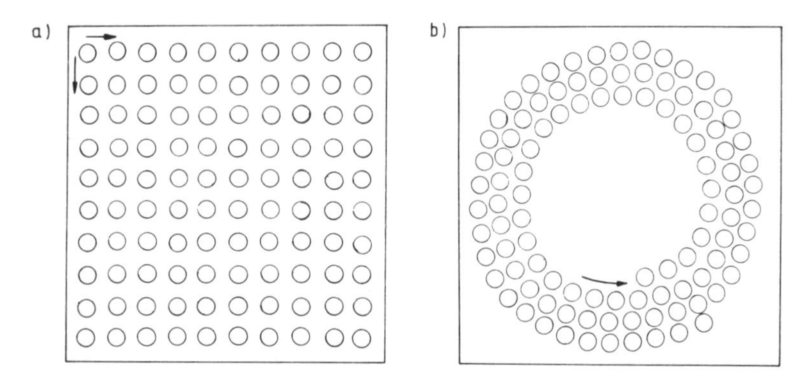


**Figure 41.** Scanning laser microanalysis: (*a*) distribution of laser crater in an $x-y$ raster, proposed in Refs. [2, 3, and 5]; (*b*) distribution of laser crater in a spirally raster, proposed in Ref. [4] reprinted with permission of Akademische Verlagsgesellschaft Geest und Portig KG, Leipzig.

power of about 1 MW was achieved with active *Q*-switching (rotating prism, 500 Hz). The analytical specimen was moved spirally on a special stage under the microscope objective (Fig. 41b). The number of steps for one spiral could be preselected at 1 to 100 (with indication of the steps and manual resetting). The length of the steps was 100 to 250 $\mu$m in four stages, and the distance between the coils of the spiral was 250 $\mu$m. Triggering of one, two, four, and eight laser shots per step was possible. To investigate the surroundings of a given position of the sample, it was possible to cause the spiral to open or close with a free internal diameter of up to 2 mm. With this stage, the following technical variants were realized:

- Local analysis
- Line analysis (circular)
- Area analysis (circle, annulus)
- Bulk or average analysis

The improved reproducibility of the spectral line intensities with 100-shot average analysis as compared with single-shot analysis according to the $\sqrt{n}$ law was not fully realized with photographic recording, since the uncertainty of measuring ratios after allowing for photographic plate background could amount to as much as $\pm 6\%$. The advantages of laser scanning average analysis are fully realized only with photoelectric recording.

A third simple technical variant was proposed by Bieber [5]: he used a mechanical stage with vernier reading and a device for accurate adjustment of the distance of the auxiliary electrodes from the surface of the specimen. This device consisted of a micrometer and a low-current auxiliary circuit with 6-V signal lamps, so that adjustment with an accuracy of 0.01 mm was possible. The optimum size of the laser craters for quantitative micro spectral analysis was between 60 and 100 $\mu$m (Fig. 41a). Bieber made both line and area analysis with photographic recording.

Moenke-Blankenburg and co-workers [2] have prepared a scanning stage which allows the following applications. In the first place, an accurate line analysis with distances of 10 to 500 $\mu$m between the measurement points and with a number of measurement points of from 2 to 25 (i.e., with a line length of 20 $\mu$m to 12.5 mm) can be made. Second, it permits an accurate area analysis of from 400 $\mu m^2$ up to a maximum of 62.5 $mm^2$ with 4 to 250 measurement points and again a distance between the measurement points of from 10 to 500 $\mu$m (Fig. 41a). Third, through a programmable summation of 2 to 10 (or even more) individual measurements, average analysis over a line up to 5 mm long and over an area of up to 2.5 $mm^2$ and therefore, depending on the depth of penetration of the laser radiation, a corresponding bulk analysis are possible.

Laser microanalysis has some unique properties not found in standard laboratory analysis methods. Bulk solid samples can be prepared rapidly either in the vapor or via an auxiliary technique. Also, species for analysis can be both prepared and excited in one step by optical emission spectroscopy [6]. Laser microanalysis with cross excitation has been applied to a wider variety of materials then any other microprobe method. These include minerals, meteorites, metals, ceramics, glasses, archaeological samples and restoration materials, biological samples, oils, welds, and various powdered materials of conducting and nonconducting matrices [4].

Despite that, there are many problems related to the reproducibility of this method, depending on the interaction of the laser light with the solid (see Chapter 3). Instrument instability can be overcome using, for example, correction methods (see Section 2.5) or using the laser double-beam microanalyzer [1, 7]. The results have a relative standard deviation of 2.5% [4], 3 to 4% [1, 7], to 39% [6, 8].

Detection limits for most elements lie in the range 0.1 to 500 pg; concentration detection limits are between 10 and 200 ppm [6]. In principle, femtogram masses of some element should be detectable, but incomplete vaporization and other plasma parameters prevent analyses at these low levels [6], (see also Chapter 3).

Routine applications of laser microanalysis with cross excitation and photographic registration have decreased in recent years together with decreased availability of commercial instruments [6]. Recently, there has been renewed interest in laser ablation for use with other forms of analysis because it exhibits several attractive features.

### 5.1.2. Construction and Mode of Action of Spectral Apparatus and Radiation Detectors

There are many ways of classifying spectroscopic instruments. They can be considered according to the physical principles that they embody; whether they utilize the principles of differential refraction, diffraction, or interference; or alternatively, according to the function that they are to carry out. From the point of view of the experimentalist, the latter classification is the most useful, since the selection will be made on the basis of function. Thus the researcher will only need to compare the merits of those instruments that are suited to the purpose.

#### *5.1.2.1. Fundamental Concepts*

Resolving power is most easily understood as it applies to monochromators. These are instruments that select and transmit a certain range of wavelengths in the spectrum while absorbing or reflecting the rest. If the range selected is from $\lambda - \Delta\lambda/2$ to $\lambda + \Delta\lambda/2$, centered on wavelength $\lambda$, the resolving power is defined as $R = \lambda/\Delta\lambda$ and is a dimensionless number. The concept can be applied to other instruments. For example, in a spectrograph, if two wavelengths $\lambda$ and $\lambda + \Delta\lambda$ appear as two separate lines on the photographic plate, the same formula applies.

A distinction must be made between theoretical and practical resolution. The theoretical resolving power of a prism spectrograph is the product of the material dispersion $\Delta n/\Delta\lambda$ and the base length of the prism $b$: $R_p = b\Delta n/\Delta\lambda$, where $n$ is the refractive index of the prism material. This is valid for an infinitely narrow slit. With incrasing slit width the practical resolving power falls below that determined theoretically.

With a grating spectrograph or spectrometer, the theoretical resolving power is equal to the product of the order number of the grating spectrum $m$ and the total number of lines in the grating $N$: $R_g = mN$. Apart from the reduction of the theoretical resolving power due to the finite slit widths and the properties of the photographic plate, with good spectrographs the practical resolving power differs insignificatly from the theoretical value. It can easily be determined by finding the wavelength difference between two spectral lines which are just separated.

What demands must be placed on a spectral apparatus suitable for laser microanalysis? For this special purpose the intensity of the light is of prime interest. Despite the use of the term luminosity in photometry, it has come into common use when speaking of the performance of an optical system. It has no connection here with the brightness of a source but is used instead to describe the "light-gathering power" of an instrument. Other terms in this sense are "etendue," "throughput," "light grasp," and "collecting power." The term denotes the amount of light that passes from the source, through the instrument, and on to the detector. It is determined by the aperture ratio $K$ of the camera objective, which is obtained from the ratio of the focal length $f$ to the diameter $d$: $K = f/d$. If the bundle of rays is not delimited by the edge of the objective but by a diaphragm, $d$ must be taken not as the diameter of the objective but as that of a circular surface having the same area as the actual diaphragm. In this case $K$ is called the "effective aperture ratio." The intensity of the light is inversely proportional to the square of the effective aperture ratio. When the plate holder is highly inclined, the angle of inclination $n$ must be included in the calculation. The luminous intensity $L$ is then $L = \cos n/K^2$. In addition to this, losses of radiation due to scattered light, absorption, and reflection must be taken into account in evaluating the intensity as a criterion for the efficiency of spectral apparatus.

The resolving power–luminosity product is commonly known as the "efficiency" of a spectrometer and is defined by $E = RL$. The chief usefulness of the term is as a means of comparing the relative merits of two competing systems. For a given instrument at a given wavelength the efficiency is fixed, but it is possible to trade resolution for luminosity, or vice versa, as required.

As a third factor one must consider the linear dispersion, $\Delta x/\Delta\lambda$. This is the ratio of the distance between neighboring spectral lines on a plate to their wavelength difference. As a rule, the reciprocal value in nm/mm is given.

Prism instruments show a pronounced wavelength dependence of the linear dispersion, while for grating instruments it is almost constant. The dispersion formula for the grating spectrum in the middle of the plate is $\Delta x/\Delta\lambda = fmA/\cos\alpha$, where $\alpha$ is the angle of incidence, at the center of the plate, equal to the diffraction angle $\beta$; $f$ is the focal length of the spectral instrument; $m$ is the order of the spectrum; and $A$ is the number of lines per millimeter. The linear dispersion of a grating spectrograph increases in proportion to the focal length, while the luminous intensity increases as the square of the focal length.

For laser microanalysis, as far as possible, a spectral apparatus of high light-gathering power and adequate dispersion is selected. It is also important, particularly for laser microspectral analysis, that the largest possible region of the spectrum be included in one spectral exposure. The main analytical lines of more than 50 elements lie between 200 and 500 nm.

#### 5.1.2.2. *Spectrographs with Photographic Recording*

For more than 25 years, photographic recording has proved satisfactory in laser microanalysis because of its properties, such as the ability to store both the position and intensity of spectral lines on long-lasting photographic material; high detection capacity over a wide spectral range, which permits the bulk of the elements of the periodic system to be recorded simultaneously; and not least, because of its simplicity of operation and relatively low purchase price and maintenance cost. But disadvantages have to be accepted in exchange, these being:

- A chemical process is necessary for the development of the photographic layer, which is unsatisfactory in relation both to time consumed and to reproducibility.
- Qualified personel and a microdensitometer with suitable accessories are necessary for the evaluation process.
- At low light levels corrections must be made in order to calculate the true intensities.

In photographic recording the intensity of the radiation is always integrated over time. The resulting blackening of the photographic layer is proportional to the amount of light (amount of light = intensity $I \times$ time $t$) [9]. The blackening corresponds to the logarithm of the reciprocal transmission of the photographic material. It is measured with a densitometer. Laser microanalysis requires a photographic detector of high sensitivity since the exposure times are very short in comparison to spectral analysis with arcs or sparks.

The task of qualitative elementary analysis is to test samples for the presence of quite distinct elements or to determine all the elements detectable in the specimen. For this purpose, spectral atlases and spectral tables containing the spectral lines and indicating the strongest lines of the elements as the analytical lines are used. Quantitative spectral analysis is based on the relationship between the intensity of the radiation of the spectral lines recorded and the concentration of the corresponding element, as described above.

### *5.1.2.3. Photoelectric Recording*

The spectrometer principle consists in arranging slits in the focal plane of a spectral apparatus that screen off the lines. Behind the slits are placed detectors such as photoelectron multiplier tubes (PMTs), which produce a photocurrent proportional to the intensity of the incident radiation. Since the intensity cannot be measured absolutely, a measurement relative to a calibration standard must be performed. The intensity of the main element of the sample can be used as the calibration standard so that an analytical line and a reference line must be measured simultaneously. In addition, a continuous background radiation value in relation to wavelength must be subtracted from the discrete line spectrum before the ratio is derived. This leads to the need for measurements in at least three channels: for an analytical or measurement line of the spectrum, for a reference or comparison line of the spectrum, and for background. Moenke-Blankenburg et al. [10] have reported the results obtained with an adapter with three channels for a spectrograph. Summarizing, the following observations were made:

- It was possible to lower the absolute limits of detection by one to two orders of magnitude.
- It was possible to reduce the relative limits of detection with small craters (a few hundred $\mu m^3$) by an order of magnitude.
- The relative standard deviation of the determination of concentration was $\leqslant 5\%$ and therefore better by a factor of about 2 than with photographic recording.
- The speed of analysis had improved manyfold.

Nevertheless, the following disadvantage remained, which favored the retention of photographic plates: Because the device is limited to fixed exit slits and photomultiplier measuring channels, the elements that could be analyzed were predetermined, or else the channels had to be adjusted such that the desired elements could be analyzed. Both conditions require that the composition of the sample be known, which is frequently not the case. The universality of the photographic recording procedure could not be achieved by photoelectric methods in this way [11, 12].

In the past 15 years there has been a steadily increasing awareness and recognition of the utility of optoelectronic image devices (OIDs) as multichannel spectrometric detectors. These devices are capable of simultaneously detecting an entire spectral region [13–24]. These detectors permit both one- and two-dimensional position coordinates, and the intensity values corresponding to them, to be determined electronically. Basically, this permits a universality analogous to that of the photographic plate, with a simultaneous manyfold increase in the speed of analysis, although still with the severe limitation that in each case, only part of the spectrum can be investigated in one working step. Two types of area detectors are of prime interest for the recording of laser micro emission spectra. The development of image-recording techniques for television was the stimulus for attempts to use vidicons for direct evaluation of spectra. Up to the beginning of the 1970s, the low sensitivity of the photosensitive layers and the small dynamic range were obstacles. Vidicons with Si multidiode targets were used by Santini in 1972 [25], Jackson in 1973 [26], Knapp et al. in 1974 [15], Milano et al. in 1974 [21], Talmi et al. in 1975–1983 [27–31], Olesik and Walters in 1984 [32], and Saisho et al. in 1985 [33].

With these detectors, the photosensitive target consists of a crystal with a

**TABLE 24. Detector Types**

| Characteristics | SIT Vidicon | SIT Vidicon | ISIT Vidicon | Diode Array |
|---|---|---|---|---|
| Photosensitive material | Si | S 20 | S 20 | Si |
| Wavelength range | 350 to 1100 nm | 350 to 840 nm[a] | 195 to 900 nm | < 180 to 100 nm (350–1100 nm with fiber optic) |
| Sensitivity (photons/count) | 2200 at 600 nm | 11 at 500 nm | 7 at 500 nm | 180 at 650 nm |
| Active elements | 499 × 256 | 499 × 256 | 499 × 256 | 499 |
| Element size | 25 μm[b] | 25 μm[b] | 25 μm[b] | 25 μm × 2.5 mm |
| Dynamic range | 4000:1 | 4000:1 | 4000:1 | 16,000:1 |
| System noise (counts rms) | < 3 | < 3 | < 3 | < 1.5 |
| Uniformity of sensitivity | ±8% | ±10% | ±12% | ±5% |
| Resolution (number of channels FWHM) | 2 | 3 | 3 | 2 |
| Persistence | 260 ms | 260 ms | 260 ms | No |
| Deviation from linear response | < 1% | < 1% | < 1% | < 1% |
| Geometric distortion | 2 channels | 3 channels | 3 channels | < 1 channel |
| Gating | No | Yes | No | No |
| Gate time | — | | — | — |
| Gating ratio | — | 1000:1 | — | — |
| Cooling | −60°C (option) | −60°C (option) | −60°C (option) | −40°C |

| Detector Type | Diode Array | Intensified Diode Array | Intensified Diode Array | Intensified Diode Array |
|---|---|---|---|---|
| Photosensitive material | Si | Si | Si | Si |
| Wavelength range | < 1800 to 1100 nm (350 to 1100 nm with fiber optic) | 350 to 900 nm[a] | 350 to 900 nm[a] | 195 to 900 nm |
| Sensitivity (photons/count) | 1800 at 650 nm | 10 at 550 nm | 10 at 550 nm | 8 at 550 nm |
| Active elements | 999 | 499 | 999 | 499 |
| Element size | 25 $\mu$m × 2.5 mm | 25 $\mu$m × 2.5 mm | 25 $\mu$m × 2.5 mm | 25 $\mu$m × 2.5 mm |
| Dynamic range | 16,000:1 | 16,000:1 | 16,000:1 | 16,000:1 |
| System noise (counts rms) | < 1.5 | < 2 | < 2 | < 2 |
| Uniformity of sensitivity | ± 5% | ± 10% | ± 10% | ± 10% |
| Resolution (number) (of channels FWHM) | 2 | 3 | 3 | 3 |
| Persistence | No | 3.2 ms | 3.2 ms | 3.2 ms |
| Deviation from linear response | < 1% | < 1% | < 1% | < 1% |
| Geometric distortion | < 1 channel | 2 channels | 2 channels | 2 channels |
| Gating | No | No | No | Yes |
| Gate time | — | — | — | 5 ns[c] |
| Gating ratio | — | — | — | $> 10^6$:1 |
| Cooling | −40°C | −40°C | −40°C | −35°C |

**TABLE 24.** *(Contd.)*

| Detector Type | Intensified Diode Array | Linear CCD | Linear CCD | Matrix CCD |
|---|---|---|---|---|
| Photosensitive material | Si | Si | Si | Si |
| Wavelength range | 195 to 900 nm | < 180 to 1100 nm | < 180 to 1100 nm | 400 to 1100 nm |
| Sensitivity (photons/count) | 8 at 550 nm | 400 at 650 nm | 400 at 650 nm[d] | 400 at 650 nm |
| Active elements | 700 | 1024 (1728) | 1728 | 576 × 384 |
| Element size | 25 μm × 2.5 mm | 13 μm × 13 μm | 13 μm × 39 μm | 23 μm × 23 μm |
| Dynamic range | 16,000:1 | 4000:1 | 4000:1 | 2500:1 |
| System noise (counts rms) | < 2 | < 2 | < 2 | < 2 |
| Uniformity of sensitivity | ± 10% | ± 5% | ± 5% | ± 5% |
| Resolution (number of channels FWHM) | 3 | 2 | 2 | 2 |
| Persistence | 3.2 ms | None | None | None |
| Deviation from linear response | < 1% | < 1% | < 1% | 1% |
| Geometric distortion | 2 channels | < 1 channel | < 1 channel | < 1 channel |
| Gating | Yes | No | No | No |
| Gate time | 5 ns[c] | — | — | — |
| Gating ratio | > $10^6$:1 | — | — | — |
| Cooling | − 35°C | − 35°C, water[e]<br>+ 10°C, air | − 35°C, water[e]<br>+ 10°C, air | − 35°C, water[e]<br>+ 10°C, air |

*Source*: After Ref. 35; reprinted by permission.

[a] UV extension with scintillator.

[b] Height size depending on readout mode.

[c] With pulse generator.

[d] Video signal increase by a factor of about 3.

[e] Temperature controller necessary.

large number of applied Si photodiodes ($\approx 10^6$). These diodes operate in the charge-storage mode and are scanned by an electron beam. The recharging current of the storage layer capacity of the Si diodes is then a measure of the charge that has been neutralized by the incident photons between two scans. The size of the recharging current is therefore proportional to the amount of light incident between two scans. An increase in sensitivity is achieved by the inclusion of an electronic image intensifier [these are silicon-intensified target (SIT) vidicons]. The spectrum falls on the photocathode of an image converter tube (UV $\rightarrow$ visible) and after conversion appears on the output screen. This pattern is stored by a vidicon. The electron beam of the vidicon scans its target. The measurement signals so obtained pass through an analog-to-digital converter to a computer. Here they are subjected to further processing. The Si multidiode target of a vidicon is located in the focal plane of a polychromator. Arranged on the target area, which is 12.5 mm long and 10 mm high, are 500 rows of diodes each 25 $\mu$m wide. The spectral range depends on the diode and the material of the target: SIT vidicon with scintillator: 195 to 900 nm, and standard vidicon: 350 to 1100 nm (see Table 24).

Through the inclusion of an image intensifier in series, the quantum yield of the SIT vidicon is two to three orders of magnitude higher than that of the standard vidicon. The sensitivity of the SIT vidicon reaches that of a photomultiplier. However, the signal-to-noise ratio is inferior. A disadvantage is the quite small size of the target, with a given resolution of about 15 lines/mm. Only one section of the spectrum can be imaged on the target.

Another type of area detector is obtained when the scanning of the silicon diodes in multiplex operation is also applied to the crystal on which the diodes are located. Such lines of photodiodes are compact microelectronic constructional elements without the disadvantages of the vacuum-tube technique, such as mechanical sensitivity, size, and vacuum tightness (see Table 24).

In 1972 and 1973, Boumans [34] reported on investigations on the use of Si diodes in spectrometry. Si diodes work at room temperature with a preamplifier, dc or ac amplifier, and a recorder. For the dynamic range, Boumans found a value of between $10^2$ and $10^3$, depending on the wavelength. The overall sensitivity was at least two orders of magnitude smaller than that of a photomultiplier. The observed noise arose from the amplifier unit. The signal-to-noise ratio was at least 100 to 300 times poorer than for the photomultiplier. The spectral resolution of the lines of diodes corresponded to the arrangement in which the exit slit had the dimensions of individual diodes.

Since 1973, and particularly in 1975 and 1976, Horlick [13] has published research results on the use of "self-scanning linear rows of Si diodes" [or self-scanned photodiode arrays (SPD)] in spectrometry. One thousand and twenty-four diodes are arranged on a support with a length of, for example, 26 mm (i.e., 40 diodes per millimeter). The scanning circuit is also located on the support. Each diode is connected with the output via a field-effect transistor (FET) switch. The FET switches are controlled successively by a counter circuit so that the measured values of the individual diodes appear at the output with a definite frequency (variable up to 10 MHz). The diodes work in the charge-storage mode; that is, the amount of incident light that has fallen on the diode between two scans is integrated. The whole circuit is included in an integrated-circuit system with 18 or 22 connections. To improve its properties and, in particular, to lengthen the time of integration, the circuit system is cooled to $-40°C$. In this way it has been possible to achieve a marked reduction in dark current.

Optoelectronic image devices (OIDs) provide an attractive alternative to the polychromator approach because they, too, provide the capability of simultaneous detection of dispersed radiation. When these detectors are interfaced with appropriate on-line computer control and data processing facilities, the signals from hundreds of virtually independent optical (spectral) channels can be integrated and digitized simultaneously. In addition, the contribution of continuum spectral background on undesirable overlapping spectral features present in the "blank" spectrum can be sub-tracted or "stripped" from the total signal for all channels. If desirable, random or fast access to the specific spectral features of interest may be employed to conserve computer memory space and time.

Other capabilities provided by these detectors include:

- Flexibility of choice of mode of signal integration and processing: real time, in-memory integration, or on-target integration
- Monitoring of ultrarapid spectroscopic events
- Compensation for source flutuations and for sensitivity and spectral response variations
- High quantum efficiency
- Linear response

A foremost limitation is set by the number of diodes in commercially available diode arrays and their physical arrangement and spatial resolution; in addition, the low amplification gain of the SPD requires longer signal-integration times; and finally, the dynamic range of the SPD is narrower than that of the PMT.

In 1978 and 1979, two series of investigations were carried out with combinations of a laser microanalyzer and two generations of optical multichannel analyzers, OMA 1 and OMA 2 from PARC [28, 30]. As connecting units the authors used polychromators with a focal distance of 0.3 m and exchangeable gratings with 600, 1200, and 2400 lines/mm from

Jarrell-Ash and McPherson. Depending on the choice of grating, it was possible to record either a large spectral range with low resolution, a medium range with medium resolution, or a small range with good resolution (or dispersion). The OMA 1 coupling was made through a 0.3-m Ebert polychromator from Jarrell-Ash. The polychromator was equipped with the user's choice of a 590- or a 2360-line/mm grating. Since the target was 12.5 mm long, the spectral range that could be covered was 70 nm or 17 nm, respectively.

In the SIT vidicon, the electron beam is directed in the form of lines from left to right in 256 tracks in such a way that in each case channel 0 the channel 499 is scanned. The scanning time amounts to 32.8 ms. In the store of the OMA control console, an intensity value is assigned to each of the 500 channels after scanning. The range of values per channel and per scan of the target is about 700 digits (or scale divisions). These values can be sampled individually at the control console and the entire contents of the store can be shown on the display simultaneously. The OMA 1 system has two stores. After the storage in each case of 500 values in stores A and B, independently of one another, comparisons of spectra and background corrections can be made by subtracting B from A. For example, the spectrum of the matrix element of the analytical sample is fed to store B and the spectrum of the analytical sample to store A. By subtracting B from A, the alloying or doping elements can be made

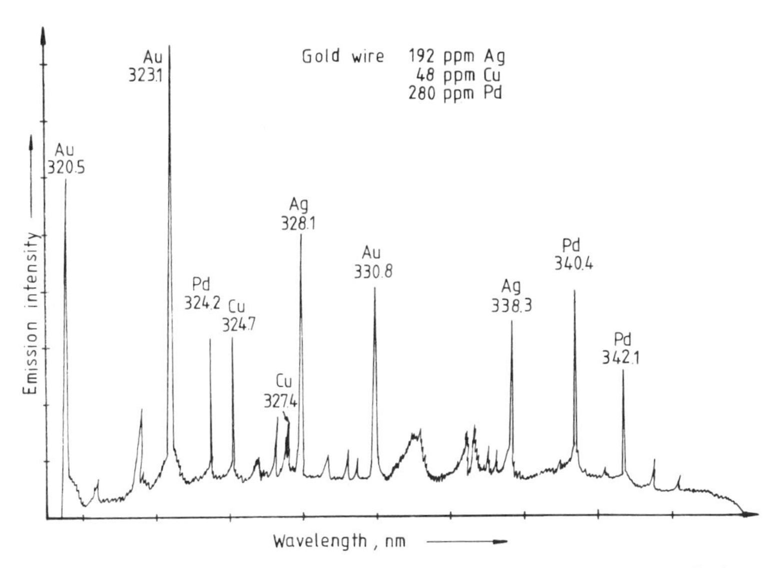


**Figure 42.** Laser micro emission spectrum from a gold wire recorded with a photodiode array in an optical multichannel analyzer.

visible on the monitor. A spectral range of 184.5 to 770 nm was used. When a length of section of the spectrum of 17 or 70 nm was not sufficient, it was necessary to investigate several sections of the spectrum successively. Under some circumstances, this could be difficult for local analyses. The results in relation to reproducibility and to absolute and relative limits of detection were comparable with those of photoelectric recording with photomultipliers.

By Talmi et al. [28, 30] the OMA 2 coupling was made through a McPherson 0.3 m Czerny-Turner vacuum polychromator, type 218, with a 1200-line/mm grating. The OMA 2 system consisted of a 1215 console, a 1216 or 1218 detector control apparatus, and a 1254 SIT vidicon which could be replaced by a silicon photodetector. The external triggering was done through the laser flash, which released a voltage pulse via the light-conducting cable of a Tektronix 604 Waveform Monitor.

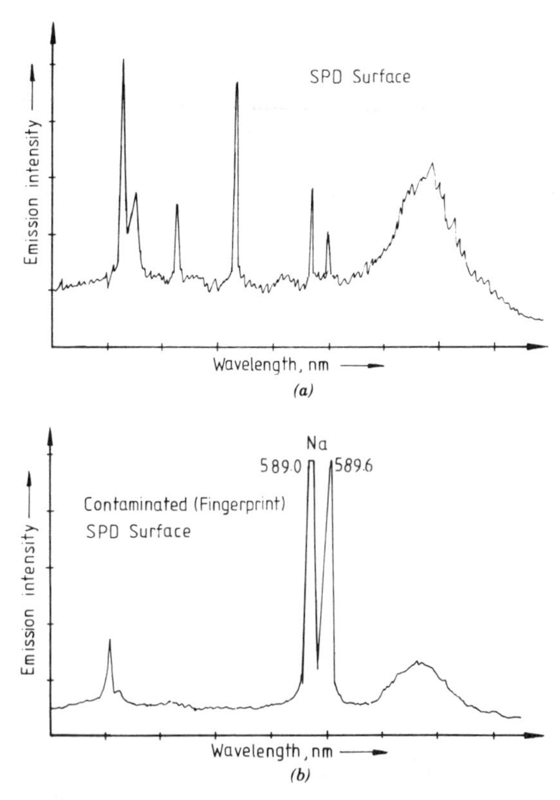


**Figure 43.** Emission spectrum from the target surface of a photodiode array: (*a*) clean surface; (*b*) fingerprint-contaminated surface. Conditions: Laser microanalyzer with laser output energy of $\approx$ 1 J, crater diameter 160 $\mu$m by 2 $\mu$m depth, wavelength range 572.8 to 600.5 nm. From Ref. [30], reprinted by permission of Elsevier Science Publishers, Amsterdam.

Using four known spectral lines of the region concerned, it is possible to program the display to show accurately the wavelength assigned to the particular channel (Fig. 42). With a frequent change in the spectral region, it is possible to work with known comparison spectra to identify lines. The OMA 2 system is capable of storing 100 spectra in a floppy disk. These spectra can be recorded at any time and shown on the display, even simultaneously.

Figure 43 shows the spectrum of the surface of a defective silicon photodiode array that was sampled and shown to contain traces of sodium. Also shown is the emission spectrum obtained from the same surface region after it had been contaminated with sodium by a fingerprint. This indicates that the combined system can be useful in surface-contamination studies.

For investigations of concentration profiles, the technical possibility of presenting the lateral cross section or the depth cross section is a great advantage. Figure 44 shows the emission spectrum of chromium obtained from a line along an electrically nonconductive dissected ruby rod. The

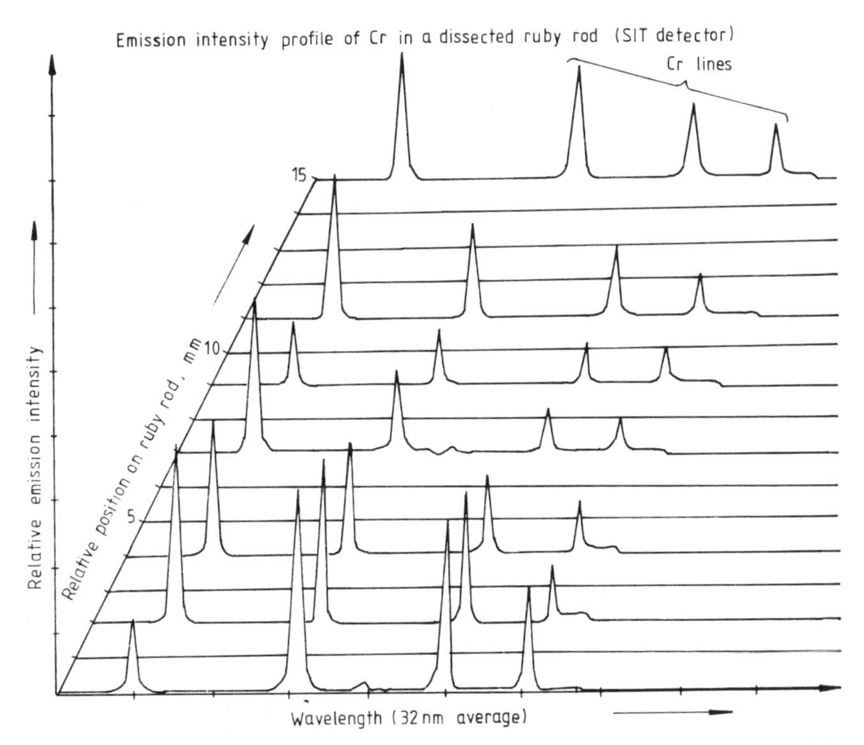


**Figure 44.** Line analysis of chromium in a ruby rod with LMA coupled with OMA with a SIT detector.

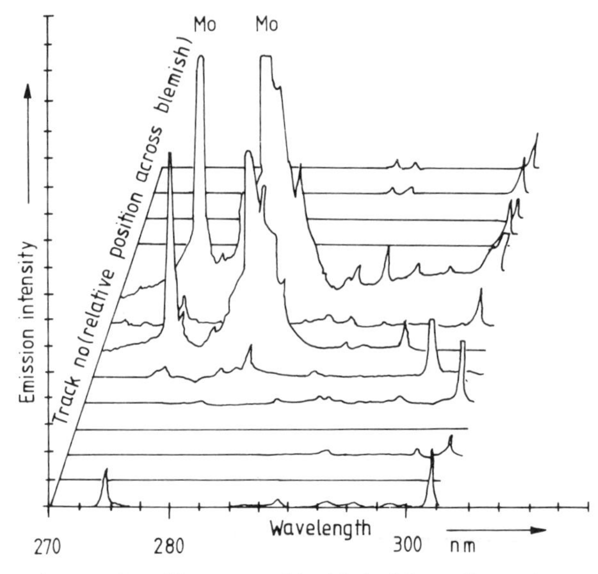


**Figure 45.** Lateral spectral profile across a blemish ($\approx 0.2$ nm diameter) on a ceramic tube showing molybdenum lines between 277 and 287 nm. From Ref. [30], reprinted by permission of Elsevier Science Publishers, Amsterdam.

intensity differences between the measuring points correspond to local variations in chronism concentrations.

Figure 45 is a similar presentation of the emission spectra obtained by lateral probing across a dark blemish found on a white ceramic tube. The emission lines have been identified as those of molybdenum, probably from contamination introduced during the fabrication process. By plotting the data with 180° rotation, it was possible to show that there were no hidden lines behind the broad molybdenum lines. The system provides a means of obtaining rapid chemical composition of lateral profiles of solid surfaces, for either qualitative or quantitative determinations [30].

Figure 46 is a depth profile (i.e., emission spectral characteristics versus probing depth) of an electrical conductor with a silver-conducting layer. Each spectrum is the result of a single laser shot. Through focusing–defocusing manipulations, and aperture, objective, and *Q*-switched stage selection, the diameter of the laser beam was gradually reduced so that different depths of the capacitor were sampled [30].

Table 25 summarizes some of the detection limits obtained with the silicon-intensifield target vidicon and the silicon photodiode array. Talmi et al. [30] reported that the detection limits of the intensified vidicon for Fe, Ni, and Mn were $1 \times 10^{-11}$, $5 \times 10^{-12}$, and $2 \times 10^{-12}$ g, respectively, compared with

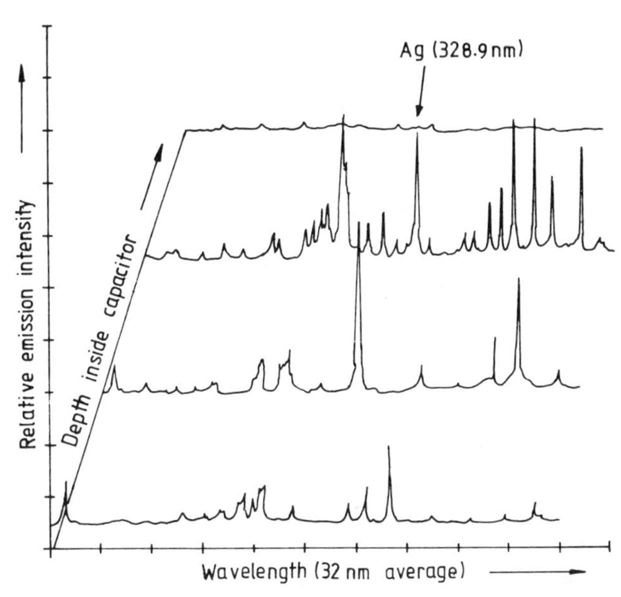


**Figure 46.** Depth spectral profile of an electric capacitor using a laser microanalyzer with an optical multichannel analyzer (SIT-vidicon). From Ref. [30], reprinted by permission of Elsevier Science Publishers, Amsterdam.

**TABLE 25. Detection Limits for Several Elements in Different Matrices with the Laser Microprobe Adapted to Imaging Detectors**

| | | | Detection Limits (ppm) | |
|---|---|---|---|---|
| Element | Matrix | Spectral Line (nm) | Intensified Vidicon | Photodiode Array |
| Ag | Copper alloy | 328.3 | <15 | |
| | Gold wire | 328.3 | <15 | |
| Cr | Copper alloy | 429.0 | 2 | 5 |
| | Aluminium alloy | 429.0 | | <5 |
| | Steel | 429.0 | <100 | <100 |
| Ga | Aluminium alloy | 417.2 | | <10 |
| Cu | Steel | 324.7 | <20 | |
| | Gold wire | 324.7 | | <10 |
| Mn | Copper alloy | 279.5 | <10 | <20 |
| Sn | Copper alloy | 284.0 | <500 | |
| Pb | Copper alloy | 280.1 | <500 | |
| Pd | Gold wire | 324.2 | | <50 |

*Source*: Refs. 28 and 30; by permission of Elsevier Science Publishers, Amsterdam.

values of $2 \times 10^{-10}$, $1 \times 10^{-11}$, and $5 \times 10^{-11}$ g for a typical photographic plate. These typical results are even more significant considering the fact that no auxiliary cross excitation was used with the vidicon.

Saisho et al. [33] coupled a laser microanalyzer with a G-500 III monochromator (Nikon), focal length 500 mm, grating 1200 gr/mm, dispersion 1.5 nm/mm, and an optical multichannel analyzer TV camera (Hamamatsu Photonics), spectral response range 120 to 850 nm, TV pickup width 7.5 nm, number of channels 1024, data processing unit PC 8801 (NEC). They got detection limits between 0.02 and 0.06% for Al and Si in ceramics and Cr, Cu, Mn, Mo, Ni, and P in steels. The reproducibility obtained by measuring the emission intensity of the Si 250.7 nm and by using a silicon wafer was a relative standard deviation of 3.3%, and the reproducibility of laser crater size was 7.3%. The relative standard deviation measured by Talmi et al. [30] lies between 5.2 and 10.2% using reference lines. Further discussions of the applications of image dissector tubes and linear diode arrays are those of de Haseth et al. [36], Santini et al. [37], and Peiko et al. [38].

The latest advances in charge-coupled device (CCD) technology have resulted in linear arrays of greater than 2048 elements and two-dimensional arrays to 380 × 488 elements. At present the dynamic range is slightly less than that obtained with diode arrays, and because the light is detected through a transparent electrode, the spectral range is more limited (Table 24).

Several commercial multichannel spectrometers are now available with a choice of SIT vidicon, ISIT, diode array, or CCD detectors, such as the above-mentioned OMA system I–III from PARC, the instrument by Hamamatsu, the OSA (Optical Spectral Analyzer) by B&M Spektronik, the Tracor Northern Rapid Scan Spectrometer, and the O-SMA (Optical Simultaneous Multichannel Analysis) by Spectroscopy Instruments. All these instruments can easily be interfaced to a computer. They offer fast performance and good linearity, sensitivity, and dynamic range and therefore are a better choice, although more expensive, than photographic emulsions. Comparisons of the suitability of multichannel detectors to spectrography include those by in Refs. 13, 27, 31, and 38 to 44.

The laser microprobe–optical multichannel analyzer system provides local and distribution analyses at a fraction of the cost of most other microprobes. Because sample preparation and chamber evacuation are not necessary, and because the spectral record is instantly digitized and stored, very rapid qualitative and semi-quantitative determinations can be performed. Quatitative determinations are possible but are subject to the availability of standard materials. Computer control greatly facilitates wavelength calibration, line identification, signal measurement, calculations, and compensation for shot-to-shot variations. At present, the limited spectral window (see Fig. 47) is the greatest disadvantage of the detection system, but computer-manipulated

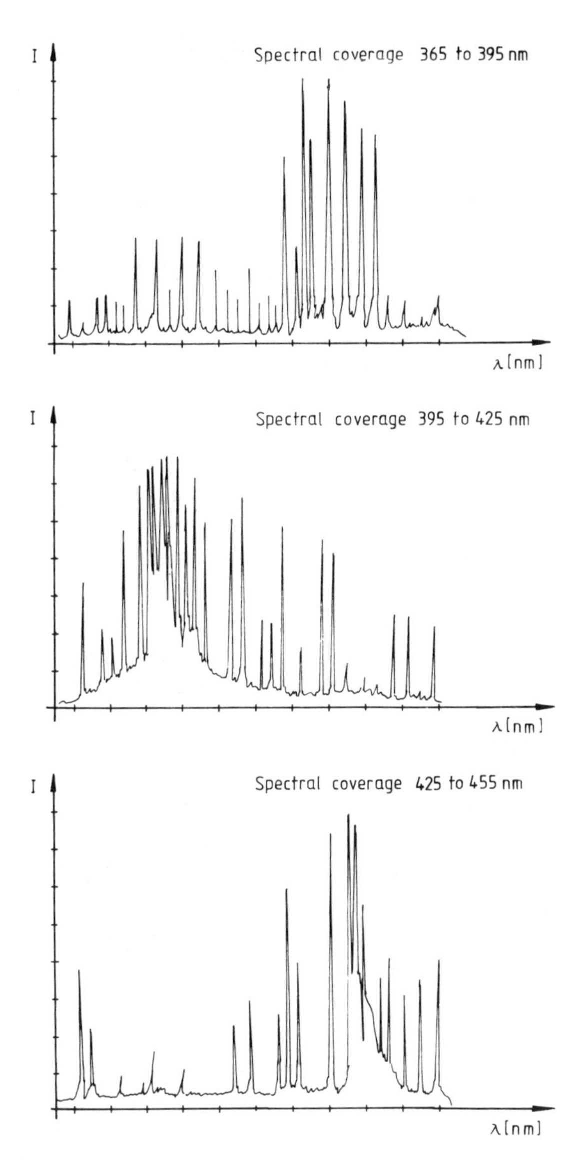


**Figure 47.** Spectral coverage of 30 nm over the range 365 to 455 nm using a laser microanalyzer together with an optical multichannel analyzer with a photodiode array.

polychromator "window" scan could greatly alleviate this shortcoming. In future, two-dimensional format spectrometers [22] could be used to expand the spectral window.

## 5.2. FIELD OF APPLICATION

Laser microanalysis have been used to determine 64 elements (see Table 26) in many types of samples. Laser microanalysis with cross excitation, especially, has been applied to a wider variety of materials than has any other microprobe technique. This includes minerals, meteorites, metals, alloys, welded joints, semiconductors, glasses, ceramics, technical silicates, reactor graphite, monocrystals, polymers, paintings, and medical and biological samples. Included in the 64 elements are light elements such as lithium, beryllium, and boron, which could not be (or not easily) determined with an electron microprobe.

The precision of about 3 to 30% is comparable with dc arc emission spectrochemical analysis of powders and is better than electron and ion microanalyses of solids.

The absolute detection limits are in the range of nanograms to picograms and therefore several orders of magnitude lower than for emission spectrochemical analysis with dc arc or spark, and approach those of atomic absorption with graphite furnace. Relative detection limits are in the range of $10^{-3}$ mass percent and therefore one order of magnitude better than the electron microprobe but one or two orders of magnitude worse than dc arc optical emission spectrography.

Table 26 shows a list of the elements of the periodic system so far determined with laser microanalysis together with the wavelengths used in optical emission spectrography and spectrometry. Detection limits are given in Table 27. The following tables summarize the application fields, their aims of analyses, and the results obtained, such as detection limits and reproducibilities.

### 5.2.1. Mineralogy, Geochemistry, and Cosmochemistry

As a rule, no preparation of the samples is necessary for either quantitative or qualitative investigations of rock-forming minerals of all kinds. The hand specimens customary in mineralogy may exhibit a maximum size of $200 \times 180 \times 70\,mm^3$ (determined by the apparatus). Polished surfaces and thin sections of the usual formats, incrustations, and coatings have a minimum size, determined by the process, of $10 \times 10 \times 2\,\mu m^3$. Mineral grains and dusts can be embedded in a layer of gelatine or fixed to adhesive tape.

The following rules apply: The sample should be free of impurities, which

TABLE 26. **List of Elements Determined with Laser Microanalysis and The Wavelengths Used in Optical Emission Spectrography and Spectrometry**

| *Z* | Element | Wavelength (nm) | References |
|---|---|---|---|
| 3 | Li | 610.4 | 4, 211 |
| | | 323.3 | 108 |
| 4 | Be | 313.0 | 4, 81, 106, 149, 176 |
| | | 313.1 | 149, 176 |
| | | 234.8 | 4, 58, 81 |
| 5 | B | 249.7 | 4, 56, 106, 108, 154 |
| | | 249.8 | 4, 56, 108, 216 |
| 6 | C | 250.9 | 4, 121, 160, 216 |
| | | 283.7 | 4, 113 |
| | | 247.8 | 4, 90 |
| 11 | Na | 589.0 | 4, 58, 82 |
| | | 589.5 | 4 |
| 12 | Mg | 279.6 | 4, 50, 53, 81, 90, 103, 106, 108, 137, 160, 168, 189, 211, 216 |
| | | 280.2 | 4, 58, 90, 108 |
| | | 285.2 | 48, 53, 108, 113, 160 |
| | | 383.8 | 138, 148 |
| | | 278.1 | 4, 106 |
| | | 277.9 | 4, 168 |
| | | 279.1 | 4, 113 |
| 13 | Al | 308.2 | 58, 106, 108, 132, 146, 189 |
| | | 396.2 | 4, 90, 168, 217, 219, 220 |
| | | 394.4 | 4, 90 |
| | | 309.3 | 108, 137 |
| 14 | Si | 288.2 | 3, 90, 108, 113, 138, 139, 148, 162, 189 |
| | | 251.6 | 4, 58, 104, 137, 139 |
| | | 252.4 | 106 |
| 15 | P | 323.4 | 4 |
| | | 255.3 | 189 |
| 16 | S | 349.7 | 4 |
| 19 | K | 766.5 | 4 |
| | | 769.9 | 4 |
| | | 404.4 | 82 |
| 20 | Ca | 393.4 | 4, 81, 90, 106, 137, 160, 211, 216 |
| | | 396.9 | 4, 53, 90, 121, 160, 216 |
| | | 315.9 | 48, 53 |
| | | 317.9 | 58 |
| 21 | Sc | 361.4 | 4, 108 |
| | | 363.1 | 4, 108 |
| | | 364.5 | 108 |

**TABLE 26.** (*Contd.*)

| Z | Element | Wavelength (nm) | References |
|---|---|---|---|
| 22 | Ti | 334.9 | 4, 106, 108, 137, 160 |
| | | 336.1 | 4, 108, 160 |
| | | 337.3 | 4, 108 |
| | | 338.4 | 4, 108, 132 |
| | | 257.3 | 168 |
| | | 278.5 | 168 |
| | | 327.2 | 189 |
| 23 | V | 311.1 | 3, 4, 58, 108, 189 |
| | | 309.3 | 4, 108, 137 |
| | | 310.2 | 4, 106, 108 |
| | | 311.8 | 4, 108 |
| 24 | Cr | 425.4 | 4, 53, 108, 139, 168 |
| | | 267.7 | 103, 106, 108, 189 |
| | | 428.9 | 4, 108, 139 |
| | | 283.6 | 4, 108, 146 |
| 25 | Mn | 257.6 | 4, 81, 106, 168, 189 |
| | | 293.3 | 50, 58, 81, 113 |
| | | 293.9 | 81, 104, 148 |
| | | 279.5 | 144, 146, 168 |
| 26 | Fe | 259.9 | 4, 58, 139, 162, 216, 217 |
| | | 259.8 | 4, 137 |
| | | 238.2 | 4, 148 |
| | | 358.1 | 99 |
| | | 302.0 | 211 |
| 27 | Co | 345.4 | 4, 58, 60, 106, 108, 189 |
| | | 340.5 | 4, 108 |
| 28 | Ni | 341.4 | 4, 53, 60, 95, 106, 108 |
| | | 349.3 | 4, 53, 106, 108 |
| | | 352.5 | 4 |
| 29 | Cu | 324.8 | 4, 50, 58, 81, 106, 138, 211, 217 |
| | | 327.4 | 3, 4, 50, 103, 108 |
| | | 282.4 | 121, 123, 138, 189 |
| | | 224.7 | 4, 168 |
| 30 | Zn | 334.5 | 4, 106, 108, 138, 189, 190, 216, 217 |
| | | 330.3 | 4, 146, 216 |
| | | 328.2 | 4, 108, 216 |
| 31 | Ga | 417.2 | 4, 108 |
| | | 403.3 | 4, 108 |
| | | 287.4 | 4, 108 |
| | | 294.4 | 4, 108 |
| 32 | Ge | 265.1 | 4, 108 |
| | | 326.9 | 4, 108 |

**TABLE 26.** (*Contd.*)

| $Z$ | Element | Wavelength (nm) | References |
|---|---|---|---|
| | | 303.9 | 108 |
| | | 312.5 | 108 |
| 33 | As | 234.9 | 4, 189 |
| | | 228.8 | 4 |
| | | 286.0 | 4 |
| | | 289.9 | 4 |
| 37 | Rb | 780.0 | 4 |
| | | 794.8 | 4 |
| 38 | Sr | 407.7 | 4, 58, 81, 82 |
| | | 421.5 | 4, 82 |
| | | 346.5 | 82 |
| 39 | Y | 371.0 | 4 |
| | | 377.4 | 4 |
| | | 378.9 | 4 |
| 40 | Zr | 342.8 | 4 |
| | | 346.3 | 4 |
| | | 347.9 | 4 |
| 41 | Nb | 405.9 | 4 |
| | | 407.9 | 4 |
| | | 266.2 | 123 |
| 42 | Mo | 281.6 | 3, 4, 106, 137 |
| | | 284.8 | 4 |
| | | 287.1 | 4 |
| 45 | Rh | 343.5 | 4 |
| | | 339.7 | 4 |
| 46 | Pd | 340.5 | 4, 108 |
| | | 342.1 | 4, 108 |
| | | 363.5 | 108 |
| 47 | Ag | 328.1 | 4, 81, 106, 108, 146 |
| | | 338.3 | 4, 50, 108, 189 |
| | | 255.3 | 190 |
| 48 | Cd | 346.8 | 4, 164 |
| | | 228.8 | 106, 108 |
| | | 226.5 | 108 |
| 49 | In | 451.1 | 4 |
| | | 244.8 | 4 |
| | | 207.9 | 4 |
| 50 | Sn | 326.2 | 4 |
| | | 317.5 | 4 |
| | | 284.0 | 146, 189 |

**TABLE 26.** (*Contd.*)

| $Z$ | Element | Wavelength (nm) | References |
|---|---|---|---|
| 51 | Sb | 259.8 | 4, 108 |
| | | 252.9 | 4, 108 |
| 52 | Te | 238.3 | 4 |
| | | 238.6 | 4 |
| 55 | Cs | 894.4 | 4 |
| | | 852.1 | 4 |
| 56 | Ba | 455.4 | 58, 83 |
| | | 416.4 | 82 |
| | | 230.4 | 4 |
| 57 | La | 394.9 | 4, 108 |
| | | 391.6 | 108 |
| | | 366.2 | 108 |
| | | 624.9 | 4 |
| 58 | Ce | 418.7 | 4, 108 |
| | | 413.4 | 4, 108 |
| | | 356.1 | 108 |
| | | 342.2 | 108 |
| 59 | Pr | 422.5 | 4 |
| | | 422.3 | 4 |
| | | 417.9 | 4 |
| 60 | Nd | 401.2 | 4, 108, 145, 164 |
| | | 410.9 | 4, 108 |
| 62 | Sm | 442.1 | 4 |
| | | 442.4 | 4 |
| | | 443.4 | 4 |
| 63 | Eu | 397.2 | 108 |
| | | 393.1 | 108 |
| | | 390.1 | 108 |
| 64 | Gd | 335.1 | 4, 108 |
| | | 335.9 | 4, 108 |
| | | 354.6 | 108 |
| | | 354.9 | 108 |
| 66 | Dy | 400.0 | 4 |
| | | 407.8 | 4 |
| 67 | Ho | 293.7 | 4 |
| 68 | Er | 349.9 | 4, 108 |
| | | 337.3 | 108 |
| | | 369.3 | 4, 108 |
| | | 361.9 | 108 |

TABLE 26. (*Contd.*)

| Z | Element | Wavelength (nm) | References |
|---|---|---|---|
| 70 | Yb | 398.8 | 4, 108 |
| | | 345.4 | 108 |
| | | 347.9 | 108 |
| 72 | Hf | 264.1 | 108 |
| | | 263.9 | 108 |
| | | 264.7 | 108 |
| 73 | Ta | 263.7 | 108 |
| | | 268.5 | 108 |
| | | 340.7 | 108 |
| 74 | W | 239.7 | 4 |
| | | 258.9 | 4 |
| | | 400.9 | 144 |
| 76 | Os | 290.9 | 4 |
| | | 305.9 | 4 |
| | | 225.6 | 4 |
| 78 | Pt | 265.9 | 4 |
| | | 306.5 | 4 |
| | | 299.8 | 4 |
| 79 | Au | 267.6 | 4, 108, 190 |
| | | 242.8 | 4, 108 |
| | | 268.8 | 123 |
| 80 | Hg | 253.6 | 4, 85, 108, 211 |
| | | 435.8 | 4, 85, 108 |
| | | 365.0 | 4, 85, 108 |
| 81 | Tl | 271.5 | 4 |
| | | 243.7 | 4 |
| | | 535.1 | 4 |
| 82 | Pb | 283.3 | 4, 106, 108, 146, 148, 189 |
| | | 405.8 | 4, 108, 211 |
| | | 280.2 | 4, 108 |
| | | 257.7 | 108, 190 |
| 83 | Bi | 306.8 | 4, 106, 108 |
| | | 289.8 | 4, 108, 190 |
| | | 262.8 | 108 |
| 90 | Th | 374.1 | 4 |
| | | 297.9 | 4 |
| 92 | U | 237.8 | 4 |
| | | 367.0 | 99 |
| | | 358.5 | 144 |

**TABLE 27. Relative and Absolute Detection Limits**

| $Z$ | Element | Relative Detection Limits (ppm) | | | | | | Absolute Detection Limits (g) | | | |
|---|---|---|---|---|---|---|---|---|---|---|---|
| | | [151]* | [179] | [8] | [100] | [58] | [108] | [104] | [81] | [152] | [100] |
| 3 | Li | | | | | | | | | | $2 \times 10^{-12}$ |
| 4 | Be | 12 | 10 | | 0.8 | 25 | | | $10^{-13}$ | | |
| 5 | B | > 32 | | | 11 | 300 | | | | | |
| 12 | Mg | | | 10 | | | 50 | | $10^{-13}$ | $3 \times 10^{-10}$ | $2 \times 10^{-15}$ |
| 13 | Al | 200 | | | | | 500 | $2.5 \times 10^{-9}$ | | | |
| 14 | Si | 270 | | 20 | | | 1,000 | $3 \times 10^{-10}$ | | $1 \times 10^{-8}$ | |
| 15 | P | > 1,100 | | | | 10 | 5,000 | | | | $1 \times 10^{-13}$ |
| 20 | Ca | | | | 24 | | 500 | | $10^{-13}$ | | $1 \times 10^{-14}$ |
| 21 | Sc | | 100 | | | | 1,000 | | | | |
| 22 | Ti | 280 | 100 | | 20 | 300 | 1,000 | $5 \times 10^{-10}$ | | $4 \times 10^{-10}$ | |
| 23 | V | 830 | 100 | | | < 90 | 5,000 | $7.5 \times 10^{-10}$ | | | |
| 24 | Cr | 60 | 100 | | 11 | 15 | | $1.5 \times 10^{-10}$ | | $2.5 \times 10^{-9}$ | |
| 25 | Mn | 40 | 100 | 20 | | 100 | 1,000 | $6 \times 10^{-11}$ | $10^{-13}$ | $5 \times 10^{-10}$ | $2 \times 10^{-15}$ |
| 26 | Fe | 240 | | 50 | 45 | | | | | | $3 \times 10^{-13}$ |
| 27 | Co | > 700 | 100 | | | 65 | 10,000 | $1 \times 10^{-9}$ | | | |
| 28 | Ni | 210 | 100 | | 25 | 40 | | | | $1 \times 10^{-9}$ | |
| 29 | Cu | | 10 | 10 | | 10 | 10 | $5 \times 10^{-11}$ | $10^{-13}$ | $5 \times 10^{-10}$ | $2 \times 10^{-15}$ |
| 30 | Zn | | | | | 100 | 1,000 | | | $5.5 \times 10^{-9}$ | $5 \times 10^{-14}$ |
| 31 | Ga | 80 | | | | 300 | 500 | | | | |
| 32 | Ge | | 10 | | | 300 | 1,000 | | | | |
| 33 | As | | | | | 3,000 | | | | | |

| | | | | | | | | | |
|---|---|---|---|---|---|---|---|---|---|
| 38 | Sr | | 100 | 10 | < 100 | 5,000 | | $10^{-13}$ | |
| 39 | Y | | 1,000 | | | 5,000 | | | |
| 40 | Zr | | 1,000 | | < 1,000 | 1,000 | | | |
| 41 | Nb | | | | | 5,000 | | | |
| 42 | Mo | | 100 | 21 | < 1,000 | | $4 \times 10^{-9}$ | | |
| 46 | Pd | | | | | 1,000 | | | |
| 47 | Ag | 10 | 10 | | 100 | 100 | | $10^{-13}$ | |
| 48 | Cd | | | | 3,000 | 500 | | | |
| 49 | In | | | | 1,000 | 500 | | | |
| 50 | Sn | 480 | 10 | 50 | < 1,000 | 100 | | | $3 \times 10^{-9}$ |
| 51 | Sb | > 500 | | | | | | | |
| 52 | Te | > 400 | | | | 5,000 | | | |
| 56 | Ba | | | | 100 | 1,000 | | | |
| 57 | La | | | | | 5,000 | | | |
| 58 | Ce | | | | | 10,000 | | | |
| 72 | Hf | | | | | 5,000 | | | |
| 73 | Ta | | | | | 50,000 | | | |
| 74 | W | > 480 | | 400 | | | | | |
| 78 | Pt | | | | | 10,000 | | | |
| 79 | Au | | | | | 1,000 | | | |
| 80 | Hg | | | | | 5,000 | | | |
| 82 | Pb | 200 | 100 | 30 | 10 | 5,000 | | | $4.5 \times 10^{-10}$ |
| 83 | Bi | 180 | | | 1,000 | 500 | | | |

*Numbers in brackets are references.

**TABLE 28. Application in Mineralogy, Geochemistry, and Cosmochemistry**

| Elements | Matrices | Object of the Analyses | Ranges of Concentrations | Limits of Detection | Reproducibility | References |
|---|---|---|---|---|---|---|
| Si, Ca, Mg, Fe, Al, Ti | Clinopyroxene | Qualitative analysis of silicate minerals | Major, minor, trace, and ultratrace components | $\approx 10^{-3}\%$ | $\pm 5\%$ | Maxwell [45] |
| Si, Mg, Ca, Fe, Mn, Ti, Al, Cr | Orthopyroxene | | | | | |
| Mg, Ca, Si, Al, B, Fe | Plagioclase | | | | | |
| Si, Na, K, Al, Mg, Ca, Fe | Feldspar in rocks | | | | | |
| Si, Ti, Mg, Fe, Al | Enstatite in meteorites | | | | | |
| Fe, Cu, Cr, Mg, Si, Ca, Ag | Bornite | Qualitative analysis of ore minerals | Major, minor, trace, and ultratrace elements | | | |
| Zn, Fe, Si, Pb, Cd, Cu, Si, Mg, Pb, Ca | Sphalerite | | | | | |
| Fe, Cu, Al, Cr, Mg, Ca | Chalcopyrite | | | | | |
| Fe, Ni, Mg, Ca, Cu, Co, Ag | Pyrite | | | | | |

| | | | | | |
|---|---|---|---|---|---|
| Fe, Ni, Mg, Cr, Ca, Ag | Pyrrhotite | | | | |
| Pb, Fe, Ag, Si, Mg, Bi, Ca | Galena | | | | |
| Ni, Fe, Cu, Zn, Co, Mg, Cu | Siegenite | | | | |
| As, Ni, Mg, Fe | Niccolite | | | | |
| As, Co, Ca, Fe, Ni, Cu, Cr | Skutterudite | | | | |
| Fe, As, Ni, Cr, Mg | Arsenopyrite | | | | |
| Fe, Mg, Si, Mn, Ca | Magnetite | | | | |
| Fe, Cr, Ti, Al, Mn, V, Mg, Cu, Ca | Chromite | | | | |
| Li, Na, Cu, Ag, Au, Be, Mg, Ca, Sr, Ba, Zn, Cd, Hg, Sc, Y, La, Se, Th, B, Al, Ga, In, Tl, U, Ti, Zr, Hf, C, Si, Ge, Sn, Pb, V, Nb, Ta, As, Sb, Cr, Mo, W, Mn, Fe, Co, Ni, Rh, Pd, Pt | Minerals, etc. | Qualitative detection of 57 chemical elements | Major components, minor components | $\leqslant 10^{-6}$ g | Berndt et al. [46] |

**TABLE 28.** (*Contd.*)

| Elements | Matrices | Object of the Analyses | Ranges of Concentrations | Limits of Detection | Reproducibility | References |
|---|---|---|---|---|---|---|
| Fe, Ca, Mg, Al, B, Ti, Zn | Dolomite, tourmaline, rutile, talc, sphalerite in quartz | Determination of crystalline inclusions in rock crystals; supplementation of laser microanalysis by IR spectrometry | Major components | | | Blankenburg et al. [47] |
| Fe, Mg, Mn, Ca | Garnets in gneisses | Quantitative analysis using standard glass of own production | Major components | | | Blackburn et al. [48] |
| Ag, Bi | Galena | Semiquantitative analysis | Major components, minor components | | | Bollingberg [49] |
| Ag, Al, Cu, Fe, Cr, Mg, Pb | Minerals | Quantitative determination; methods for increasing sensitivity: special | Major components | | ±8% | Georgieva and Petrakiev [50] |

| | | | | | |
|---|---|---|---|---|---|
| | | treatment of photographic plates, external magnetic field, time-resolved spectra | | | |
| Zr, Si, Al, Ca, Zn, Cr, Fe, Mg, Y, Mn, Ti, Ce, Sc, Th, Nb, Ba, Ta | Zircon, orthite, columbite | Determination of minerals in pegmatites | Major and minor components | | Kirchner et al. [51] |
| Ag, Mg, Cu | Galena | Increase in sensitivity of microspectral analysis by an external magnetic field | Minor components, trace elements | | Petrakiev et al. [52] |
| Ni, Fe, Cr, Cu, Mg, Ca, Cr, Li | Meteorites | Local analysis of microareas to distinguish two meteorites | | | Nikolov et al. [53] |
| Al, Fe, Ca, Co, Si, Mg, Mn, Cu, Na, Ni, Ti, Cr | Meteorite "pawel" (chondrules, matrix, crust) | Semiquantitative analysis | Major, minor, and trace elements | $\approx 10^{-3}\%$ | Petrakiev et al. [54] |
| Ti, Fe, Al, Cu, Na, Ca, Ba | Quartz | Qualitative analysis with $Q$-switched laser | Major and minor elements | | Moenke-Blankenburg and Mohr [55] |
| Ca, Mg, Fe | Beryl | | | | |

**TABLE 28.** *(Contd.)*

| Elements | Matrices | Object of the Analyses | Ranges of Concentrations | Limits of Detection | Reproducibility | References |
|---|---|---|---|---|---|---|
| Si, Mg, Fe, Ca Ba Mn Cu Sr B, Fe, Mn, Mg | Silicates, calcite witherite, rhodochrosite, malachite, azurite, celestite, boracite | Quantitative analysis | Major components | | | Karyakin et al. [56] |
| Li, Cd, Pt, Ag, Sm, Bi, Mo, Ga, Ce, Nd, Co, Zn, Sr, Hg, Sn, Au, In, Pd, U, Ba, Ti, Sb, Ag, B, Ge, Pb, K | Meteorites | Semiquantative determination of the composition of silicate and ore minerals and also of metallic inclusions in five meteorites for the purposes of differentiation | Major components, minor components, traces | | | Dimov et al. [57] |
| 32 elements | Minerals | Qualitative investigation | Major components, minor components, traces | | | Schrön [58] |

| | | | | | | |
|---|---|---|---|---|---|---|
| Mn, Ba, Be, Cu, Sr, V, Ca, Na | Synthetic samples | Production of calibration samples as pressings and glasses, construction of calibration curves. | 3 to 300 ppm | | | Moenke-Blankenburg [59] |
| 39 elements | Diverse samples | Determination of the relative limits of detection for 39 elements and the absolute values for five elements | | 10–3000 ppm | | |
| five elements | | | | $10^{-10}$ to $10^{-11}$ g | | |
| Mg, Al, Si, Ca, Ti, Cr, Mn, Co, Ni, Cu, Zn, Ag, Pb | Pyrite | Differentiation of pyrites on the basis of change in the Co/Ni ratio | Minor components, traces | | 15 to 30% | Smirnov et al. [60] |
| Ag, Cu, Fe, Pb, V, Ge | Chalcopyrite, bornite, chalcosite, covelline, disperse ore fraction | Qualitative and semiquantitative analysis of the constituents of copper-bearing marl, local and distribution analysis | Major components, minor components, traces | | | Idzikowski and Schrön [61] |

**TABLE 28.** *(Contd.)*

| Elements | Matrices | Object of the Analyses | Ranges of Concentrations | Limits of Detection | Reproducibility | References |
|---|---|---|---|---|---|---|
| Mn, Cr, Cu | Pressings and glasses from rock powders, calibration crystals | Calibration curves for quantitative analysis | 10 to $10^4$ ppm | $10^{-3}$ to $10^{-2}$% | ±7 to ±28% | Moenke-Blankenburg et al. [62] |
| Ba, Mn, Sr, Cu, Si, K, Fe, Mg, Sn | Aragonites | Increase in sensitivity by working in an atmosphere of argon; semiquantitative local analysis | Minor components, traces | | | Dimitrov et al. [63, 64] |
| Ti, Cu, Ag, Sn, Zn | Inclusions of glasses and liquids in quartzes from rocks | Investigation on the origin of ore bodies and distribution; qualitative analysis | | | | Takenouchi and Imai [65] |
| La, Ce, Y, Ca, Sr, Ba | Minerals of rare earths | Quantitative analysis with calibration samples of own production | > 2% | | | Idzikowski [66] |

|  |  |  |  |  |  |  |
|---|---|---|---|---|---|---|
| Fe, Si, Mn, Al, Mg, Cu, Ag, Ti, Si | Carbonates | Semiquantitative analysis of layers in stalactites to investigate the process of their origin | Minor components, traces |  |  | Dimitrov et al. [67] |
| Ga, Ni | Rock powder pressings BM, GM, TB[a] | Method of data processing for quantitative microanalysis | $10^{-4}$ to $10^{-3}$% |  | ± 19% | Hüfner [68] |
| Si, Ca, Na, Ti, Al, Fe, Mg, Mn, K | Rock powder pressings BM, GM, TB, TS[a] | Quantitative analysis of microamounts with high concentrations of elements | $10^{-2}$ to 100% |  | ± 5% at a concentration of 10% | Quillfeldt and Berka [69] |
| Al | Standard rock samples (based on silicates) in the form of KBr pressings $Al_2O_3$, Si | Working procedure for quantitative spectrochemical analysis with the LMA 10 | 2 to 15% |  |  | VEB Carl Zeiss [70] |
| Fe |  |  | 1 to 7% |  |  |  |
| Mn |  |  | 0.03 to 0.5% |  |  |  |
| Ti |  |  | 0.1 to 1% |  |  |  |
| Bi, Ca, Cr, In, Mg, Na, Ni, Sn | Glass | Quantitative analysis; bulk | Minor components | $10^{-2}$% |  | Felske et al. [71] |

**TABLE 28.** *(Contd.)*

| Elements | Matrices | Object of the Analyses | Ranges of Concentrations | Limits of Detection | Reproducibility | References |
|---|---|---|---|---|---|---|
| Co, Cr, Cu, Fe, Mn, Mo, Ni, Ti | $BaF_2$, $Al_2O_3$, $Yb_2O_3$, $WO_3$ | Quantitative analysis; bulk | | $10^{-3}$% | | Brech [72] |
| Ag, Al, Au, Bi, Co, Cu, Fe, Mg, Mn, Ni, Pb, Si, Ti, Zn | Pyrites | Semiquantitative analysis, bulk | | | | Eremin and Kelch [73] |
| Al, Ca, Co, Cr, Cu, Fe, Mg, Mn, Na, Ni, Si, Ti | Meteorites | Semiquantitative analysis, microregions | Minor components and trace elements | $10^{-3}$% | | Petrakiev et al. [74] |
| Al, Ba, Ca, Fe, Mg, Mn, Si, Ti | Minerals, oxides | Qualitative analysis, local distribution | Major components and minor components | | | Kozak et al. [75] |
| Al, Bi, Ca, Cu, Fe, Mn, Sn, Ti, W | Fluid inclusions in minerals | Qualitative analysis, local | | | | Bennett and Grant [76] |
| Al, Co, Fe, Ca, Mg, Mn, Ni, Pr, Si | Pyrrhotite, pyrite, chalcopyrite, apatite | Quantitative analysis, local distribution | Major components and minor components | $10^{-2}$% | | Atamanova [77] |
| B, Cr, Fe, Tl, Zn | Crystals, glasses | Quantitative analysis, bulk | Trace elements | $10^{-3}$% | | Panteleev et al. [78] |

| | | | | | | |
|---|---|---|---|---|---|---|
| Si, Fe, Ti, Al, K, V, Cu, Ni | Basaltic pillows | References to geotectonic position and alteration processes | Major and trace elements | | | Kramer et al. [79] |
| | Lunar material: silicates, basalts | Using the laser for simulating of impact processes | Fracture phenomena, fused welds, glasseous particles | | | Kramer et al. [80] |
| Mn | Li-Fe-Mn phosphates, Mg-Fe-Mn borates | Quantitative analyses with pressed standard | Major constituent | | | Matsushita et al. [81] |
| Mn, Be | Silicates | powders | Minor and trace elements | $10^{-13}$ g | | |
| Na, K | Nitrates, carbonates | Fundamental analyses | | | | Debras-Guédon and Liodec [82] |
| Mn, Fe, Ba | Manganese-iron accumulates | Distribution analyses in growth rings, middle transect, at the water side and at the sediment side of an accumulate (see Figs. 48 and 49) | Major and minor elements; Ba: $10^4$ ppm | | Ba: 9.5% | Jahn et al. [83, 84] |

**TABLE 28.** (*Contd.*)

| Elements | Matrices | Object of the Analyses | Ranges of Concentrations | Limits of Detection | Reproducibility | References |
|---|---|---|---|---|---|---|
| Hg | Cinnabar | Determination of detection limit, optimization with Plackett–Burman program | Major component | $3 \times 10^{-8}$ g | 4.5 to 13.3% | Landmann [85] |
| | Minerals | Crater-forming processes | | | | Feklitschew et al. [86] |
| | Diverse geological sample types | Photodiode array spectroscopic methods for the qualitative and quantitative analysis | | | | McGeorge et al. [87] |
| Al, Fe, Ca, Si, Mg, Na, Ti, K, Mn, Cr | Coal, char, mineral matter in the combustion zone | Simultaneous measurements of size and composition of individual coal particles (see Fig. 50) | $\geq$ 1000 ppm<br>100 to 1000 ppm<br>$\leq$ 100 ppm | | 10 to 20% | Otteson and Radziemski [88] |

*Source*: Reprinted with permission from *Prog. Analyt. Spectrosc. 9*, L. Moenke, copyright 1986, Pergamon Journals Ltd.

[a]BM, GM, TB, TS are standard samples obtained from the Central Geological Institute, Berlin, GDR.

Further application examples are given in Ref. 4.

could falsify the results. The object of analysis must lie on the surface (in the case of thin sections, the coverslip must be removed). The light path to the spectrograph must not be hindered by unevenness of the sample.

Since laser microanalysis is a relative process, for quantitative determinations standard or calibration samples with known concentrations of the

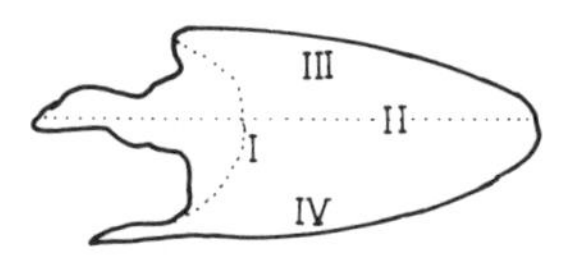


**Figure 48.** Local distribution analyses of major and minor elements in ferromanganese concretions by laser microanalysis based on optical emission spectrography; distribution patterns of laser shots at a growth ring (I), at the middle transect (II), at the water side of the accumulate (III), and at the sediment side of the accumulate. From Refs. [83, 84] reprinted by permission of VEB Gustav Fischer Verlag, Jena.

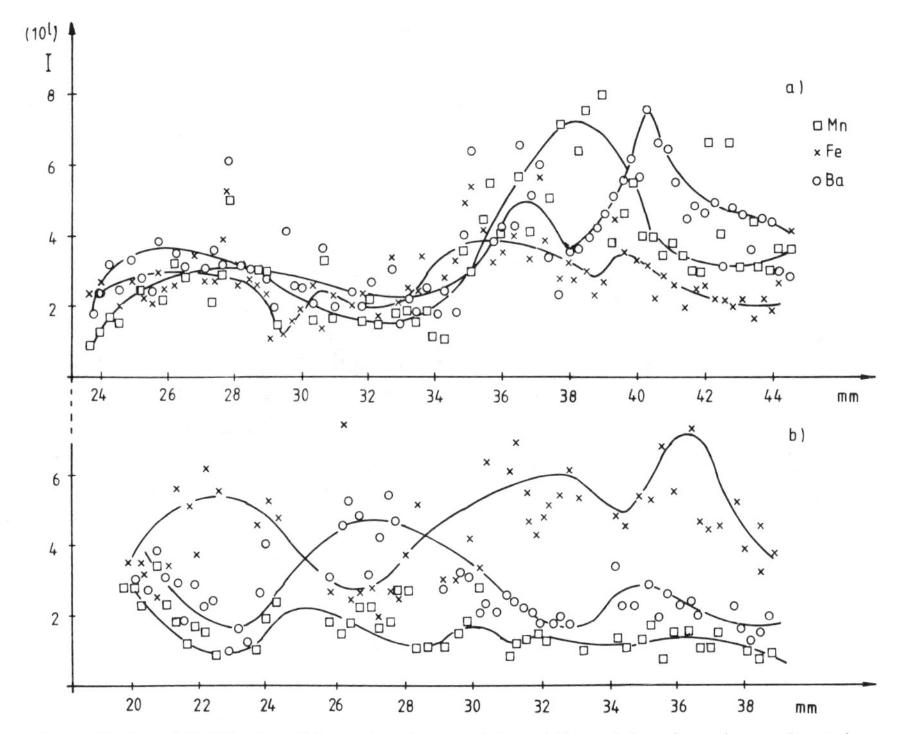


**Figure 49.** Local distribution of the major elements Mn and Fe, and the minor element Ba at the water side of an accumulate (*a*) and at the sediment side of the same accumulate (*b*); linear correlation factors: (*a*) Ba/Fe, +0.6; Ba/Mn, +0.68; Fe/Mn; +0.5; (*b*) Ba/Fe; −0.44; Ba/Mn, +0.54; Fe/Mn, −0.45. From Refs. [83, 84], reprinted by permission of VEB Gustav Fischer Verlag, Jena.

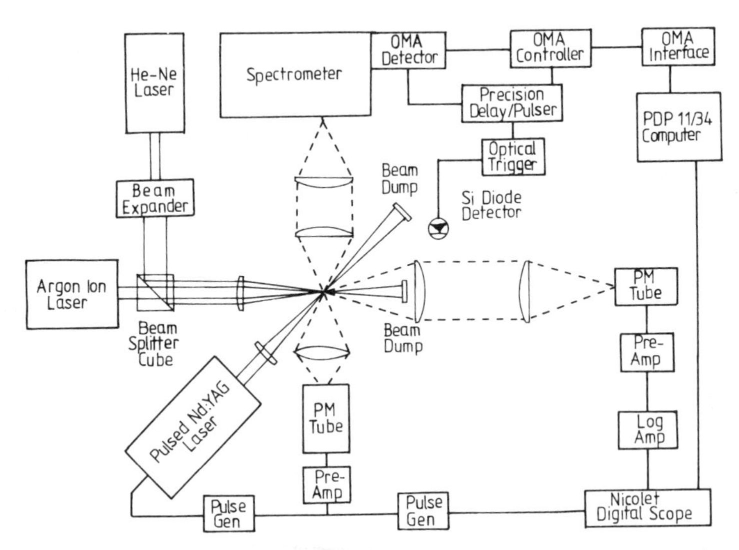


**Figure 50.** Schematic diagram of particle sizing and laser spark spectroscopy diagnostic system, used by Otteson and Radziemski [88] for simultaneous measurement of size and composition of coal particles (70 μm in diameter). The *Q*-switched Nd:YAG laser operates at 532 nm with a pulse width of 7 ns and an energy of about 35 mJ. The emission lines were viewed with a 0.5-m monochromator with a 700-element intensified diode array detector. Reprinted by permission of the authors.

elements are required for drawing up calibration curves. The preparation of suitable calibration samples requires a considerable amount of preparative work, since the samples must substantially resemble the objects of analysis in composition and consistency. Furthermore, high demands are placed on their homogeneity, depending on the nature of the microanalysis. Such calibration samples can be made from pure starting materials by mixing and pressing, sintering, or fusing. As starting materials, we have used powders of minerals or rocks subjected to accurate chemical analysis and available commercially. Table 28 gives examples of qualitative and quantitative microchemical analysis of isolated or single minerals and of local and distributed components in minerals and rocks.

### 5.2.2. Metallurgy and Related Fields

Metals, alloys and semiconductors, carbides, sulfides, nitrides, silicates, and oxides as constituents of inclusions or slags require qualitative and quantitative analyses as well in bulk as in microregions. Local analyses of inclusions or heterogeneities on the order of magnitude of 10 to 100 μm are possible.

Distribution analyses are usual if the lateral resolution limit and a precision of about 5 to 20% are sufficient. That is often the fact when testing welds produced by different technologies or when damage occurs. Furthermore, surface and layer analyses could be carried out down to layer thicknesses of $\geqslant 1\ \mu m$. The detection limits for a lot of elements are in the range $10^{-2}$ to $10^{-3}\%$.

Since nonconducting objects can also be investigated directly, in contrast to electron-beam microanalysis, there is no necessity for vapor coating with conducting substances and the formation of a good electrical contact with the preparation container, which in the case of laser analyses, is the ordinary microscope stage. The surface of a sample need not be flat and should, if possible, not be polished, since if it is polished, part of the laser radiation is reflected (i.e., it is lost in terms of its task of vaporizing and exciting the substance of the object). A reduction of the sample to a maximum size of $60 \times 60 \times 50$ mm was necessary for working in a vacuum or to $100 \times 120 \times 64$ mm for working in a gaseous atmosphere (see Section 3.6). Otherwise, the maximum sample dimensions determined by the apparatus are as given in the guidelines in Table 23. A sample preparation for nonconducting calibration samples similar to that for mineral calibration samples is necessary. Metallic calibration samples are available commercially in a wide variety. Homogeneity tests of suitability for laser microanalysis are sometimes necessary (see Section 4.2). Table 29 gives a review of the possibilities of qualitative and quantitative laser microanalysis as a microchemical method and as a method for local and distribution analysis. References 2, 4, 6, 11, 12, and 89 to 94 are monographs and review articles with relevant application chapters.

### 5.2.3. Silicate Technique, Crystal Synthesis, Chemistry, Forensic Science, Archaeology, Restoration of Art Objects, Medicine, and Biology

Laser microanalysis has been applied in many diverse fields. In glass manufacturing and glassworking industries, trial melts and faults must be examined. Of interest are the still unmolten ingredients (melt residues), which convey information on the course of the melting process and allow conclusions to be drawn on the behavior of the raw materials. Devitrification products—so-called stones, knots, and crystalline segregates—are analyzed; the same is true of foreign bodies such as nonmolten particles of the material of the melt container and the tools. Similar problems can also arise in ceramic research and production. Distribution analyses and homogeneity tests are usual.

In crime detection often only small objects or minute quantities of material are at the disposal of the investigator. Common materials for analysis include paint, lacquer, soil traces, combustion residues, metal traces at the point of

**TABLE 29. Applications in Metallurgy and Related Fields**

| Elements | Matrices | Aim of the Analysis | Range of Concentration | Limits of Detection | Reproducibility | References |
|---|---|---|---|---|---|---|
| Cr, Ni | Steel standard samples | Quantitative analysis with intensity ratios Cr/Fe, Ni/Fe | Major components | | Cr: 3.82% Ni: 5.26% | Runge et al. [95] |
| Cr, Ni | Steel standard samples | Quantitative analysis | Cr: 0.02 to 2%. Ni: 0.05 to 1% | | 0.04 (s.d. according to 0.92% Cr) | Vilnat et al. [96] |
| Al, Ti, Si, Ca, Mn, Fe, Mg, Zr | Steels, slags, refractory | Qualitative analyses of inclusions | Major components | | | Ryan and Cunningham [97] |
| V, Cu, Mn, Si, Mo, Ni, Cr | Steel | Scanning of a sample surface with a sequence of laser pulses | Major and minor components | | 8 to 10% | Felske et al. [3] |
| Ni, In, Cr | $Al_2O_3$ | Scanning of a sample surface with a sequence of laser pulses | Major and minor components | | | |
| Au, Ni | Gold–nickel welding seam | Line analysis of the distribution of the components perpendicular to the joint | Major components | | | Moenke-Blankenburg [98] |

| | | | | | | |
|---|---|---|---|---|---|---|
| Sn, Fe, Cr, Ni, Al, B, Cd, C, Co, Cu, Hf, Mn, Si, Ti, U | Zirconium alloys, steel, sintered $UO_2$, Al–Mn alloys | Quantitative investigations of the influence of the matrix effect | Major and minor components; traces 0.03 to 1.2%, 0.5 to 270 ppm | | | Cerrai and Trucco [99] |
| Co, Fe, Mn, Mo, Ni, B, Be, Cr, Pb, Sn, Sr, W | $BaF_2$, $Al_2O_3$, $Yb_2O_3$, $WO_3$ | Trace analysis in powdered material; reproducibility, limits of detection, matrix effects | $10^{-5}$ to $10^{-2}$% | 0.4 to 200 ppm | 4 to 28% | Whitehead and Heady [100] |
| B | Steel | Quantitative analysis; distribution; micro | $10^{-10}$ to $10^{-3}$% | | $10^{-3}$% | Webb and Cotterill [10] |
| Al, Cu | Cu–Al alloys | Quantitative analysis; bulk | 5 to 10% | | | Kasatkin and Taganov [102] |
| Cr, Fe, Cu, Mg, Si, Al | Steel, Al alloys | Time-resolved spectra, changes of intensity; microphotograms of spectra; calibration curves | | | | Petrakiev and Dimitrov [103] |
| Fe, Si, Mn, Cr, Ni, Ti, Al, Mg, Ca | Steels | Qualitative analysis of inclusions | Major components, minor components | | | Müller-Uri and Volkmann [104] |

**TABLE 29.** *(Contd.)*

| Elements | Matrices | Aim of the Analysis | Range of Concentration | Limits of Detection | Reproducibility | References |
|---|---|---|---|---|---|---|
| Al, Zn | Copper, ferrite materials | Investigation of the diffusion of metals | Minor components | | | Gegus [105] |
| Mg, Be, Ca, Mn, Cu, Ag, Sr | Steel, quartz | Quantitative analysis; preparation of standard samples, determination of the detection limit and of reproducibility | Traces | $10^{-14}$ to $10^{-13}$ g; $10^{-5}$ to $10^{-4}$% | ±13% | Yamane [106] |
| Si, Mg, Mn, Cr, Fe, Ni | Metals, alloys | Microanalysis in a gas atmosphere | Major and minor components | | | Petrakiev and Dimitrov [107] |
| 51 elements | Oxide and metallic materials | Semiquantitative determination | $10^{-3}$ to $10^{-1}$% | | | Schroth [108] |
| Ca, Mg, Si, V | Cr–Mo–V steel | Investigation of fissures in welded seams | Major and minor components | | | Tassewa and Petrakiev [109] |
| Ag, Al, Bi, Co, Cr, Cu, Fe, Mn, Nb, Ni, Pb, Si, Sn, Ti, V, Zn | Bronze, Fe alloys, Al alloys, Cu alloys | Semiquantitative analysis; bulk | 100 to $10^{-1}$% | | | Webb and Webb [110] |

| | | | | | | |
|---|---|---|---|---|---|---|
| Mg, Al, Ni, Cu, Mo, Ag | Metal alloys, pressings from oxide and sulfide powders | Quantitative microanalysis | 0.1 to 20.5% | | | Schroth [111] |
| Zn | Cu alloys | Quantitative analysis; bulk | 10 to 20% | | | Michnov et al. [112] |
| Mn, Cr, V, Ti, Mg, Cu, Mo, Ni, Co | Steel, E series (Hungary) | Quantitative average analysis in microregions; scanning process | Major and minor components | | ±10% | Moenke-Blankenburg [59] |
| V, Mn, Si, Mg, Ti | Borax beads with oxide admixtures, steel | Coupling of the methods of laser microanalysis and IR and atomic absorption spectrometry | | $10^{-10}$ g, $10^{-2}$% | 10 to 15% | Ohls [113] |
| Cr | Steel | Local analysis in some two-phase structures; change in chromium content | | | | Janošikova [114] |
| Si, Al, Fe, Mn, Cr, Mg | | Inclusion analysis | | $10^{-1}$ to $10^{-2}$% | ±10% | |

**TABLE 29.** *(Contd.)*

| Elements | Matrices | Aim of the Analysis | Range of Concentration | Limits of Detection | Reproducibility | References |
|---|---|---|---|---|---|---|
| Mn, Cr, Si | Steel | Quantitative spectroanalytical scanning microanalysis to investigate changes in concentration in segregations, inclusions, or structural phases—average analysis | 0.9 to 3.6% Cr; 0.1 to 0.7% Mn; 0.07 to 0.8% Si | | | Bieber [5] |
| Si | Spheroidal graphite cast iron | Concentration differences in ferritic and pearlitic structures | 1.69 to 2.58% Si | | | |
| Cr, Cu, Mn, Mo, Ni, Si, V, W | Low-alloy steels | Working procedures for quantitative analysis | 0.02 to 5% | | | VEB Carl Zeiss [115] |
| Cr, Mn, Mo, Ni | Low- to high-alloy steels | | 0.3 to 3.6% | | | |
| C, Co, Mn, P, Si | Gray cast iron | | 0.09 to 3.6% | | | |

| | | | | | | |
|---|---|---|---|---|---|---|
| Co, Hf, Mo, Nb, Ta, V | Ni alloys | | 1 to 15% | | | |
| Cu, Fe, Mg, Mn, Si, Ti, Zn | Al alloys | | 0.04 to 2% | | | |
| Al, Fe, Mn, Ni, Pb, Sn, Zn | Cu alloys | | 0.008 to 10% | | | |
| Ag, Bi, Cu, Fe, Ni, Sb, Sn, Zn | Pb–Sb alloys | | 0.0001 to 14.2% | | | |
| Cr, Fe | Metals, oxides | Quantitative analysis; bulk | 2 to 10% | | | Krivchikova and Toganov [116] |
| Si, B, Al, Cu | Semiconductors | Detection of $SiB_6$ segregation, analysis of conductor paths on chips, local analysis of parts and intermediate layers of an Si diode | | | | Mohaupt and Pätzmann [117] |
| Si | GaAs | Quantitative analysis | | 2.3 ng | ±45% | |
| | | Possibilities of application in metallurgical analysis; examples | | | | Ohls et al. [118] |
| Mn | Welding seam | Concentration profile | 1.04 to 1.18% | | | |

**TABLE 29.** (*Contd.*)

| Elements | Matrices | Aim of the Analysis | Range of Concentration | Limits of Detection | Reproducibility | References |
|---|---|---|---|---|---|---|
| C, S, P | Steel | Microanalysis in vacuum and in a protective gas atmosphere | | | | Mohr [119] |
| Cr, Ni, Mo, V, Mn, Ti, W, Si, S, P | Steels | Photoelectric recording of LMA pulses | 0.003 to 27.8% | $10^{-3}$ to $10^{-2}$% $10^{-10}$ to $10^{-12}$ g | ±3 to ±7.4% | Moenke-Blankenburg et al. [120] |
| Cu, Fe, Pb, Zn | Brass | Influence of gas medium of excitation processes | | | | Dimitrov and Petrakiev [121] |
| Fe, V, Cr | Steel | Quantitative analysis, calibration | Major components | | 23 to 28% | Schrön [122] |
| Au, Cu, Al, Fe, Nb, Co, Cu, Al | Single-metal crystals | Investigation of the relationship between crystal structure and vaporization caused by laser radiation | | | ±4 to ±25% | Kirchheim et al. [123] |
| Ni | Steel, E series (Hungary) | Data processing procedure for quantitative microanalysis | $10^{-3}$% | | ±12% | Hüfner [124] |

| | | | | | | |
|---|---|---|---|---|---|---|
| Mn | Al alloys, HC and HP series (Hungary) | Data processing procedure | $10^{-4}$ to $10^{-3}$% | | ±7 to ±14% | |
| Cu, Fe, Mn, Zn | Brass | Qualitative analysis; bulk | | | | Petuch et al. [125] |
| Ag, Au, Ni, Zn | Alloys | Quantitative analysis; local, micro | To $10^{-8}$ g | | | Mankevich et al. [126] |
| Sn, V | Cu alloys | Quantitative analysis; bulk | To $10^{-2}$% | | | Moenke-Blankenburg et al. [127] |
| Cr, Ni, Mn, Si, Ti | Steel | Distribution analysis | Minor elements | | 10 to 13% | Maksimov et al. [128] |
| Au, Ag, Cu, Ni, Zn | Gold | Quantitative analysis of micro amounts of high concentration | 14 to 24 carats | | ±5% | Quillfeldt and Berka [129] |
| Au | Au | Correction method for quantitative analysis | 14 to 24 carats | | 1.6 to 3.5% | Moenke-Blankenburg and Quillfeldt [130] |
| Si, V, Mn, Cr, Ti, W, Ni, Cu | Steel E series (Hungary | Homogeneity investigations in microregions | | | | Maul and Quillfeldt [131] |

**TABLE 29.** (*Contd.*)

| Elements | Matrices | Aim of the Analysis | Range of Concentration | Limits of Detection | Reproducibility | References |
|---|---|---|---|---|---|---|
| Fe, Cr, Co, Ti, Al, Si | Alloys | Investigation on the corrosion of gas turbines; quantitative local analysis | 0.27 to 18.3% | | | Floris and Muttoni [132] |
| Cr, Ni, Mn, Mo, Cu, Al, Ti, V, W | Steel | Combination of atomic emission and absorption; limits of detection | $10^{-3}$ to 100% | | | Quillfeldt [133] |
| Pb, Sn, Fe, Mn, Al, Ni, Mg, Si | Cu alloys | | | | | |
| Fe, Cu, Mg, Si, Ti, Mn, Cr, Ga | Al alloys | | | | | |
| Sn, Fe, Ni, Zn, Mn | Copper | Photoelectric recording of LMA pulses with optical multichannel analyzer I | | $10^{-11}$ to $10^{-12}$ g | | Moenke-Blankenburg et al. [134] |
| Cu, Pb | $Sb_2Te_3$; $Bi_2Te_3$ | Homogeneity test of doped materials of $V_2$-$VI_3$ type | Trace elements | | 15% | Priemuth [135] |

| | | | | | | |
|---|---|---|---|---|---|---|
| Si | Steel | LTE model, electron temprature (~8500 K) and peak electron densities (2.2 × $10^{23}$ $m^{-3}$) were measured | 0.55% | 0.03% | 30% | Adrain et al. [136] |
| Ni, Cr, Mn, Si, Ti | Steel | Analysis in gas medium He | 0.4 to 21% | | | Maksimov et al. [137] |
| Ag, Cr, Mn, | Copper | Photoelectric recording with optical multichannel analyzers I and II; limits of detection | | (Ag) 15 ppm; (Cr) 2 ppm; (Mn) 10 ppm; | 5.2 to 10.2% | Talmi et al. [138, 143] |
| Pt | Razor-blade steel | | | (Pt) $4 \times 10^{-11}$ g | | |
| Si, Cr | Steel | Quantitative analysis, comparison between $Q$-switched and normal ruby laser operation | Minor components | 0.01% | < 10% | Adrain et al. [139] |

**TABLE 29.** *(Contd.)*

| Elements | Matrices | Aim of the Analysis | Range of Concentration | Limits of Detection | Reproducibility | References |
|---|---|---|---|---|---|---|
| Si | Mild steel | Measurements of electron temperature and electron density (LTE) | 0.01 to 1% | | | Adrain et al [140] |
| Al, Pb, Sn | Copper | Improvement of reproducibility and sensitivity | 0.01 to 1.5% | | 9.5 to 19.4% | Dimitrov and Maximova [141] |
| Al, Cr, Mn, Ni, Si, Ti, V | Low-alloy steels | Quantitative analysis; bulk | 1 to $10^{-2}$% | | | Nikitina et al. [142] |
| Cr | Steel | Quantitative analysis | 0.1 to 1% | See Table 25 | | Talmi [143] |
| Ag, Cu, Pd | Gold ribbon | Concentration profile | | | | |
| Ag | Capacitor | Layer analysis, depth profile | | | | |
| Co, Mn, Mo, Ti, U, W | Metallic oxides | Hypothesis of dynamic equilibrium in the analytical zone of a laser torch | $10^{-4}$ to 20% | | | Wulfson et al. [144] |
| Mn | Al standards | Improvement by using a cylindrical lens in front of the photoplate | 0.78% | | 0.047 | Nikolova and Krasnobaeva [145] |

| | | | | | | |
|---|---|---|---|---|---|---|
| Ag, Bi, Cu, Cr, Al, Mn, Fe | Metals, alloys | Collector for laser erosion products | Minor elements | 0.01% | 20% | Rudnevsky et al. [146] |
| Cr, Si, Mn, Cu, Ti, Ni, V, Nb, Mo, Al, Zr, Co | Steel, cast iron | Distribution analysis of welds | Minor elements | 0.01 to 0.05% | | Buzási et al. [147] |
| Cr, Cu, Mn, Mo, Ni, P, Si | Steel | Fundamental studies with a SIT vidicon | Minor elements | 0.02 to 0.05% | 3.3 to 7.3% | Saisho et al. [33] |
| Fe, Pb, Mn, Cu, Zn, Mg, Sn, Si | Metallic standard samples | Decreasing the laser radiation absorption in the plasma plume and improving the sensitivity in quantitative LMA | 0.04 to 1.55% | | 9.5 to 19.4% | Dimitrov and Zheleva [148] |
| Be | Be-Cu alloys | Time-resolved laser-induced breakdown | Be: 0.0001 to 0.22% | 2 ppm | 7% | Millard et al. [149] |
| Ge, Si, Cu, Se | Glassy chalcogenide semiconductors | Kinetics of microplasma, interaction processes between laser plume and sample | | | | Armenski and Dimitrov [150] |

*Source*: Reprinted with permission from *Prog. Analyt. Spectrosc. 9*, L. Moenke, copyright 1986, Pergamon Journals Ltd.

TABLE 30. Laser Microanalysis Predominantly of Nonconducting Materials Such As Glasses, Ceramics, Enamels, and Crystals As Well As Miscellaneous Samples Such as Graphite, Oil, Dust, Varnishes, Films, and Fibers

| Elements | Matrices | Aim of the Analysis | Range of Concentrations | Limits of Detection | Repro-ducibility | References |
|---|---|---|---|---|---|---|
| Al | $Al_2O_3$ | Quantitative distribution of the element in an $Al_2O_3$ knot from the center of the knot radially to the surrounding glass | 2.6 to 10% | | | Ryan et al. [153] |
| Cr, Al, Fe, Mg | Chrome spinel | Quantitative distribution of the elements in a $Cr_2O_3$-rich stone and the surrounding glass | 0.1 to 35% | | | |
| Fe, Si, Ca, Cu, Mg, Al, Cr, Mn, B, Ba | Glass, etc. | Qualitative analysis of silica glass frits with inclusions of foreign glass and metals | Main components, secondary components | | | Moenke and Moenke-Blankenburg [4] |
| Zn, Si, Ca | Glass | Qualitative analysis of a special glass | Main components, secondary | | | |

| | | | | | | |
|---|---|---|---|---|---|---|
| | | with drop-shaped inclusions | components | | | |
| Zr, Si, Al | Baryta crown glass | Quantitative analysis of a glass streak | 0.046 to 0.053% Zr | | | Maul and Quillfeldt [131] |
| Sn | Borosilicate glass | Calibration curve | 0.04 to $10^{-4}$% | | ±10.7% | |
| B | Borosilicate glass | Comparison of quantitative analysis of glasses with arc-OES and LMA | | | | Birkenfeld and Moenke-Blankenburg [154] |
| Al, Fe, Mn, Ti | Glass powder | Working procedure on quantitative analysis | 0.001 to 0.3% | | | VEB Carl Zeiss [115] |
| Al, Zr, Si | Borosilicate glass | Semiquantitative analysis of streaks | High concentrations | | | Horst et al. [155] |
| Co, Mn, Fe | Borosilicate | Homogeneity tests (see Section 4.2) | 61% $B_2O_3$; 8% $SiO_2$; 30% $Na_2O$; 0.7% Co, Mn, Fe | | | Pacher [156]; Moenke-Blankenburg and Pacher [157] |
| | $\|SiO_4\|^{-n}$ $\|BO_3\|^{-n'}$ $\|GeO_4\|^{-'n}$ $\|PO_4\|^{-n'}$ | Relation between the chemically inhomogeneous glass structure and the laser damage threshold | Major components | $10^{-3}$ to $10^{-5}$% | | Karapetyan and Maksimov [158] |

**TABLE 30.** *(Contd.)*

| Elements | Matrices | Aim of the Analysis | Range of Concentrations | Limits of Detection | Repro-ducibility | References |
|---|---|---|---|---|---|---|
| Fe, Ti, Si, Al, Ca, Zr, Cr | Refractory material | Qualitative analysis of inclusions in a a fragment of a pot | Major components, minor components | | | Moenke and Moenke-Blankenburg [4,159] |
| Al, Si, Zr, Mg | Refractory materials | Qualitative analysis of punctate inclusions in Bakor 33 | Major components, minor components | | | |
| Be | Fine ceramics | Analysis of acicular crystals in a steatite product | Minor components | | | |
| Mg, Ca, Ti | Ceramics $MgO-CaO-TiO_2$ system | Investigation of change in microstructure at different temperatures of synthesis; semiquantitative analysis of light, gray, and dark phases | Major components | | | Dimitrov et al. [160] |

| | | | | | | |
|---|---|---|---|---|---|---|
| Al, Si, B | Ceramics, silicon wafer | Fundamental studies with a SIT vidicon | Minor elements | 0.05% | 3.3% | Saisho et al. [33] |
| Al, B, Ca, Fe, Mg, Si, Ti | Masrock ($SiO_2$), Durital ($SiO_2$, $Al_2O_3$) | Analysis of inhomogeneities | 0.05 to 45% (oxides) | | 20 to 25% | Siechau et al. [161] |
| Fe, Mg, Ti, Ca, Si, B, Be, Cr, Cu, Mn, Mo, Nb, Ni, Pb, Sn, Zn, Ba | Durital E 90 and M 70, Masrock | Major, minor, and trace analyses of compositional variations from grain to grain | 0.01 to 23.05% (oxides) | | | Siechau et al. [162] |
| Fe, Cu, Si, Ag, Ti | Ceramics with metal coating and protective varnish | Removal of multiple layers and their laser microanalysis by defocusing the laser beam | Major elements | | | Moenke-Blankenburg and Quillfeldt [163] |
| Cd | $\beta$-$Al_2O_3$ | Program algorithms for processing of data from LMA with photographic registration | Minor elements | | | Nikolova and Krasnobaeva [164] |
| La | $BaTiO_3$ ceramics, doped with La | Quantitative determination | 0.2 at % | | 13% | Zander and Moenke-Blankenburg [165] |

**TABLE 30.** *(Contd.)*

| Elements | Matrices | Aim of the Analysis | Range of Concentrations | Limits of Detection | Reproducibility | References |
|---|---|---|---|---|---|---|
| Sn | $BaTiO_3$ ceramics, doped with Sn and La | Local and distribution analysis | 1 to 30 at 3% | | 14 to 8% | Zander and Moenke-Blankenburg [166] |
| Mg, Be, Ca, Mn, Cu, Ag, Sr, B, Cd, V, Fe, Cr, Zr, Sn, Bi, Mo, Ti, Zn, Co, Al, Ni, Pb | Nonconducting materials | Determination of detection limits | | $10^{-1}$ to $10^{5}$%, $10^{-10}$ to $10^{-14}$ g | | Yamane [106] |
| Fe, Ca, Al, Mg, Si, Ti, Ni, Sn, Zr, Sr | NaI | Qualitative analysis of particles in single crystals | Minor components | | | Yamane and Yamada [167] |
| Cr, Mn, Cu, Mg, Al | Mono crystals: $Ca_2V_2O_7$, $CaV_2O_6$, $CeO_2$, $Nb_2O_5$, $TiO_2$ | Two-stage method of OES with laser sampling | Impurity concentrations: 0.005 to 1.5% | $10^{-2}$ to $10^{-3}$% | 15 to 35% | Boitsov and Zilberstein [168] |
| Al, Ba, Ca, Cr, Fe, Mg, Mo, Ni, Si, Ti, V | Graphite | Quantitative micro-analysis of impurities in reactor graphite | Minor components, traces | 0.1 to 23 ppm, $10^{-12}$ to $10^{-10}$ g | | Nickel et al. [169, 170] |
| Al, Cr, Ni | Graphite | Influences of discharge atmospheres | Trace analysis | 4 to 8 ppm | | Dörge et al. [171] |

| | | | | | | |
|---|---|---|---|---|---|---|
| Be, Ti, Y, Zr | Graphite | LMA under reactor conditions | 25 to 80 ppm | | | Stupp [172] Stupp and Overhoff [173] |
| Ca, Zn, Ba, Mg, P | Lubricating oil | LMA for classification of oils with the help of Karhunen–Loeve transformation | Additives (traces) | | | Borszéki et al. [174] |
| Cr, Ti, Al | Corrosion layer on gas turbine materials | Distribution of elements in the corrosion layer | Low concentration | | 15% | Floris and Muttoni [132] |
| Si, Mg, Ca, Fe, Mn, Al, Ti, Cu, Zn | Dust | Qualitative analysis | Major and minor components | | | Litz [175] |
| Be | On filters | Quantitative analysis of particles on filters with a cylindrical lens and by rotation of the filter | 3.6 ng or 0.45 ng/cm$^2$ | | 4% | Cremers and Radziemski [176] |
| Miscellaneous | Nylon and other fibers, films | Qualitative analysis of impurities | | 0.1 to 5%, 10 to 500 ppm (depending on apparatus) | | McGillivray [177] |

*Source*: Reprinted with permission from *Prog. Analyt. Spectrosc. 9*, L. Moenke, copyright 1986, Pergamon Journals Ltd.

**TABLE 31. Application in Forensic Sciences, Archaeology, and Restoration of Art Objects[a]**

| Elements | Matrices | Aim of the Analysis | Range of Concentrations | Limits of Detection | References |
|---|---|---|---|---|---|
| Miscellaneous | Paints, varnishes, metals, alloys, tools | Local analysis, analysis of trace amounts, layer analysis; qualitative | Major and minor components | | Schicht [178] |
| 60 elements | Bullet holes in fabrics, paints, glasses, metals, nonmetals, hair, bones, nails | Local analysis, trace analysis, layer analysis; qualitative | Major and minor components | | Moenke and Moenke-Blankenburg [4] |
| Miscellaneous | Motor vehicle paint | Comparative investigations for identity of material (e.g., after hit-and-run traffic accidents) | Major and minor components | $10^{-8}$ to $10^{-12}$ g | Neuninger [179] |
| Au, Ag, Cu, Zn, Rh | White gold, brass, rhodium layer | Nondestructive testing of valuable objects; layer analysis | Major components | | |
| Fe, Si, Mn, Pb, Ti, Zn | Steel, vanish | Investigation of molten metallic beads (proof of breakage into strongroom) | Major components | | |

| | | | | | |
|---|---|---|---|---|---|
| Cu | Copper cable, knife | Investigation of foreign metal rubbed of onto a knife (proof of theft) | Major components | | |
| Miscellaneous | Glass | Investigation of small particles of glass (e.g., of headlamp fragments in traffic accidents) | | | |
| Miscellaneous | Metals and alloys | Investigation of inhomogeneities in metals (failures of material as causes of accidents) | Major components | | |
| Y, Zr | Minerals | Analysis of small amounts of mineral materials | Traces | About 1000 ppm | |
| Co, Cr, Mn, Mo, Ni, Pb, Se, Sr, Ti, V | | Limits of detection | | About 100 ppm | |
| Ag, Be, Cu, Ge, Sn | | | | About 10 ppm | |
| Cu, Zn, Al, Pb | Steel | Detection of foreign metals rubbed off onto tools, such as steel bolts, shears, screwdrivers | Major and minor components | | Rudolph [180] |

**TABLE 31.** *(Contd.)*

| Elements | Matrices | Aim of the Analysis | Range of Concentrations | Limits of Detection | References |
|---|---|---|---|---|---|
| Fe, Cr, Mn, Ni, Si, Co, Mo, P, W | Steel | Investigation of the inclusion of a foreign body in a steel pipe (elucidation of the cause of damage) | Major and minor components | | Neuninger [181] |
| Cu, Zn | Copper | Investigation of melting on and around electrical conductors (cause of fire) | Major components | | |
| Si, Al | Glass | Investigation of foreign body inclusions | Major components | | |
| Cu | Azurite | Identification of pigments in multilayer wall paintings | Major components | | Petrakiev et al. [182] |
| Al | Ultramarine | | | | |
| Pb | Minium | | | | |
| Hg, S | Cinnabar | | | | |
| Fe | Venetian red | | | | |
| Fe, Mn | Umber | | | | |
| Pb | White lead | | | | |

| | | | | |
|---|---|---|---|---|
| P | Bone black | | | |
| Cu | Malachite | | | |
| Cu, Sn, Ag, As, Fe, Ni, Pb, Sb, Au, Co | Bronze | Chemical investigation of a bronze cup for determining the deposit from which the copper was obtained | Major and minor components, traces | Neuninger and Sauter [183] |
| Mo, Cr, Fe, Pb, Cu | Inks | Qualitative analysis for forensic reports | Major and minor components | Mehrotra and Sidhana [184] |
| Si, Mg, Ca, Fe, Al, Ti, Cu, Zn, Mn | Dusts | Qualitative analysis of trace amounts | Major and minor components | Litz [175] |
| Ba, Ca, Ti, Cu, Al, Sb, Mg, Pb, Fe, Si, Zn, Ag | Paints, metals fragments | Forensic analysis | Major and minor components | Baisane et al. [185] |
| Pb, Ba, Sb, Cu, Ca, Mg | Bullet holes | | | |
| Ca, Pb, Ti, Mg, Fe, Si | Headlamp glass | | | |
| Hg | Hair | | | |
| Pb, Sn, Cu | Coins | Distribution of alloying elements can give information about origin and age | Minor components | Bajnóczi and Major [186] |

**TABLE 31.** *(Contd.)*

| Elements | Matrices | Aim of the Analysis | Range of Concentrations | Limits of Detection | References |
|---|---|---|---|---|---|
| Au, Pb, Sn, Bi, Ni, Zn | Roman coins | Classification of coins with pattern recognition methods | Minor and trace components | | Borszéki et al. [187, 188] |
| Cu, Zn, Sn, Pb, P, Ag, Au, Sb, TL, V, Cr, Fe, Co, Ni, Mg, Al, Si, Bi, Mn | Roman coins | Semiquantitative analysis | Major, minor and $10^{-4}$% trace elements | | Bakos and Gegus [189] |
| Pb, Zn, Sn, Bi, Sb, Ag, Ni and Cu, Pb, Bi | Archaeological bronze and silver findings | Semiquantitative determinations | 0.003 to 40% | | Gegus [190] |
| Ag, Cu, Hg, Zn | Brass, copper, silver | Qualitative analysis of findings | Major and minor components | | Geisler [191] |
| Several elements | Art objects | Qualitative analysis of paints, glasses, metals | | | Schreiner et al. [192] |
| Cd, Zn, Cr, Pb, Fe | Pigments in artistic paintings | Restoration of paintings | Major components | | Landmann [193] |
| Na, Cu, Ca | Ultramarine | Qualitative analysis for restoration of paintings | Major, minor, and trace elements | | Landmann [194] |

| | | | | |
|---|---|---|---|---|
| Au, Ag, Cu, Zn, Sb, Ca, Mg, Al, Si, Fe | Gold | Qualitative analysis of leaf gold | Major, minor, and trace elements | Landmann [194] |
| Cu, Sn, Fe, Mn, Ca, Si, Mg | Copper alloys | Explanation of origin and method of its metallurgy in Bronze Age | Major, minor, and trace elements | |
| Hg, Pb | Pigments: cinnabar/vermilion, white lead | Quantitative analysis of ratios with standard deviations of 0.03 to 0.07 | 5 to 60% cinnabar in white lead | |
| Cu | Azurite, verdigris, malachite in white lead | Quantitative analysis with addition method for calibration | Major components | |
| Ag, Cu, Pb, Sn, Fe, Zn, Si, Mg | Coins | Distinction between Byzantine coins and their Latin and "Bulgarien" imitations | Ag: 0.3 to 6%. Sn: 0.01 to 5%. Pb: 0.1 to 5%. Zn: 0.1 to 1%. Fe: 0.01 to 1%. Si: 0.01 to 3%. Mg: 0.005 to 0.1%. | Papazova et al. [195] |

**TABLE 31.** *(Contd.)*

| Elements | Matrices | Aim of the Analysis | Range of Concentrations | Limits of Detection | References |
|---|---|---|---|---|---|
| Cu, Sn, Pb, Ni, Zn, Ag, As, Sb, Tl, Cd, Fe, Si | Bronze | Determine origin of used ores, method of manufacture of objects of Bronze Age | | | Gäckle et al. [196] |
| Ag, Al, Au, Be, Bi, Ca, Cu, Fe, Mg, Mn, Ni, Pb, Pd, Pt, Si, Sn, Zn | Gold, bronze, staghorn, clay, red terra, sigillata, brass, glass | Analyses of ancient Roman products provide clues to origins, manufacturing places, workshops, and trade routes | Major, minor, and components | | Feustel and Maul [197] |

*Source*: Reprinted with permission from *Prog. Analyt. Spectrosc. 9*, L. Moenke, copyright 1986, Pergamon Journals Ltd.

[a] Further application examples are given in Refs. 198 to 201; see also Ref. 216 in Table 32.

**TABLE 32. Application in Medicine and Biology[a]**

| Elements | Matrices | Aim of the Analysis | Range of Concentration | Limits of Detection | Reproducibility | References |
|---|---|---|---|---|---|---|
| Mn, Fe, Be, Cu, Zn, Ca | Tissues from brain, pancreas, ligaments, and calculi | Comparative analysis of sections of tissue | | $10^{-10}$ mol | | Rosan et al. [202] |
| Mg, Al, Si, Ca, P, Fe, Cu, Zn | Calcified tissues | Semiquantitative determination of inorganic components in calcified tissue | 0.001 to 10% | | | Goldman et al. [203] |
| Ca, Mg | Freeze-dried sections of human pylorus | Development of an internal calibration method for quantitative histochemical investigation | | $10^{-7}$ to $10^{-12}$ mol | ±20% | Rosan et al. [204] |
| Mg, Al, Fe, Zn | Teeth | Analysis of the distribution of the elements in tooth sections | Main component (up to 12.7%), secondary components, traces | $10^{-13}$ mol | | |
| Ca, P, Mg, Si, Fe, Al, Ca, Ti, Zn | Tooth constituents, calculi, bones | Semiquantitative local analysis | 0.01 to 10% | $10^{-10}$ mol | | Sherman et al. [205] |

**TABLE 32.** (*Contd.*)

| Elements | Matrices | Aim of the Analysis | Range of Concentration | Limits of Detection | Reproducibility | References |
|---|---|---|---|---|---|---|
| Ca, P, Mg, Na, Al, Fe, Cu | Skin | Semiquantitative local analysis during and after a skin disease | 0.0001 to 10% | | | Wilson et al. [206] |
| Ca, Mg, Zn, Cu, Mn | Liver; kidney | Analysis of inorganic constituents in freeze-dried sections | | $10^{-4}$ to $10^{-16}$ g | | Beatrice et al. [207] |
| Fe | Red blood corpuscles | | | $10^{-13}$ g | | |
| Zn | Rat sperm | | | $10^{-15}$ g | | |
| Ti, Fe, Ca, B, Mg, Si, Cu, Mn, Al | Butterfly wings | Investigation of pigmentation | 0.05 to 1.8% | | | Dimitrov et al. [208] |
| Fe, Si, Mn, Mg, Al, Ca, Zn, Ti, B, P | Houseflies | Investigation of various parts of the body | 0.018 to 1.5% | | | |
| Fe, Mg, Ca, Cu, Al | Rat kidneys | Histotopographical investigations | Secondary components, traces | | | Kozik et al. [209] |
| Pb | Brain | Quantitative determination | Traces | | 3.45 to 7.7% and 14 to 20% | Kozik et al. [210] |

| | | | | | |
|---|---|---|---|---|---|
| | | activity on the basis of PbS concentrations; Ag of photographic plate as internal standard | | | |
| Li, Fe, Cu, Zn, Hg, Pb, Mg, Ca | Dried samples of gelatin | Determination of the limits of detection | | Li: 0.2 pg. Mg: 0.002 pg. Ca: 0.01 pg. Fe: 0.3 pg. Cu: 0.02 pg. Zn: 0.05 pg. Hg: 0.3 pg. Pb: 0.1 pg. | Treytl et al. [211] |
| | Fibromas, skin tissue, heart, cancerous tissue | Quantitative histochemistry with a laser | | $10^{-12}$ to $10^{-15}$ g | Glick and Marich [212] |
| | Skin of a finger and of an ear | In vivo analysis of human and animal tissue | | | |
| Ca, Mg, Na, F, Fe, Si, Ti, Cu | Urinary calculi | Distribution of trace elements in 10 phosphate-containing concretions | | | Dimitrov and Marinov [213] |
| Mn, Al, Cu, B, Fe, Cr, Ti, Zn, Ca, Mg | Pancreas, islets of Langerhans, cells A and B | Qualitative distribution analysis | Minor and trace elements | | Galabova et al. [214] |

**TABLE 32.** (*Contd.*)

| Elements | Matrices | Aim of the Analysis | Range of Concentration | Limits of Detection | Reproducibility | References |
|---|---|---|---|---|---|---|
| Ag, Al, Pb, Ca, Cd, Cu, Cr, Fe, Mg, Mn, Ni, P, Si, Ti | Urinary calculi | Analysis of the distribution of the elements in individual regions of urinary calculi | | | | Schulz et al. [215] |
| Ca, P, Mg, Cu, Mn, Fe, Si, Al, Na, Ti | Tooth stone | Semiquantitative analyses; comparison | Major, minor, and trace elements | | | Dietz et al. [216] |
| Ca, Mg, Cu, Fe | Dried residue of blood serum | Quantitative analysis | $10^{-2}$ to $10^{-4}$% or $10^{-9}$ to $10^{-11}$ g | | 6.9 to 18.6% | Marinov and Dimitrov (217) |
| Mg, Cu, Zn, Ca, Fe, P, Na, K | Pancreas of rats | Comparison of analyses with laser microanalysis, electron microprobe, electron microscope | | | | Jablenska et al. (218) |
| Al | Lumbar vertebrae of rats | Distribution analysis in relation to dose, time and location | 22.5 to 2300 ppm | | 3.5 to 46.3% | Günther [219] and Schmidt et al. [220] |

| | | | | | |
|---|---|---|---|---|---|
| Zn | Lumbar vertebrae of rats | Distribution analysis: negative result; therefore, analysis with LM-ICP (see Table 33) | $< 500$ ppm | | Al Hamadi and Hegewald (221) |

*Source*: Reprinted with permission from *Prog. Analyt. Spectrosc. 9* L. Moenke, copyright 1986, Pergamon Journals Ltd.

[a]Literature reviews up to 1971 and 1973, respectively, are given by Harding-Barlow [222] and Goldman [223]. For further literature, see Ref. 224.

impact of a bullet, splinters, chips, fragments of metal, glass, ceramics, and stones.

The effectively slight destructive feature of LMA is indispensible for the analysis of art objects and archaeological treasures. Analysis of paint pigments, for example, can help in dating and establishing the authenticity of works of art and can assist in finding original methods for restoration.

As early as 1963, medical and biological investigations were being carried out for the determination of toxic elements in different tracts of rat brain, in islets, acini, pancreas, elastin, and in concrements with an error margin of about 20%. The first reports of analytical work on the living body have been published. In the field of dentistry the main interest has been in analyzing teeth or tooth fillings. An overview of all application fields is to be found in Refs. 2, 4, 6, 11, 12, 89–94, 151, and 152. Detailed information is given in Tables 30 to 32.

## REFERENCES

1. A. Strasheim, *Fresenius Z. Anal. Chem.*, **324**, 793 (1986).
2. L. Moenke-Blankenburg, *Prog. Anal. Spectrosc.*, **9**, 335 (1986).
3. A. Felske, W. D. Hagenah, and K. Laqua, *Z. Anal. Chem.*, **216**, 50 (1966).
4. H. Moenke and L. Moenke-Blankenburg, *Einführung in die Laser-Mikro-Emissionsspektralanalyse*, 2nd Ed., Akademische Verlagsgesellschaft Geest und Portig KG, Leipzig, GDR, 1968; 4th Ed., *Laser Microspectrochemical Analysis*, Adam Hilger Ltd., London, England, 1973, and Crane, Russak & Co., Inc., New York, 1973.
5. B. Bieber, *Jena Rev.*, 248 (1974).
6. L. J. Radziemski, R. W. Solarz, and J. A. Paisner, *Laser Spectroscopy and Its Applications*, Marcel Dekker, Inc., New York, 1987.
7. Yanbin Zhu, *Proc. Pittsburgh Conf. and Exposition*, 1984, Abstract 207.
8. S. D. Rasberry, B. F. Scribner, and M. Margoshes, *Appl. Opt.*, **6**, 87 (1967).
9. W. Schrön and L. Rost, *Atom-Spektralanalyse*, VEB Deutscher Verlag für Grundstoffindustrie, Leipzig, GDR, 1969.
10. L. Moenke-Blankenburg, J. Mohr, and W. Quillfeldt, Wissenschaftlich-Technische Information des Zentralen Geologischen Instituts, GDR, **17**, 343 (1976).
11. K. Laqua, in *Analytical Laser Spectroscopy*, N. Omenetto, Ed., John Wiley & Sons, Inc., New York, 1979.
12. L. Moenke-Blankenburg, in *Advances in Optical and Electron Microscopy*, Vol. 9, R. Barer and V. E. Cosslett, Eds., Academic Press, Inc., London, 1984, pp. 243–322.
13. G. Horlick, *Anal. Chem.*, **45**, 1490 (1973); **46**, 133 (1974); *Appl. Spectrosc.*, **29**, 48, 167 (1975); **30**, 113 (1976).
14. J. W. Busch, N. G. Howell, and G. H. Morrison, *Anal. Chem.*, **46**, 575, 2074 (1974).

15. D. O. Knapp, N. Omenetto, L. P. Hart, F. W. Plankey, and J. D. Winefordner, *Anal. Chim. Acta*, **69**, 455 (1974).
16. F. L. Fricke, O. Rhodes, and J. Caruso, *Anal. Chem.*, **47**, 2018 (1975).
17. H. Haraguchi, F. O. Fowler, J. D. Johnson, and J. D. Winefordner, *Spectrochim. Acta*, **32A**, 1539 (1976).
18. Y. Talmi, D. C. Baker, J. R. Jadamec, and W. A. Sauer, *Anal. Chem.*, **50**, 936A (1978).
19. W. H. Woodruff and G. H. Atikinson, *Anal. Chem.*, **48**, 186 (1976).
20. A. Danielsson, P. Lindblom, and E. Soderman, *Chem. Scr.*, **6**, 5 (1974).
21. M. J. Milano, H. L. Pardue, T. E. Cook, R. E. Santini, D. W. Margerum, and J. M. T. Raycheba, *Anal. Chem.*, **46**, 374 (1974).
22. H. L. Felkel, Jr., and H. L. Pardue, *Anal. Chem.*, **49**, 1112 (1977); **50**, 603 (1978).
23. H. L. Felkel, Jr., and H. L. Pardue, in *Multichannel Image Detectors*, Y. Talmi, Ed., ACS Symposium Series 102, American Chemical Society, Washington, D.C., 1978, p. 59.
24. K. W. Busch, B. Malloy, and Y. Talmi, *Anal. Chem.*, **51**, 670 (1979).
25. R. E. Santini, *Anal. Chem.*, **44**, 826 (1972).
26. K. W. Jackson, *Spectrosc. Lett.*, **6**, 315 (1973).
27. Y. Talmi, *Anal. Chem.*, **47**, 658A, 699A (1975).
28. Y. Talmi, H.-P. Sieper, L. Moenke-Blankenburg, and W. Quillfeldt, *Proc. 21st. Colloquium Spectroscopicum Internationale*, Cambridge, 1979, Abstracts 40.
29. Y. Talmi and R. Simpson, *Appl. Opt.*, **19**, 1401 (1980).
30. Y. Talmi, H.-P. Sieper, and L. Moenke-Blankenburg, *Anal. Chim. Acta*, **127**, 71 (1981).
31. Y. Talmi, Ed., *Multichannel Image Detectors*, ACS Monograph Series 102, American Chemical Society, Washington, D.C., 1979; ACS Symposium Series 236, American Chemical Society, Washington, D.C., 1983.
32. J. W. Olesik and J. P. Walters, *Appl. Spectrosc.*, **38**, 578 (1984).
33. H. Saisho, K. Sushida, H. Hashimoto, and E. Nakamura, *Proc. 24th Colloquium Spectroscopicum Internationale*, Garmisch-Partenkirchen, West Germany, 1985, p. 502.
34. P. W. J. M. Boumans et al., *Spectrochim. Acta*, **27B**, 247 (1972); **28B**, 277 (1973).
35. Prospectus B&M Spektronik, *Solid State Multichannel Detectors.*
36. J. A. de Haseth, W. S. Woodward, and T. L. Isenhour, *Anal. Chem.*, **48**, 1513 (1976).
37. R. E. Santini, M. J. Milano, and H. L. Pardue, *Anal. Chem.*, **45**, 915A (1973).
38. L. Peiko, J. Haas, and D. Osten, *Proc. Soc. Photo-Optical Instrumentation Engineers*, Vol. 116, *Solid State Imaging Devices*, SPIE, 1977, p. 56.
39. R. A. Saroyan, *Rep. UCID-16497*, Lawrence Livermore Laboratory, 1974.
40. K. J. Timmins and D. W. Snow, *Proc. 25th CSI*, Toronto, Canada, 1987, Abstract H4.2, p. 137.

41. J. Tabani and A. Montaser, *Proc. 25th CSI*, Toronto, Canada, 1987, Abstract H4.6, p. 139.
42. M. B. Denton, H. A. Lewis, and G. R. Sims, in *Multichannel Image Detectors*, Vol. 2, Y. Talmi, Ed., ACS symposium Series 236, American Chemical Society, Washington, D.C., 1983.
43. M. B. Denton, R. B. Bilhorn, and G. R. Sims, *Proc. Pittsburgh Conf.*, Atlantic City, N.J., 1986, No. 265.
44. B. Willman, S. Lee, A. G. Douglas, and R. R. Alfano, *Int. Lab.*, 18 (March 1988).
45. J. A. Maxwell, *Chem. Can.*, **15**, 10 (1963); *Can. Mineral.*, **7**, 727 (1963).
46. M. Berndt, H. Krause, L. Moenke-Blankenburg, and H. Moenke, *Jenaer Jahrb.*, 45 (1965).
47. H.-J. Blankenburg, L. Moenke-Blankenburg, H. Moenke, and K. Wehrberger, *Krist. Tech.*, **1**, 351 (1966).
48. W. H. Blackburn, Y. J. A. Pelletier, and W. H. Dennen, *Appl. Spectrosc.*, **22**, 278 (1968).
49. H. J. Bollingberg, *Spectrochim. Acta*, **27B**, 247 (1969).
50. L. Georgieva and A. Petrakiev, *Bol. Geol. Min.*, **80**, 491 (1969).
51. E. Kirchner, W. Meditz, and H. Neuninger, *Ann. Naturhist. Mus. Wien*, **73**, 37 (1969).
52. A. Petrakiev, L. Georgieva, and G. Dimitrov, *C. R. Acad. Bulg. Sci.*, **22**(9), 983 (1969).
53. N. Nikolov, A. P. Petkov, G. Dimitrov, and D. Dimov, *Congr. Colloq. Univ. Liège*, **59**, 267 (1970).
54. A. Petrakiev, G. Dimitrov, S. Beltschew, and N. Nikolov, *Izv. Fiz. Inst. ANEB Bulg. Akad. Nauk.*, **21**, 195 (1971).
55. L. Moenke-Blankenburg and J. Mohr, *Jenaer Jahrb.*, 195 (1970).
56. A. V. Karyakin, M. V. Achmanova, and V. A. Kaigorodov, *Proc. 12th CSI*, Exeter, England, 1965, p. 353.
57. D. Dimov, A. Petrakiew, G. Dimitrov, and A. Ivanov, *Bol. Geol. Min.* (*Bulg.*), **84**, 637 (1972).
58. W. Schrön, *Z. Angew. Geol.*, **18**, 350 (1972).
59. L. Moenke-Blankenburg, *Proc. 7th Hüttenmännische Materialprüfertagung*, Balatonszeplak, Hungary, 1973.
60. V. I. Smirnov, N. I. Eremin, V. E. Kelch, and D. R. Sakija, *Jena Rev.*, 16 (1972); V. I. Smirnov, N. I. Eremin, V. E. Kelch, V. M. Okrugin, and N. E. Sergeeva, *Jena Rev.*, 240 (1974).
61. A. Idzikowski and W. Schrön, *Z. Angew. Geol.*, **20**, 256 (1974).
62. L. Moenke-Blankenburg, H. Moenke, J. Mohr, W. Quillfeldt, W. Grassme, and W. Schrön, *Spectrochim. Acta*, **30B**, 227 (1975).
63. G. Dimitrov, D. Dimov, and A. Paneva, *Proc. EUROANALYSIS II*, Budapest, Hungary, 1975, Abstracts 36.
64. G. Dimitrov and M. Marinov, *Chemia Analytyczna* (*Bulg.*), **20**, 715 (1975).

65. S. Takenouchi and H. Imai, *Econ. Geol.*, **70**, 750 (1975).

66. A. Idzikowski, *Proc. 9th Conf. Institute for Inorganic Chemistry*, Polytechnical University, Wroclaw, Poland, 1975, pp. 37–55.

67. G. Dimitrov, D. Dimov, and A. Lazarova, *Proc. 7th National Conf. Spectroscopy*, Slanchev Bryag, Bulgaria, 1976, Abstracts D7, p. 102.

68. M. Hüfner, *Jena Rev.*, 312 (1976).

69. W. Quillfeldt and K. Berka, *Jena Rev.*, 302 (1977).

70. VEB Carl Zeiss, Jena, GDR, *Brochure 32-A-373-1*.

71. A. Felske, W.-D. Hagenah, and K. Laqua, *Spectrochim. Acta*, **27B**, 295 (1972).

72. F. Brech, *Analysis Instrumentation*, Vol. 6, Plenum Press, New York, 1969, p. 215.

73. N. Eremin and V. Kelch, *Dokl. Akad. Nauk SSSR*, **191**, 166 (1970).

74. A. Petrakiev, G. Dimitrov, S. Belchev, and N. Nikolov, *Jena Rev.*, 21 (1971).

75. J. Kozak, L. Pavel, and S. Kristoufka, *Pol. J. Soil Sci.*, **12**, 19 (1979).

76. J. N. Bennett and J. N. Grant, *Mineral. Mag.*, **43**, 945 (1980).

77. S. P. Atamanova, *Zh. Prikl. Spektrosk.*, **32**, 202 (1980).

78. V. V. Panteleev, V. A. Rozantzev, and A. A. Jankovsky, *Zh. Prikl. Spektrosk.*, **38**, 357 (1983).

79. W. Kramer, E. Kramer, H. J. Rösler, W. Klemm, W. Schrön, R. Starke, and P. Lange, *Z. Geol. Wiss. (Berlin)*, **8**, 1403 1(980).

80. W. Kramer, P. Lange, and W. Schrön, *Z. Geol. Wiss. (Berlin)*, **9**, 195 (1981).

81. J. Matsushita, T. Yamane, and S. Suzuki, *Proc. Conf. Analytical Measuring Methods in Short-Time Spectroscopy*, Tokyo, 1969, Abstract 20; *Bunko Kenkyu*, **19**, 147 (1970).

82. J. Debras-Guédon and N. Liodec, *Bull. Soc. Fr. Ceram*, **61**, 61 (1963).

83. K. Jahn, Untersuchungen von Mn-Fe-Akkumulaten aus der westlichen Ostsee hinsichtlich ihrer qualitativen und quantitativen Zusammensetzung, Diploma thesis, Martin Luther University, Halle, GDR, 1986.

84. L. Moenke-Blankenburg, K. Jahn, and L. Brügmann, Chemie der Erde, 1989 (in press).

85. M. Landmann, Studium der Plasmaeigenschaften und Erarbeitung optimaler Bedingungen für eine gezielte Anwendung der Laser-Mikro-Spektralanalyse zur substanziellen Kunstgutuntersuchung, Doctoral thesis, Martin Luther University, Halle, GDR, 1988.

86. W. G. Feklitschew, D. K. Scherbatschew, Yu. I. Iljaschenko, and W. W. Rjuchin, *Nov. Dannye Mineralach (SSSR)*, **32**, 171 (1985).

87. S. W. McGeorge, J. B. Falconer, and R. L. Lyke, *Proc. 25th CSI*, Toronto, Canada, 1987, Abstract H4.5, p. 139.

88. D. K. Ottesen and L. J. Radziemski, *Proc. Conf. Laser Applications to Chemical Analysis*, Lake Tahoe, Nev., 1987, Abstract TU A 6-2, p. 67.

89. R. H. Scott and A. Strasheim, *Laser Emission Excitation and Spectroscopy*, Vol. 1 in *Applied Atomic Spectroscopy*, E. L. Grove, Ed., Plenum Press, New York, 1978, pp. 73–118.

90. N. Omenetto, *Analytical Laser Spectroscopy*, Vol. 50 in *Chemical Analysis*. P. J.

Elving, J. D. Winefordner, and I. M. Kolthoff, Eds., John Wiley & Sons, Inc., New York, 1979.

91. K. Dittrich and R. Wennrich, *Prog. Anal. At. Spectrosc.*, **7**, 139 (1984).
92. R. S. Adrain and J. Watson, *J. Phys. D Appl. Phys.*, **17**, 1915 (1984).
93. R. S. Adrain, in *Industrial Applications of Lasers*, H. Koebner, Ed., John Wiley & Sons, Inc., New York, 1984, pp. 135–175.
94. E. H. Piepmeier, Ed., *Analytical Applications of Lasers*, Vol. 87 in *Chemical Analysis*, P. J. Elving, J. D. Winefordner, and I. M. Kolthoff, Eds., John Wiley & Sons, Inc., New York, 1986.
95. E. F. Runge, R. W. Minck, and F. R. Bryan, *Spectrochim. Acta*, **20**, 733 (1964).
96. J. Vilnat, N. Liodec, and J. Debraś-Guédon, *Proc. 12th CSI*, Exeter, England, 1965, Abstract p. 343.
97. J. R. Ryan and J. L. Cunningham, *Met. Prog.* 98 (1966).
98. L. Moenke-Blankenburg, *Tech. Rundschau* (*Bern*), **49**, 6 (1967).
99. E. Cerrai and R. Trucco, *Metall. Ital.*, **59**, 615 (1967); *Energ. Nucl.* (*Milan*), **15**, 581 (1968).
100. A. B. Whitehead and H. H. Heady, *Appl. Spectrosc.*, **22**, 7 (1968).
101. M. S. W. Webb and J. C. Cotterill, *Anal. Chim. Acta*, **43**, 351 (1968).
102. V. I. Kasatkin and R. I. Taganov, *Zh. Prikl. Spektrosk.*, **8**, 223 (1968).
103. A. Petrakiev and G. Dimitrov, *Proc. 15th CSI*, Madrid, 1969, Abstracts, p. 267.
104. G. Müller-Uri and P. Volkmann, *Jenaer Jahrb.*, 225 (1969–70).
105. E. Gegus, *Proc. 3rd. Conf. High Purity Materials Science and Technology.* Dresden, East Germany, 1970, Abstract 180.
106. T. Yamane, *Seramikkusu*, **6**, 680 (1971); *Bunseki Kagaku* (*Jpn. Anal.*), **20**, 1202 (1971).
107. A. Petrakiev and G. Dimitrov, *Proc. 16th CSI*, Heidelberg, West Germany, 1971, Vol. 1, p. 186.
108. H. Schroth, *Z. Anal. Chem.*, **253**, 7 (1971); **255**, 257 (1971).
109. S. Tassewa and A. Petrakiev, *Proc. 16th CSI*, Heidelberg, West Germany, 1971, Vol. 1, p. 191.
110. M. S. W. Webb and R. J. Webb, *Anal. Chim. Acta*, **55**, 67 (1971).
111. H. Schroth, *Z. Anal. Chem.*, **261**, 21 (1972).
112. S. A. Michnov, V. V. Panteleev, V. S. Strichner, and A. A. Jankovsky, *Zh. Prikl. Spektrosk.*, **17**, 394 (1972); see also A. A. Jankovsky, *Quantum Electronics and Laser Spectroscopy* (in Russian), Nauka i Technika, Mink, Russia, 1979.
113. K. Ohls, *Proc. 17th CSI*, Florence, Italy, 1973, Part I, p. 376; see also *Z. Anal. Chem.*, **264**, 97 (1973).
114. V. Janošikova, *Proc. 17th. CSI*, Florence, Italy, 1973, Part II, p. 766.
115. Kombinat VEB Carl Zeiss, Jena, GDR, *Publication 32-A-373-1.*
116. E. P. Krivchikova and R. I. Toganov, *Zh. Prikl. Spektrosk.*, **19**, 601 (1973).
117. G. Mohaupt and G. Pätzmann, *Jena Rev.*, 252 (1974).
118. K. Ohls, G. Becker, and H. Grote, *Jena Rev.*, 245 (1974).

119. J. Mohr, *Exp. Tech. Phys.*, **22**, 327 (1974); *Hena Rev.*, 245 (1979).
120. L. Moenke-Blankenburg, J. Mohr, and W. Quillfeldt, *Proc. 10th. Spektrometertagung*, The Hague, The Netherlands, 1974.
121. G. Dimitrov and A. Petrakiev, *Annu. Univ. Sofia Fac. Phys.*, **66**, 205 (1974).
122. W. Schrön, *Jena Rev.*, 112 (1975).
123. R. Kirchheim, M. Nagorny, K. Maier, and G. Tölg, *Anal. Chem.*, **48**, 1505 (1976).
124. M. Hüfner, *Jena Rev.*, 312 (1976).
125. M. L. Petuch, W. D. Satzunkevich, and A. A. Jankovsky, *Zh. Prikl. Spektrosk.*, **25**, 33 (1976); **29**, 1109 (1978); **32**, 414 (1980).
126. W. N. Mankevich, A. T. Nepotschoitschi, D. A. Skiwa, and V. V. Vasilieva, *Zh. Prikl. Spektrosk.*, **25**, 719 (1976).
127. L. Moenke-Blankenburg, D. Böwe, and W. Quillfeldt, *Proc. 7th National Conf. Spectroscopy*, Varna, Bulgaria, 1976.
128. D. E. Maksimov, N. K. Rudnevsky, W. P. Rabtschikova, C. M. Tschechowin, I. W. Schlapnikov, and I. C. Schklaeva, *Zavod. Lab.*, **4**, 445 (1977).
129. W. Quillfeldt and K. Berka, *Jena Rev.*, 302 (1977).
130. L. Moenke-Blankenburg and W. Quillfeldt, Proc. 20th CSI, Prague, Czechoslovakia, 1977.
131. W. Maul and W. Quillfeldt, *Jena Rev.*, 234 (1977).
132. F. Floris and E. Muttoni, *Jena Rev.*, 230 (1977).
133. W. Quillfeldt, *Jena Rev.*, 226 (1978).
134. L. Moenke-Blankenburg, W. Quillfeldt, and R. Spitzer, *Proc. 8th National Conference on Atomic Spectroscopy*. Varna-Druschba, Bulgaria, 1978, p. 244.
135. A. Priemuth, Doctoral thesis, Martin–Luther–University, Halle–Wittenberg, Halle, German Democratic Republic, 1978; *Proc. 5th Conf. Mikrosonde*, Leipzig, GDR, 1981, p. 276.
136. R. S. Adrain, D. R. Airey, and E. J. Ormerod, *Proc. 5th International Gas Discharge Conf. IEE*, 1978, Publ. 165, p. 70.
137. D. E. Maksimov, N. K. Rudnevsky, W. P. Rabtschikova, and E. N. Pranitschnikova, *Zavod, Lab.*, **4**, 333 (1979).
138. Y. Talmi, H.-P. Sieper, L. Moenke-Blankenburg, and W. Quillfeldt, *Proc. 21st CSI*, Cambridge, England, 1979, Abstracts 40.
139. R. S. Adrain, J. Watson, P. H. Richards, and A. Maitland, *Opt. Laser Technol.*, **12**, 137 (1980).
140. R. S. Adrain, D. R. Airey, R. C. Klewe, and E. J. Ormerod, *Proc. 5th IEE Conf. Gas Discharge*, 1978, Publ. 165, p. 70; *Proc. 7th IEE Conf. Gas Discharge*, 1980, Publ. 189, p. 231.
141. G. Dimitrov and T. Maximova, *Spectrosc. Lett.*, **14**, 737 (1981).
142. O. I. Nikitina, N. K. Ivanova, L. A. Slinko, and L. N. Sacharchenko, *Zh. Pirkl. Spektrosk.*, **34**, 197 (1981).
143. Y. Talmi, H.-P. Sieper, and L. Moenke-Blankenburg, *Anal. Chim. Acta*, **127**, 71 (1981).

144. E. K. Wulfson, W. I. Dborkin, A. W. Karjakin, and P. J. Musakov, *Zh. Prikl. Spektrosk.*, **35**(3), 415 (1981).
145. L. Nikolova and N. Krasnobaeva, *Isw. Chim.* (*Bulg.*), **17**(3), 274 (1984).
146. N. K. Rudnevsky, A. N. Tumanova, and E. V. Maximova, *Spectrochim. Acta*, **39B**, 5 (1984).
147. A. Buzási, M. Györfi, M. Pankotai, and I. Kosma, *Proc. 5th. EUROANALYSIS*, Krakow, Poland, 1984.
148. G. Dimitrov and T. Zheleva, *Spectrochim. Acta*, **39B**, 1209 (1984).
149. J. A. Millard, R. H. Dalling, and L. J. Radziemski, *Appl. Spectrosc.*, **40**(4), 491 (1986).
150. St. Armenski and G. Dimitrov, *Phys. Status Solidi*, 1989 (in press).
151. K. G. Snetsinger and K. Keil, *Am. Mineral.*, **52**, 1842 (1967).
152. Y. Katsuno, Morinaka, and S. Hiroshi, *Japan Analyst*, **17**, 376 (1968).
153. J. R. Ryan, E. Ruh, and C. B. Clark, *Ceram. Bull.*, **45**, 260 (1966).
154. M. Birkenfeld and L. Moenke-Blankenburg, Laboratory report, Martin Luther University, Halle, GDR, 1986.
155. H. J. Horst, N. Koch, and E. Bader, *Silikattechnik*, **33**, 22 (1982).
156. M. Pacher, Untersuchungen zur quantitativen Lasermikrospektralanalyse: Anwendung und Vergleich mehrerer Modelle, Diploma thesis, Martin–Luther–University, Halle, GDR, 1985.
157. L. Moenke-Blankenburg and M. Pacher, *Acta Chim.* (*Hung.*), 1989 (in press).
158. G. O. Karapetyan and L. V. Maksimov, *Kvantovaya Elektroni.* (*Moscow*), **11**(9), 1839 (1984).
159. H. Moenke and L. Moenke-Blankenburg, *Sprechsaal Keram. Glas Email Silik.*, **100**, 112 (1967).
160. J. Dimitrov, G. Dimitrov, and A. Petrakiev, *Mat. Sci. Eng.*, **5**, 334 (1970).
161. R. Siechau, M. Mazurkiewicz, and H. Nickel, *Berichte der Kernforschungsanalge*, Jülich, FRG, Jül-1834, 1983.
162. R. Siechau, K. Bürgers, M. Mazurkiewicz, and H. Nickel, *Fresenius Z. Anal. Chem.*, **314**, 221 (1983).
163. L. Moenke-Blankenburg and W. Quillfeldt, *Jena Rev.* 1972, p. 91.
164. L. Nikolova and N. Krasnobaeva, *Isw. Chim. Bulg.* **20**(3), 349 (1987).
165. U. Zander and L. Moenke-Blankenburg, Laboratory report, Martin–Luther–University, Halle, GDR, 1987.
166. U. Zander and L. Moenke-Blankenburg, Laboratory report, Martin–Luther–University, Halle, GDR, 1987.
167. T. Yamane and N. Yamada, *Spectrochim. Acta*, **30B**, 305 (1975).
168. A. A. Boitsov and K. I. Zilberstein, *Spectrochim. Acta*, **36B**, 1201 (1981).
169. H. Nickel, F. A. Peuser, and M. Mazurkiewicz, *Spectrochim. Acta*, **33B**, 675 (1978).
170. H. Nickel, F. A. Peuser, M. Mazurkiewicz, and W. Dörge, *Jena Rev.*, **24**, 199 (1979).
171. W. Dörge, M. Mazurkiewicz, and H. Nickel, *Berichte der Kernforschungsanlage*, Jülich, FRG, Jül-1439, 1977.

172. H. J. Stupp, *Berichte der Kernforschungsanlage*, Jülich, FRG, Jül-933-RG, 1973.
173. H. J. Stupp and T. Overhoff, *Spectrochim. Acta*, **29B**, 77 (1974); **30B**, 77 (1975).
174. J. Borszéki, L. Koltai, L. Bartha, J. Inczédy, and E. Gegus, *Hung. J. Ind. Chem (Veszprem)*, **12**, 193 (1984).
175. E. Litz, *Jena Rev.*, 237 (1977).
176. D. A. Cremers and L. J. Radziemski, *Appl. Spectrosc.*, **39**(1), 57 (1985).
177. R. McGillivray, *Proc. Anal. Div. Chem. Soc.*, **12**, 232 (1975).
178. H. Schicht, *Forum Kriminal.*, **9**, 11 (1968).
179. H. Neuninger, *Arch. Kriminol.*, **144**, 121 (1969); also private communication.
180. E. Rudolph, Nachweis von Fremdmetallabrieb auf Stahlwerkzeugen mit Hilfe der Laser-Mikrospektralanalyse, Engineering thesis, Ingenicur Hochschule Köthen, Köthen, GDR, 1969.
181. H. Neuninger, *Mikrochim. Acta, Suppl. IV*, 239 (1970).
182. A. Petrakiev, A. Samov, and G. Dimitrov, *Jena Rev.*, 250 (1971).
183. H. Neuninger and F. Sauter, *Forsch. Stillfried*, **1**, 69 (1974).
184. V. K. Mehrotra and S. K. Sidhana, *Forensic Sci.*, **9**, 1 (1977).
185. S. O. Baisane, V. S. Chincholkar, and B. N. Mattao, *Jena Rev.*, 206 (1979); *Ind. Acad. Forensic Sci.*, **17**, 31 (1978).
186. G. Bajnószi and R. Major, *Magy. Kem. Lapja*, **37**, 385 (1982).
187. J. Borszéki, J. Inczédy, E. Gegus, and F. Ovári, *Fresenius Z. Anal. Chem.*, **314**, 410 (1983).
188. J. Borszéki, J. Inczédy, F. Ovári, and E. Gegus, *Acta Archaeol. Acad. Sci. Hung.*, **35**, 1 (1983).
189. M. Bakos and E. Gegus, *Acta Archaeol. Acad. Sci. Hung.*, **31**, 1 (1979).
190. E. Gegus, *Proc. 24th CSI*, Garmisch-Partenkirchen, West Germany, 1985, p. 1.
191. H. Geisler, *Produktivkräfte und Produktionsverhältnisse*, Berlin, GDR, 1985, p. 271.
192. M. Schreiner, P. Dolezal, G. Banik, and F. Mairinger, *Proc. 24th CSI*, Garmisch-Partenkirchen, FRG, 1985, p. 508.
193. M. Landmann, *Neue Museumskd*, **1**, 51 (1985).
194. M. Landmann, Studium der Plasmaeigenschaften und Erarbeitung optimaler Bedingungen für eine gezielte Anwendung der Laser-Mikro-Spektralanalyse zur substantiellen Kunstgutuntersuchung. Doctoral thesis, Martin–Luther–University, Halle, German Democratic Republic, 1988.
195. M. Papazova, L. Georgieva, and A. Petrakiev, *Izv. Natl. Istor, Musei (Bulg.)*, *VI*, 96 (1986).
196. M. Gäckle, W. Nitzschke, and K. Wagner, *Jahresschr. Mitteldtsch. Vorgesch.*, **71**, 57 (1988).
197. R. Feustel and W. Maul, *Jena Rev.*, 1987, p. 131.
198. R. Ashok, *Nat. Galery Tech. Bull. (London)*, **3**, 43 (1979).
199. R. P. Bleck, E. Litz, W. Maul, and W. Timpel, *Ausgrabungen Funde*, **36**, 223 (1981).
200. W. Schrön, *Neue Museumskd.*, **1**, 67 (1982).
201. M. Schreiner, G. Stingeder, and M. Grasserbauer, *Fresenius Z. Anal. Chem.*, **319**, 600 (1984).

202. R. C. Rosan, M. K. Healy, and W. F. McNarry, Jr., *Science*, **142**, 236 (1963).

203. H. M. Goldman, M. P. Ruben, and D. B. Sherman, *Oral. Surg. Oral. Med. Oral. Pathol.*, **17**, 102 (1964).

204. R. C. Rosan, F. Brech, and D. Glick, *Fed. Proc.* (*USA*) *Suppl.*, **24**, 126, 542 (1965).

205. D. B. Sherman, M. P. Ruben, H. M. Goldman, and F. Brech, *Science*, **122**, 767 (1965).

206. R. G. Wilson, L. Goldman, and F. Brech, *Arch. Dermatol.*, **95**, 491 (1967).

207. E. S. Beatrice, I. Harding-Barlow, and D. Glick, *Proc. 18th Annual Meeting of the Histochemistry Society*, Chicago, 1967, paper 3.

208. G. Dimitrov, I. Vassilev, and A. Petrakiev, *Annu. Univ. Sofia*, **63**, 123 (1968–69).

209. M. Kozik, B. Arcimovicz, and J. Dembczynski, *Acta Histochem.*, **37**, 203 (1970).

210. M. B. Kozik, J. Wachol, and B. Arcimovicz, *Histochemie*, **24**, 245 (1970); **26**, 24 (1971); **26**, 212 (1971); M. B. Kozik, *Folia Histochem Cytochem.*, **16**, 31 (1978); **17**, 153 (1979).

211. W. J. Treytl, J. B. Orenberg, A. J. Saffir, and D. Glick, *Anal. Chem.*, **44**, 1903 (1972).

212. D. Glick and K. W. Marich, *Proc. European Histochemistry Meeting*, Royal Microscopy Society, Nottingham, England, 1975, Abstract 281.

213. G. Dimitrov and M. Marinov, *Chem. Anal.* (*Warsaw*) **20**, 715 (1975).

214. R. Galabova, A. Petrakiev, A. Paneva, and P. Petkov, *Acta Histochem.*, **54**, 66 (1975).

215. E. Schulz, W. Berg, R. H. Börner, E. Litz, and U. Maier, *Jena Rev.*, 97 (1981).

216. G. Dietz, G. Klinger, E. Litz, and U. Maier, *Jena Rev.*, **99** (1981).

217. M. Marinov and G. Dimitrov, *Proc. 24th CSI*, Garmisch-Partenkirchen, FRG, 1985, p. 78.

218. R. Jablenska, A. Petrakiev, Z. Kamenova, L. Georgieva, V. Ogneva, and P. Petkov, *Proc. 31st. International Congress on Pure and Applied Chemistry*, Sofia, Bulgaria, 1987, Abstracts.

219. D. Günther, Die lasermikrospektralanalytische Bestimmung der Verteilung von Aluminium in Lendenwirbeln von Ratten, Diploma thesis, Martin–Luther–University, Halle, GDR, 1987.

220. R. Schmidt, L. Moenke-Blankenburg, and D. Günther, *Acta Histochem.*, 1989 (in press).

221. K. Al Hamadi and S. Hagewald, Zn-Bestimmungen in Lendenwirbeln von Ratten mit LM-ICP-Spektrometrie nach intraperitonealer, dosis- und zeitabhängiger Applikation von Zinkchloridlösungen, Diploma thesis, Martin–Luther–University, Halle, GDR, 1988.

222. I. Harding-Barlow, K. G. Snetsinger, and K. Keil, in *Microprobe Analysis*, C. A. Andersen, Ed., John Wiley & Sons, Inc., New York, 1971.

223. L. Goldman, in *Modern Techniques in Physiological Science*, J. F. Gross, R. Kaufmann, and E. Wetterer, Eds., Academic Press, Inc., London and New York, 1973.

224. R. Schmidt and L. Moenke-Blankenburg, *Acta Histochem.*, **80**, 205 (1986).

CHAPTER

6

# LASER MICROVAPORIZATION COMBINED WITH PLASMA EXCITATION BASED ON OPTICAL EMISSION SPECTROMETRY

In ICP the heating of the atomic reservoir is through induction heating (which is based on the well-known effect of electromagnetic induction discovered by Faraday in 1831). Inside the coil, made of copper tubing with two or three windings, a torch consisting of three concentric quartz tubes is arranged. Argon or nitrogen is delivered through the outer tubes and cools the outside of the plasma and protects the torch. The intermediate gas flow may be omitted, can propagate the plasma, or can be the plasma gas. The central tube is used for the injector gas and consists of argon plus the sample as an aerosol. The radio-frequency generator produces between 2 and 30 kW forward power at between 5 and 50 MHz, typically a few kilowatts at 27.12 MHz. When the power is on the coil, an ac magnetic field is generated axially through the coil. When the argon gas is flowing, the plasma is ignited using a spark. The electrons rapidly reach ionizing energies. The neutral argon is ionized by collisional energy exchange. The temperature of the plasma is in the region of 8000 K, which gives an excellent excitation source for emission spectrometry as well as for laser micro emission spectrometry. The advantage of the torch configuration is that the sample atoms remain concentrated in a narrow channel in the middle of the plasma plume, so that an approximation of the ideal point source is reached. Second, the sample atoms experience higher temperatures, thus emitting more light and becoming less sensitive to interference. Its precison is about 1 to 5%. The dynamic range runs from three to five decades and starts to resemble the possibilities in modern electronic equipment, which is indeed needed to take advantage of this feature. The detection limits reported in the literature lie in the ppb to ppm range.

## 6.1. ICP SPECTROMETRY OF SOLIDS

ICP spectrometry is a standard method for the analysis of liquid samples and for solids after sample dissolution. However, two disadvantages are given for solid sample dissolution: the risk of contamination and the dilution of the

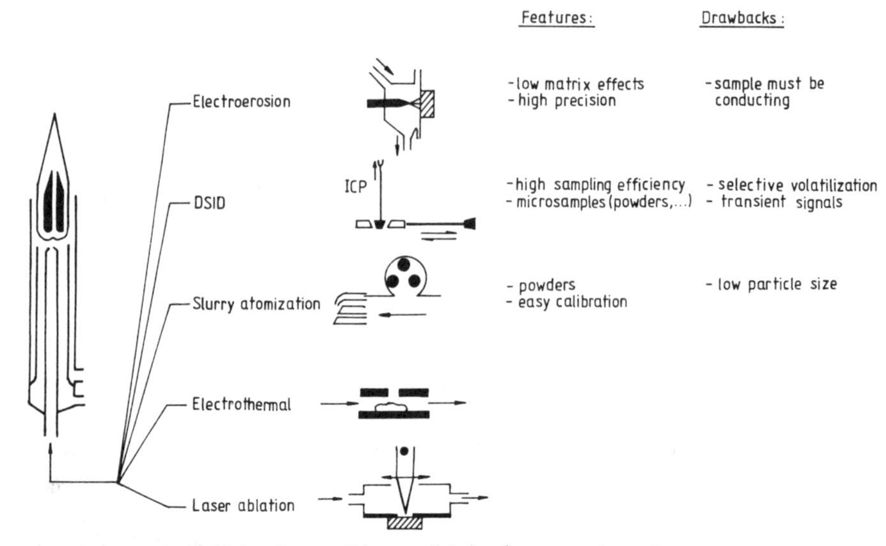


**Figure 51.** Techniques for direct solids sampling in plasma spectrometry.
DSID, direct sample insertion device. From Ref. [2], reprinted by permission of Pergamon Journals Ltd.

analyte due to the solvent. Furthermore, chemical procedures are necessary and a lot of samples, including silicates, ceramics, and special alloys, are difficult to dissolve. Therefore, direct solid sampling will be of interest and has been proposed since the early work in ICP spectrometry [1, 2]. Aims that should be attained include preserving good analytical precision, low matrix effects, and easy calibration, well known from work with solutions. Figure 51

**TABLE 33. Requirements of an Ideal Technique for Direct Solids Sampling in Plasma Spectrometry**

| |
|---|
| *Sample Volatilization* |
| High volatilization rate |
| Independence of sample nature |
| *Analyte Transport* |
| High removal efficiency |
| Low memory effects |
| Requiring only low transport gas |
| *Analyte Aerosol* |
| Low particle size |
| Composition of volatilized material must equate that of the sample |

*Source*: Ref. 2; published with permission of Pergamon Journals Ltd.

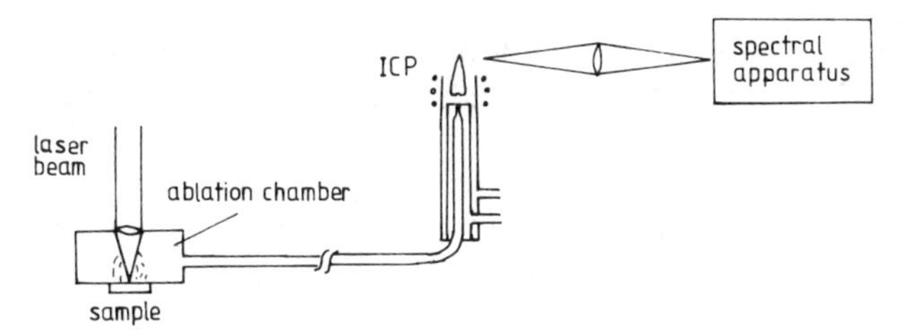


**Figure 52.** Principle of LM-ICP-OES. The laser-ablated material is transferred into an inductively coupled plasma.

shows five techniques for direct solids sampling which can be used in ICP spectrometry. Three of these techniques (electroerosion, electrothermal technique, and the direct sample insertion device) are investigated by Broekaert et al. [2]. All techniques must fulfill some conditions, which are dictated by the processes of sample ablation, analyte transport, and plasma insertion (see Table 33).

Laser ablation offers a significant potential for the direct introduction of solids into the ICP. First attempts to combine laser ablation with inductively coupled plasma excitation were done by Abercrombie in 1977 [3], Salin et al. in 1979 [4], Carr and Horlick in 1980 [5], Thompson et al. in 1981 [6], and others [7–11].

The principle most commonly used is shown in Fig. 52. The sample is set up in a laser ablation cell, and the sample surface is shot by a laser beam. The sample vapor produced from the surface by a laser shot is introduced into the ICP by the flow of argon carrier gas. The emission signals of various elements can be measured with different kinds of spectral apparatus.

### 6.1.1. Instrumentation of LM-ICP-OES

A first equipment combining laser vaporization with ICP excitation was described by Abercrombie et al. [3] in 1977. They used a pulsed carbon dioxide laser and a fully automated ICP spectrometer which could analyze 25 elements simultaneously every 10s. The sample particles were collected and fixed on a paper tape.

Salin et al. [4] and Carr and Horlick [5, 8] used a ruby laser in the free-running mode as well as in the $Q$-switched mode. $Q$-switching was carried out using a KDP Pockels cell. The energy output of the laser was limited at 1 to 2 J. In the free-running mode this energy was spent in about 1 ms as more than 100 spikes, each of 10 to 20 ns duration. The output power of the $Q$-switched laser was about 100 MW in a single pulse of 10 to 20 ns duration. The maximum repetition rate of the laser was about 1 pulse every 4 s.

A RF-ICP was used with 2.5-kW power and 27.12-MHz radio frequency. The plasma torch geometry corresponds to that used by [12]. The flow rates of the coolant, auxiliary, and carrier gases (Ar) were 18, 0.1, and 0.8 to 1.2 liters/min, respectively. The normal power setting for the ICP was 1.5 kW.

The connection from the sampling cell to the plasma torch was kept as short as possible. Therefore, the sample chambers were placed directly below the torch. Figures 53 and 54 show two sample chambers, designed by Carr and Horlick [8] to accept different sample geometries. The chambers are built to handle samples in the form of rods ($\frac{3}{8}$ in. in diameter and $\frac{1}{2}$ to 4 in. in length) and in the form of disks ($\frac{1}{2}$ to 5 in. in diameter and $\frac{1}{32}$ to $2\frac{1}{2}$ in. in thickness). The samples can be moved. The laser beam is focused on the sample surface using a short-focus (5 cm) spherical lens.

Spectral data were obtained with a 1024-element photodiode array spectrometer, developed by Horlick [13]. Such a simultaneous multichannel

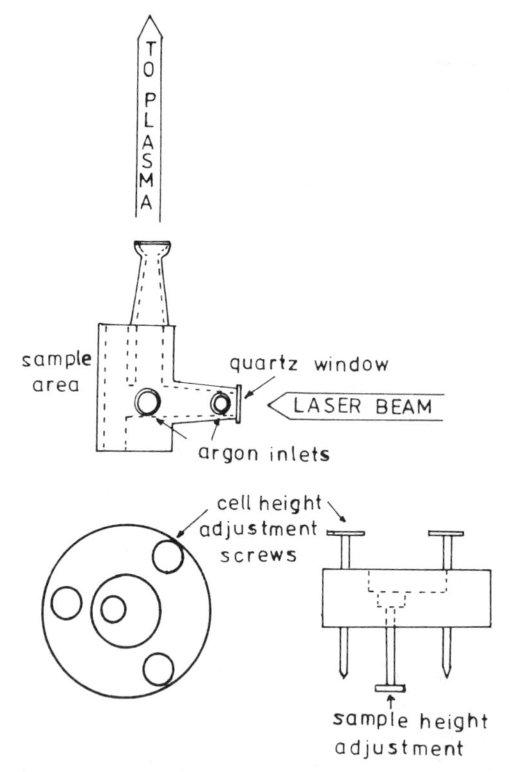


**Figure 53.** Schematic of rod-type sample chamber. From Ref. [8], reprinted by permission of Pergamon Journals Ltd.

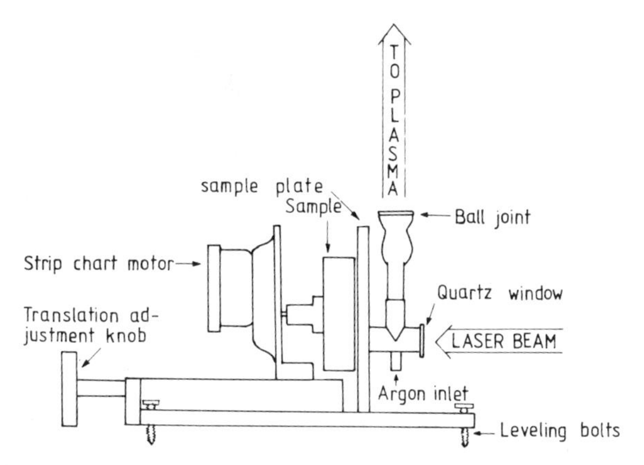


**Figure 54.** Schematic of disk-type sample chamber. From Ref. [8], reprinted by permission of Pergamon Journals Ltd.

spectrometer is a critical requirement for registration of transient emission signals generated by LM-ICP. A limitation is introduced by the 50 nm spectral window, which divides the entire spectral range from 180 to 1000 nm in about 17 parts. The resolution is sufficient with 0.05 nm per diode.

Most workers dealing with LM-ICP-OES [6, 7, 9–11] use the coupling principle shown in Fig. 52. In 1981, Thompson et al. [6] first coupled a commercially available laser microanalyzer (laser energy about 1 J) with an ICP spectrometer, also commercially available. Cells of various diameters from 20 to 50 nm were used to accommodate different sample sizes. The cells were open-bottomed, sealing being achieved by mounting the sample and cell on an inverted rubber bung. The laser window was a 35-mm disk of 2-mm-thick optical silica. The height of the cell above the surface of the sample was limited to $\leqslant 17$ mm, depending on the focal distance of the objective. To synchronize the moment of vaporization by laser beam interaction with the sample and the initiation of the integration cycle of the spectrometer, a light-actuated switch was provided. The light pulse was transmitted to the switch by an optic fiber. The optic fiber was fixed at the glassy wall of the chamber, directed toward the focal point of the laser beam. The switch was connected in parallel with the normal start known of the spectrometer and initiated a momentary short circuit in response to an increase in light intensity.

Figure 55 is a schematic of the equipment used by Ishizuka and Uwamino [9]. The maximum energy provided by the ruby laser run to about 3 J in $Q$-switched mode and to 30 J in normal mode. The laser radiation was focused using a lens with a focal distance of 50 mm. The sample chamber was made of stainless steel walls with an inner diameter of 13 mm and a height of 35 mm,

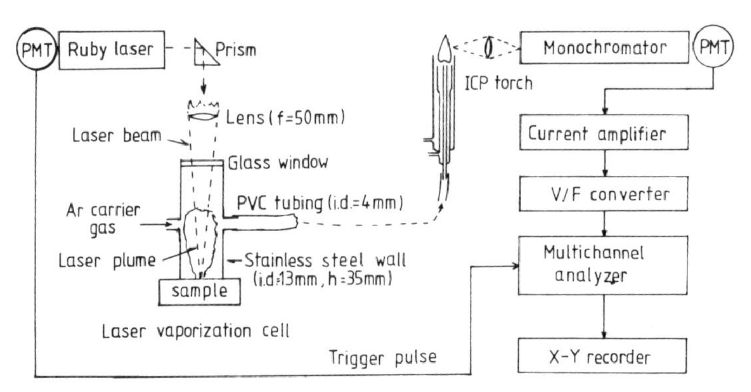


**Figure 55.** Schematic of the laser-ICP system used by Ishizuka and Uwamino. From Ref. [9], reprinted by permission of Pergamon Journals Ltd.

and a Pyrex window. The length of the PVC tubing varied between 40 and 200 cm. The RF generator had a frequency 27 MHz and a maximum power output of 2 kW, of which 1.2 kW was used. The argon carrier gas flow rate amounts 1.1 liters/min, the coolant gas flow rate 14 liters/min, and the plasma gas flow rate 1 liters/min. The spectrometer was of a 1-m mounting Czerny–Turner type with a 2400-line/mm grating and a reciprocal dispersion of 0.4 nm. The emission signal from the photomultiplier was amplified with a current amplifier. After converting the voltage to frequency, it was counted with a multichannel analyzer. The analyzer was triggered by the pulse of the laser flashlamp.

Dittrich et al. [10] used a commercially available laser microanalyzer (developed by [14]; the same as that used by Thompson et al. [6] and by Günther and Gäckle [11]) their own ICP torch, and a time-consuming spectrograph with a photoplate as detector or with a PMT adaptation for sequential analyses.

Figures 56 and 57 shows schematically the equipment used by Günther and Gäckle [11]. The output energy of the ruby laser is a maximum of 1 J in the normal mode and 0.1 to 1 J in the giant pulse or semi-$Q$-switched mode (see Table 2 and Table 23, first column). The sample chamber (scheme shown in Fig. 57) is patented by Schrön [15, 16]. Two types of chambers were developed by the authors [11]. The ICP spectrometer [17] can provide sequential *and* simultaneous analyses. The technical data are: radio frequency 27 MHz, working power 2.5 kW, plasma gas 11 liters/min, auxiliary gas 0.5 liter/min, and carrier gas 4.5 liters/min. The detectors are PMTs.

A new spectrometer system usable for ICP spectrometry has recently been described [18–20]. This system is comprised of a reprogrammable preselection polychromator coupled to an echelle-based high-resolution spectrometer

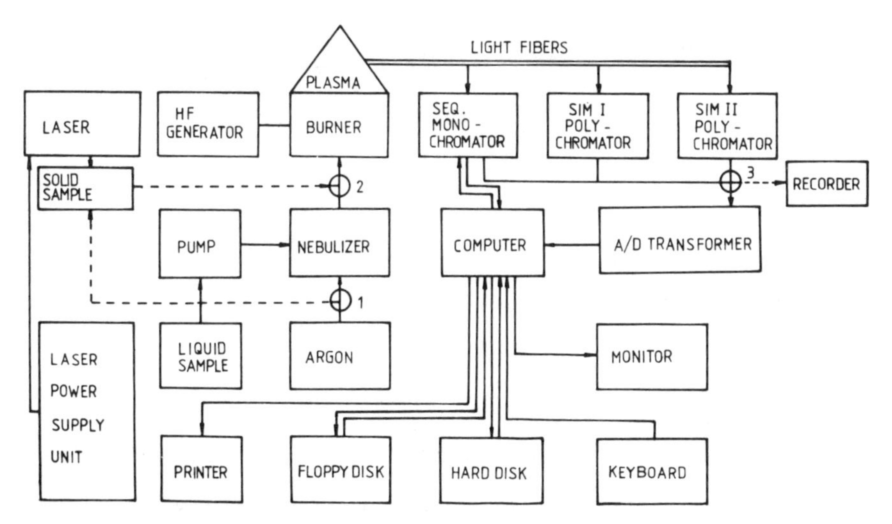


**Figure 56.** Block diagram of LM-ICP-OES used by L. Moenke-Blankenburg and co-workers. From Ref. [11], reprinted by permission of VEB Grundstoffverlag, Leipzig.

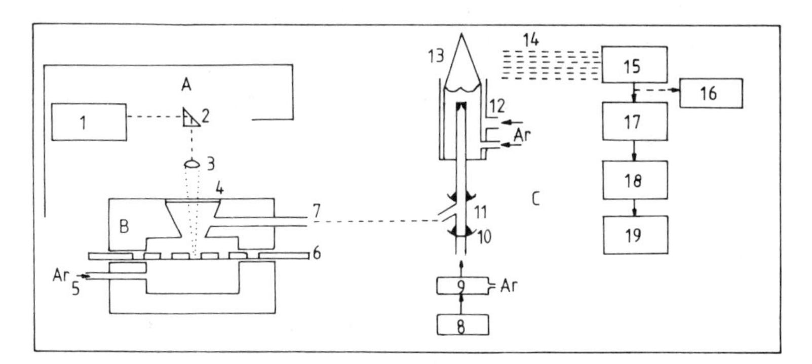


**Figure 57.** Detail from Fig. 56 showing sample chamber set up according to Refs. [15, 16]. (*A*) Microscope: 1, laser, 2, prism, 3, objective. (*B*) Sample chamber: 4, quartz window; 5, argon input; 6, sample store; 7, tubing. (*C*) ICP spectrometer: 8, pump; 9, nebulizer; 10, valve; 11, joining piece for tubing; 12, torch; 13, plasma; 14, fiber optic; 15, spectrometer optic; 16, recorder; 17, AD transformer; 18, computer; 19, monitor. From Ref. [11], reprinted by permission of VEB Grundstoffverlag, Leipzig.

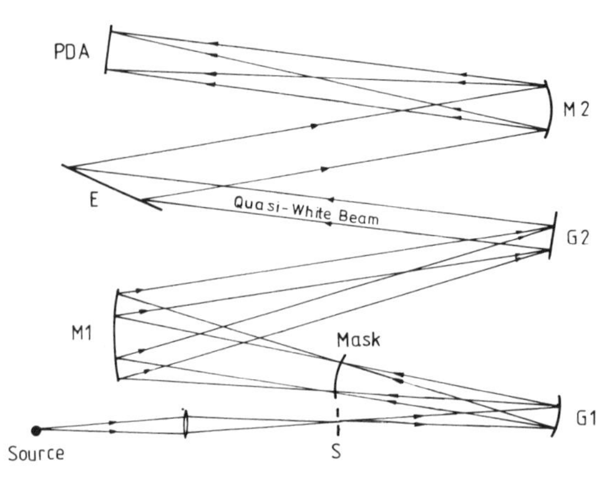


**Figure 58.** Optical scheme of plasmaarray spectrometer. S, entrance slit; G1, concave grating; M1, corrective mirror; G2, plane grating, ; E, echelle grating; M2, camera mirror; PDA, photodiode array detector. From Ref. [19], reprinted by permission of Pergamon Journals Ltd.

incorporating a linear photodiode array detector (1024-element linear self-scanning PDA). The optical scheme of the so-called "plasmaarray" spectrometer [19] is shown in Fig. 58. This system is capable of simultaneous multielement analysis for analytical wavelengths ranging from 190 to 415 nm with a high resolution of $\leqslant 0.01$ nm and a reciprocal linear dispersion of 0.08 nm/mm. One of the advantages for using this type of spectrometer for LM-ICP-OES is the fact that it does not require any optical components to be moved for wavelength scanning. However, the flexibility of the spectrometer is dependent on the ability to place different masks at the Rowland circle of G 1 without significantly attenuating the desired line radiation that would normally pass through the preselection polychromator focal plane. The maximum number of slots that can be cut into a single mask is for a set of 10 to 20 elements.

### 6.1.2. Theoretic and Methodic Investigations

The effects of laser target interaction and vaporization of solid materials on laser parameters are discussed in Chapter 3. The effect of sample chamber design, its volume and aerodynamics, of length of PVC tubing, and of the transport behavior of the sample gas are now of interest.

The effect of distance between the laser ablation cell and the ICP torch on the emission intensity of Mn (0.27%) in steel was investigated [9] by varying the length of the PVC tubing, from 40 to 70 to 100 to 200 cm. Figure 59 shows the emission signal profiles obtained by using a *Q*-switched laser of

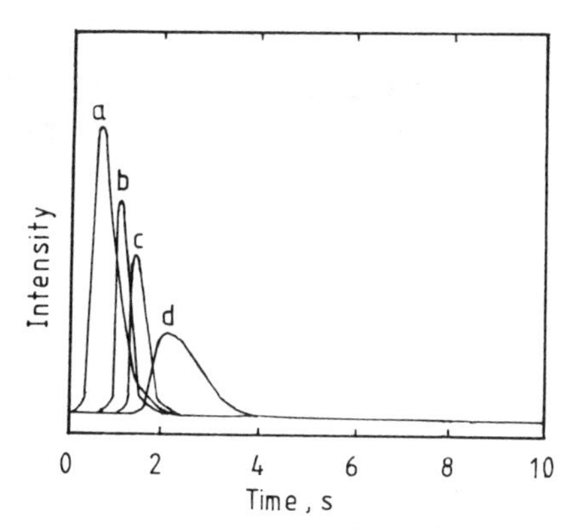


**Figure 59.** Profiles of emission signals for Mn (0.27%) in steel obtained by varying the length of PVC tubing: (*a*) 40 cm; (*b*) 70 cm; (*c*) 100 cm; (*d*) 200 cm. The laser was used in the *Q*-switched mode. From Ref. [9], represented by permission of Pergamon Journals Ltd.

about 2 J output. The peak height decreased gradually when the length was increased, and peak broadening resulted with increased length. A comparison of peak height and peak area, and of the influence of a *Q*-switched and a normal-mode laser, is given in Fig. 60. With the lasers in each mode, the degree of the decrease in intensity with increasing length was larger for the peak height value than for the peak area value, and was larger in the *Q*-switched mode than in the free-running mode. Analogous results were obtained by Günther and Gäckle [21] for lengths of tubing between 0.7 and 2.7 m.

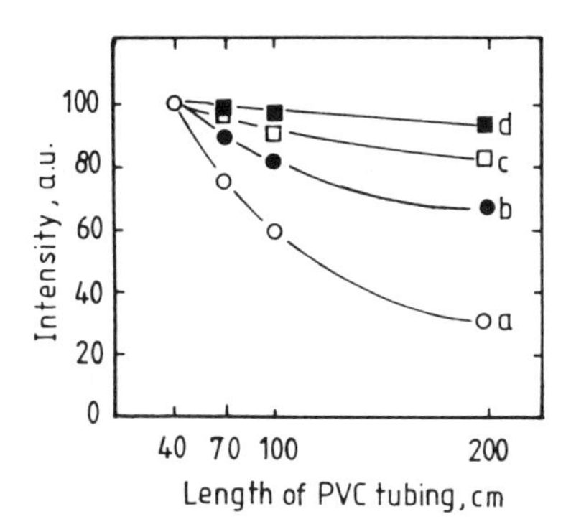


**Figure 60.** Effect of length of PVC tubing on the emission intensity of Mn (0.27%) in steel. (*a*) Peak height and (*b*) peak area were obtained by the *Q*-switched laser; (*c*) peak height and (*d*) peak area were obtained by the normal laser. From Ref. [9], reprinted by permission of Pergamon Journals Ltd.

According to Ramsing et al. [22] and Matschiner et al. [23], Günther and Gäckle [11] proposed using the formula for dispersion $D$ for flow injection analysis for the transport of sample gas in an Ar stream from the laser ablation cell to the plasma torch via PVC tubing:

$$D = \frac{c_0}{c_{\max}} = \frac{k_1 H_0}{k_2 H_{\max}} = 2\pi^{3/2} r^2 \delta^{1/2} F^{1/2} L^{1/2} T^{1/2} s_v^{-1} \tag{57}$$

where $D$ = dispersion
$c_0$ = starting concentration
$c_{\max}$ = maximum concentration of the dispersed sample
$H_0$ = peak height at $c_0$
$H_{\max}$ = peak height at $c_{\max}$
$k_1, k_2$ = factors
$r$ = radius of the flow system
$\delta$ = dispersion number, to be calculated (see [24]).
$F$ = flow velocity
$L$ = length of tubing
$T$ = residence time
$s_v$ = sample volume

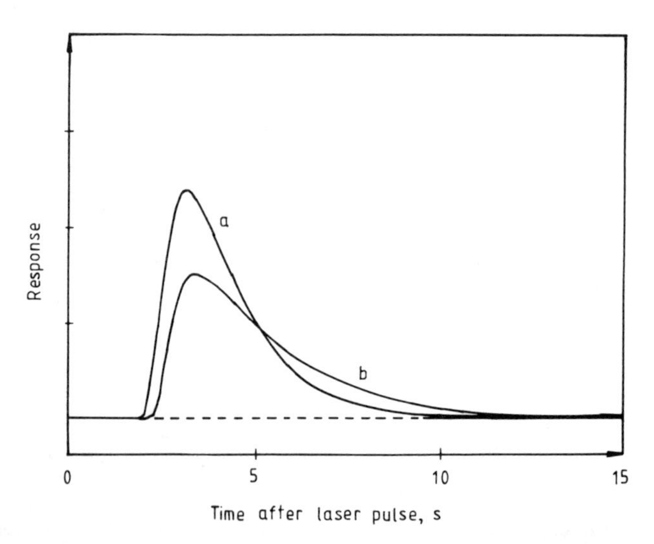


**Figure 61.** Instantaneous response as a function of time on the iron channel after laser ablation of a sample of steel, in chambers of different volumes: (*a*) about 5 cm³; (*b*) about 30 cm³. From Ref. [6], reprinted by permission of the Royal Society of Chemistry.

The dispersion of the gaseous sample in the carrier gas (Ar) increases with the square root of the length $L$ of the transport way and the flow velocity $F$, respectively. The dispersion is directly proportional to the square of the radius $r$ of the tubing and reversed in proportion from the sample volume.

The main problem with LM-ICP coupling is efficient generation and maintenance of a reasonable population of the desired chemical species (i.e., excited atomic vapor for the emission system). The laser plume contains significant amounts of solid particles, which means that both laser ablation and laser vaporization take place. Therefore, effective systems need the capability of further vaporizing the solid fraction of the ablated material. The high temperature and a more-or-less long residence time of introduced species of ICP should be quite effective for this purpose.

Figure 61 shows a signal obtained from an iron channel after a sample of mild steel was subjected to a laser pulse. The signal is unaffected for the first 2 s after the pulse, than rises rapidly to a maximum in about 1 s, and decays to the level with a half-life of about 1.2 s. Trace constituents show the same decay profile as those of the major constituents. Metallic and nonmetallic samples gave the same results, so there seems to be no differentiation of the vaporized sample as a fraction of time.

### 6.1.3. Fields of Application

Thompson et al. [6] investigated the components Co, Cr, Mn, Mo, V, P, S, and Si in steel. They compared some estimated detection limits of LM-ICP-OES

**TABLE 34. Some Estimated Detection Limits for the ICP-Laser Microprobe Compared with Nebulization on the Same ICP Instrument**

| | Concentration Detection Limit for Steel ($\mu$g/g) | | Absolute Detection Limit (pg) | |
|---|---|---|---|---|
| Element | Laser | Nebulizer[a] | Laser[b] | Nebulizer[c] |
| Cr | 15 | 2 | 15 | 25 |
| S | 15 | 45 | 15 | 250 |
| P | 10 | 35 | 10 | 150 |
| Mn | 80 | 2 | 80 | 5 |
| Mo | 60 | 4.5 | 60 | 20 |
| Cu | 20 | 1.5 | 20 | 12 |
| V | 10 | 1.5 | 10 | 20 |
| Ni | 70 | 1.0 | 70 | 70 |

*Source*: After Ref. 6; reprinted by permission.

[a] Assumed 1% solution of the steel.

[b] Assumed 1 $\mu$g of sample ablated.

[c] Assumed 2% efficiency at 0.7 ml/m uptake.

**TABLE 35. Reproducibility of the Laser ICP System on Steel Standard BCS 402[a]**

| Statistic | Fe | Cu | Ni | Cr | Si | Ni/Fe |
|---|---|---|---|---|---|---|
| % in BCS 402 | 96.3 | 0.23 | 0.73 | 0.55 | 0.27 | |
| Mean response (mV) | 440.33 | 26.56 | 25.31 | 74.81 | 53.85 | 0.0573 |
| Standard deviation | 8.07 | 1.26 | 0.39 | 7.00 | 13.26 | $6.4 \times 10^{-4}$ |
| Covariance (%) | 1.8 | 4.7 | 1.5 | 9.0 | 24.6 | 1.1 |

*Source*: After Ref. 6; reprinted by permission.

[a]The statistics are based on 10 replicate single laser pulses, blank corrected.

with ICP-OES after nebulization of the soluted steel sample (see Table 34). The reproducibility of the LM-ICP-OES on steel standard samples used by Thompson et al. [6] is shown in Table 35.

Carr and Horlick [8] analyzed high- and low-alloy aluminum samples and NBS naval brass standards using a photodiode array spectrometer by recording a series of spectra at 0.8-s intervals after firing the laser. A typical data set is shown in Fig. 62. After 0.8 s, emission starts and after about 4.8 s, all

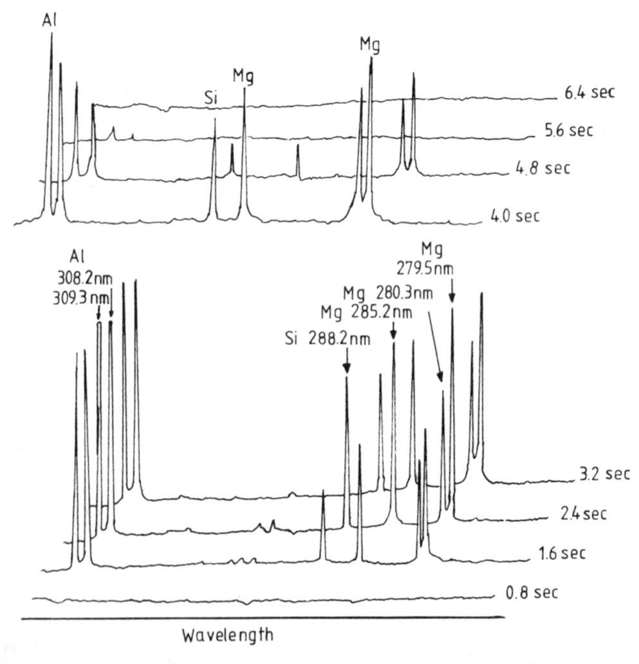


**Figure 62.** Time behavior of ICP emission signals from a high-alloy aluminum sample. From Ref. [8], reprinted by permission of Pergamon Journals Ltd.

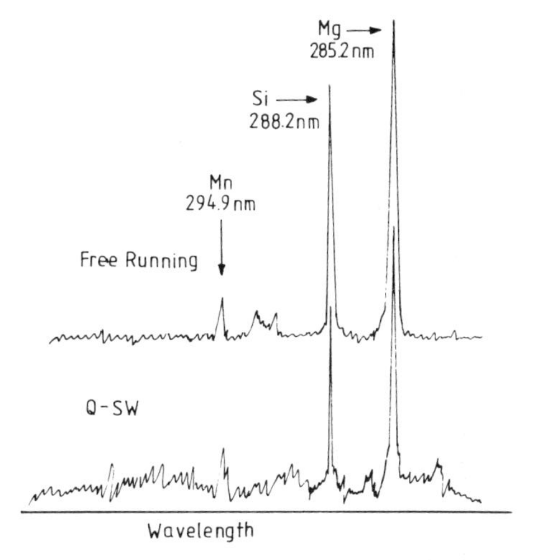


**Figure 63.** Comparison on the emission signals for free-running and *Q*-switched modes of laser operation. From Ref. [8], reprinted by permission of Pergamon Journals Ltd.

emission has finished. The wavelength range simultaneously observed included 315 to 270 nm. Changing the wavelength region or the type of sample did not influence the overall time behavior.

Spectra produced by a *Q*-switched and a free-running laser are compared in Fig. 63. The low-alloy aluminum sample contains 0.1% Si, 0.076% Mn, and 0.038% Mg. The evaporated material was 25 $\mu$g in the *Q*-switched mode and 500 $\mu$g in normal-mode laser action, which means for the *Q*-switched spectrum only about 10 ng of Mg presents quite a strong Mg line. For the free-running mode the relative standard deviation ranges from 1.7 to 9.4%, with an average of about 5% [8].

Ishizuka and Uwamino [9] analyzed standard samples of steel, brass, aluminum alloy, and titanium-based alloy. The results are given in Table 36. Dittrich et al. [10] analyzed copper standards, glasses [solid and powdered (KBr pellets)], and copper slates. Table 37 shows some results obtained with copper standard samples.

In glass-KBr (1:1) pellets the elements As, B, Ba, Ca, Cu, Fe, Mn, Pb, Zn, and Sb were determined in the ppm range (the relative standard deviation; see Table 38).

Cluster analyses of 26 elements in copper slates as well as distribution analyses of drill cores were carried out by Karjakin et al. [10].

Table 39 shows the determined elements, matrices, their concentration ranges, and relative standard deviations obtained by Günther and Gäckle [11, 21].

**TABLE 36. Precision Data for Various Elements in Steel, Brass, Al Alloy, Ti Alloy, and Detection Limits for Various Elements in Steel Obtained by *Q*-Switched and Normal Lasers**

| | RSD (%) | | | *Q*-Switched Laser | | Normal Laser | |
|---|---|---|---|---|---|---|---|
| Element Concentration (%) | *Q*-Switched Laser | Normal Laser | Analytical Line (nm) | Conc. (ppm) | Weight (pg) | Conc. (ppm) | Weight (pg) |
| Al 0.021 | 2.6 | 4.6 | I 396.2 | 20 | 20 | 2 | 60 |
| Co 0.15 | 10 | 3.1 | II 228.6 | 8 | 8 | 0.6 | 20 |
| Cr 0.69 | 6.8 | 3.8 | II 267.7 | 10 | 10 | 1 | 30 |
| Cu 0.098 | 5.7 | 8.6 | I 324.8 | 9 | 9 | 0.3 | 9 |
| Mn 0.25 | 3.7 | 4.6 | II 257.6 | 3 | 3 | 0.3 | 9 |
| Ni 0.32 | 2.5 | 3.3 | II 231.6 | 20 | 20 | 1 | 30 |
| V 0.31 | 8.3 | 11 | II 310.2 | 20 | 20 | 1 | 30 |

*Source*: After Ref. 9; published with permission of Pergamon Press Ltd.

**TABLE 37. Analytical Results of Laser-ICP-OES in Comparison with Laser-AAS-ETA**

| Method | Number of Shots | Element | Wavelength (nm) | Detection Limits Absolute (pg) | Detection Limits Relative (ppm) | RSD (%) |
|---|---|---|---|---|---|---|
| ICP-OES photographical | 10 | Zn | 334.5 | 1500 | 75 | 6 |
| | 10 | Mn | 260.6 | 200 | 10 | 5 |
| | 10 | Pb | 283.3 | 2500 | 125 | 6 |
| | 10 | Ag | 328.1 | 300 | 15 | 2 |
| ICP-OES photoelectrical | 1 | Ag | 328.1 | 54 | 27 | 20 |
| AAS-ETA | 1 | Zn | 213.9 | 1 | 0.5 | 20 |
| | 1 | Mn | 279.5 | 2 | 1.0 | 23 |
| | 1 | Pb | 283.3 | 13 | 6.5 | 18 |
| | 1 | Ag | 328.1 | 4 | 2.0 | 20 |

*Source*: After Ref. 10; reprinted by permission.

**TABLE 38. Relative Standard Deviation of LM-ICP-OES of Glass-KBr (1:1) Pellets with and without an Internal Standard**

| | R.S.D. (%) | |
|---|---|---|
| Element | Without Internal Standard | With Internal Standard |
| B | 6 | 2 |
| Ba | 12 | 4 |
| Sb | 20 | 3 |

*Source*: After Ref. 10; reprinted by permission.

**TABLE 39. Results of LM-ICP-OES**

| Element | Matrices | Wavelength (nm) | Concentration Range (at %) | R.S.D. (%) |
|---|---|---|---|---|
| La | $BaTiO_3$[a] | 379.4 | 0.05 – 0.5 | 3[b] |
| Ca | $BaTiO_3$ | 317.9 | 4 –16 | 2.3[b] |
| Sn | $BaTiO_3$ | 283.9 | 5 –30 | 9 |
| Pb | $Bi_2Te_3$ | 261.4 | 0.001– 2 | 12.5 |
| Pb | $Bi_2Te_3$ | 261.4 | 0.005– 4 | 11 |
| Fe | Bronze | 238.2 | 0.02 – 0.52 | 10 |
| Ni | Bronze | 221.6 | 0.032– 0.4 | 7.5 |
| Zn | Bronze | 213.8 | 1.02 –36.5 | 9.8 |
| Sn | Bronze | 283.9 | 0.83 –10.42 | 11.3 |
| Zn | Bone[c] | 213.8 | 0.005– 0.18 | 14.2 |

*Source*: After Günther and Gäckle [11]; published with the permission of VEB grundstoff-verlag, Leipzig.

[a] See also Ref. 25.

[b] With internal standard (Ba).

[c] See also Ref. 26.

## 6.2. MIP SPECTROMETRY OF SOLIDS

Soon after the first coupling of laser vaporization with ICP excitation, the same principle was realized with a microwave-induced plasma (MIP) by Leis and Laqua [27]. This procedure was applied to the determination of some elements in aluminum and zinc samples. The limits of detection when using photographic and photoelectric measurements are presented in Table 40.

**TABLE 40. Detection Limits of Some Elements in Zinc with Photographic and Photoelectric Recording**

| Photographic | | | | | Photoelectric Recording | | | | |
|---|---|---|---|---|---|---|---|---|---|
| | | | | | | | | Detection Limit | |
| Element | Wavelength (nm) | Excitation Energy (eV) | Detection Limit (ppm) | Calculated Detection Limit (ppm) | Element | Wavelength (nm) | Excitation Energy (eV) | Relative (ppm) | Absolute (ng) |
| Al I | 309.3 | 4.0 | 18 | 2 | Ai I | 396.2 | 3.1 | 250 | 0.5 |
| Cd I | 228.8 | 5.4 | 5 | 0.4 | Cd I | 228.8 | 5.4 | 20 | 0.04 |
| Mg II | 279.6 | 4.4 | 8 | 0.4 | Mg II | 279.6 | 4.4 | 13 | 0.03 |
| Pb I | 283.3 | 4.4 | 22 | 1 | Pb I | 368.4 | 4.3 | 150 | 0.3 |
| Sn I | 317.5 | 4.3 | 15 | 3 | Cu I | 324.8 | 3.8 | 150 | 0.3 |

*Source*: After Ref. 27; published with permission of Pergamon Press Ltd.

**TABLE 41. Precision Data, Detection Limits, and Concentration Ranges in Working Curves for Various Elements**

| | | | | Precision[a] | | Detection Limit | | | | |
|---|---|---|---|---|---|---|---|---|---|---|
| | | | | R.S.D. (%) | | Peak Height | | Peak Area | | |
| Element | Sample | Analytical Line (nm) | Conc. (%) | Peak Height | Peak Area | Conc. (ppm) | Weight (pg) | Conc. (ppm) | Weight (pg) | Conc. Range in Working curve (%) |
| Al | Steel | 396.15 | 0.07 | 13.8 | 8.5 | 9.3 | 8.4 | 14 | 13 | |
| Cr | Al alloy | 357.87 | 0.031 | 7.2 | 8.0 | 13 | 10 | 17 | 14 | |
| Cu | Steel | 324.75 | 0.16 | 4.3 | 2.3 | 2.4 | 2.2 | 4.2 | 3.8 | 0.017–0.16 |
| | Al alloy | 324.75 | 0.20 | 10.4 | 4.2 | 2.1 | 1.7 | 2.3 | 1.8 | 0.04–1.01 |
| Fe | Brass | 371.99 | 0.088 | 3.5 | 5.9 | 2.7 | 2.7 | 3.8 | 3.8 | 0.044–0.26 |
| | Al alloy | 371.99 | 0.20 | 6.2 | 4.1 | 12 | 9.6 | 13 | 10 | 0.075–0.97 |
| Mn | Al alloy | 403.08 | 0.13 | 2.4 | 4.4 | 5.4 | 4.3 | 4.7 | 3.8 | 0.026–0.30 |
| Mo | Steel | 386.41 | | | | 22 | 20 | 25 | 23 | |
| Ni | Brass | 352.45 | 0.07 | 1.2 | 2.9 | 3.8 | 3.8 | 3.9 | 3.9 | 0.043–0.16 |
| | Steel | 341.48 | 0.24 | 6.6 | 6.8 | 12 | 11 | 16 | 14 | 0.015–0.24 |
| | Al alloy | 352.45 | 0.20 | 3.8 | 5.4 | 13 | 10 | 9.0 | 7.2 | 0.012–2.01 |
| Pb | Brass | 283.31 | 2.0 | 5.5 | 4.5 | | | | | |
| | Al alloy | 283.31 | | | | 12 | 9.6 | 21 | 17 | |
| Ti | Al alloy | 365.35 | 0.15 | 8.4 | 12.1 | 22 | 18 | 15 | 12 | |
| Zn | Al alloy | 213.86 | 0.035 | 7.3 | 6.3 | 0.9 | 0.7 | 1.2 | 1.0 | 0.035–0.20 |

*Source*: Ishizuka and Uwamino [28]; published with permission of Pergamon Press Ltd.

[a]The precision was evaluated from the data obtained at five points which were each shot five times with the laser beam.

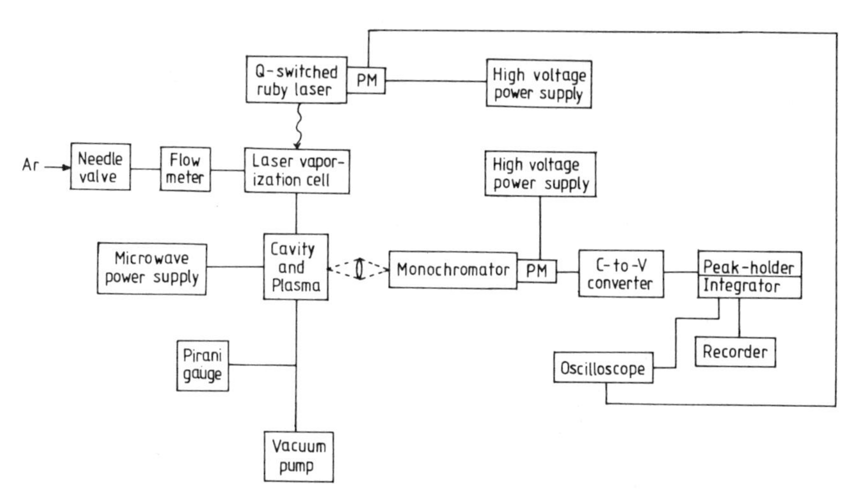


**Figure 64.** Block diagram of laser-MIP apparatus. Reprinted with permission from T. Ishizuka and Y. Uwamino, *Anal., Chem., 52*, 125. Copyright 1980, American Chemical Society.

A laser vaporization/microwave-induced plasma system was used for the determination of Al, Cr, Cu, Fe, Mn, Mo, Ni, Pb, Ti, and Zn in brass, steel, and aluminum alloy by Ishizuka and Uwamino [28]. For the apparatus and other details, see Figs. 64 to 69 and Table 41. The precision attained was 1.2 to 13.8% for the peak height method and 2.3 to 12.1% for the peak area method. The detection limits in the solid samples ranged from 0.5 ppm (0.7 pg) for Zn in aluminum alloy to 22 ppm (20 pg) for Mo in steel.

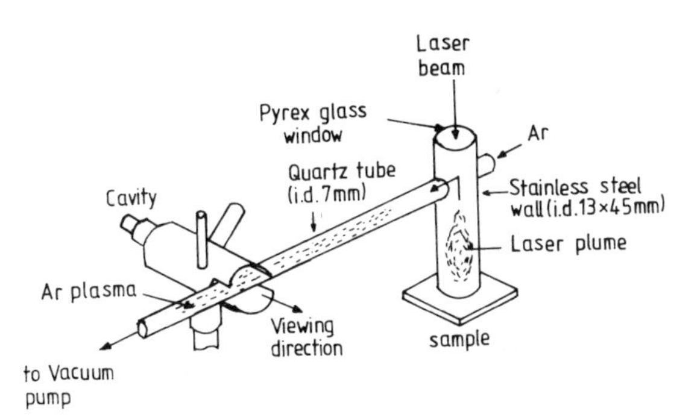


**Figure 65.** Schematic diagram of laser vaporization/microwave-induced plasma cell system. Reprinted with permission from T. Ishizuka and Y. Uwamino, *Anal. Chem., 52*, 125. Copyright 1980, American Chemical Society.

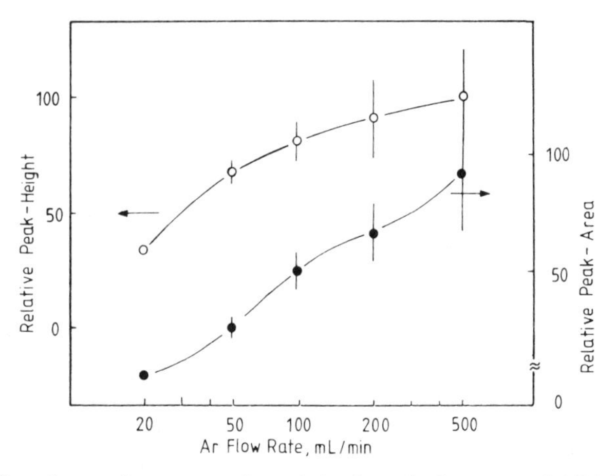


**Figure 66.** Effect of argon flow rate on the emission intensity for copper (0.04%) in aluminum alloy. Reprinted with permission from T. Ishizuka and Y. Uwamino, *Anal. Chem.*, *52*, 125. Copyright 1980, American Chemical Society.

## 6.3. DCP SPECTROMETRY OF SOLIDS

Direct determination of copper in solids by laser ablation–dc argon plasma emission spectrometry is described by Mitchell et al. [29, 30]. The block diagram of the equipment is shown in Fig. 70. Powdered copper ore and a cellulose binder were mixed in a ball mill and pelletized in a press. A special

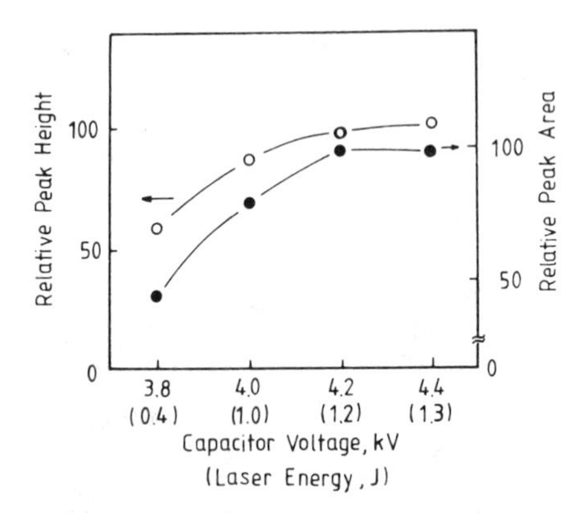


**Figure 67.** Effect of laser energy (capacitor voltage) on the emission intensity for iron (0.26%) in brass. Reprinted with permission from T. Ishizuka and Y. Uwamino, *Anal. Chem.*, *52*, 125. Copyright 1980, American Chemical Society.

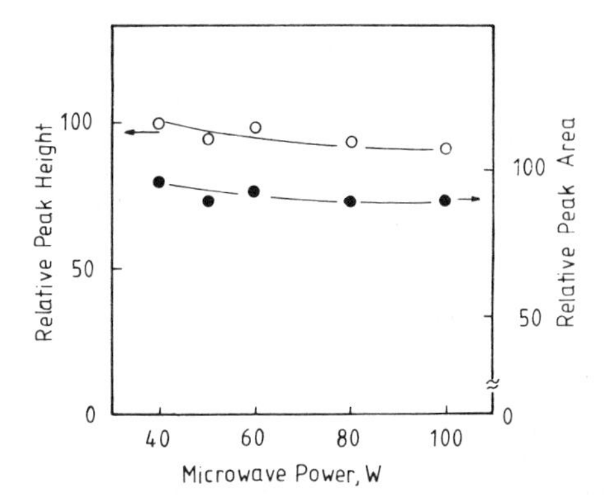


**Figure 68.** Effect of microwave power on the emission intensity for copper (0.055%) in aluminum alloy. Reprinted with permission from T. Ishizuka and Y. Uwamino, *Anal. Chem., 52*, 125. Copyright 1980, American Chemical Society.

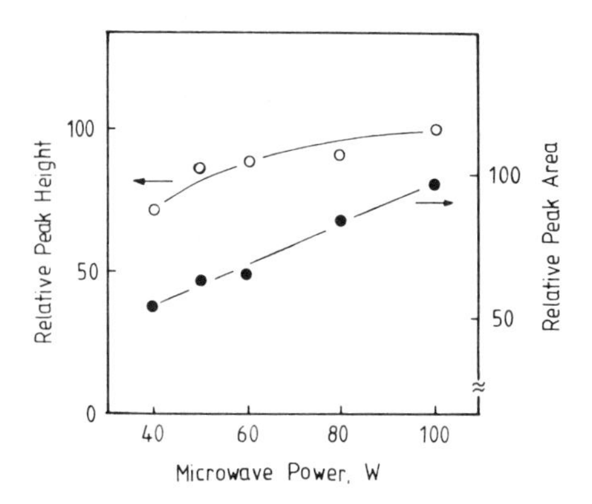


**Figure 69.** Effect of microwave power on the emission intensity for molybdenum (0.085%) in steel. Reprinted with permission from T. Ishizuka and Y. Uwamino, *Anal. Chem., 52*, 125. Copyright 1980, American Chemical Society.

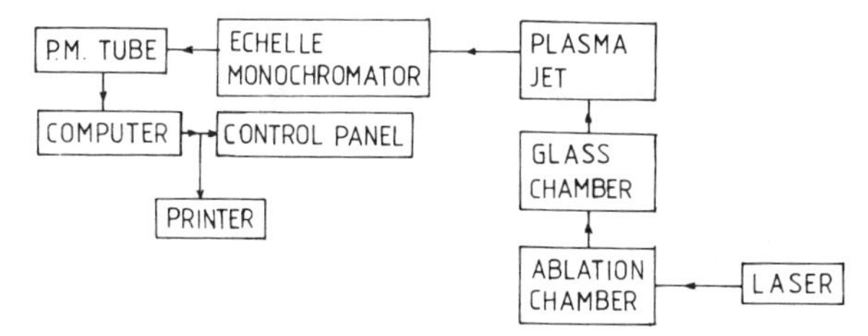


**Figure 70.** Block diagram of laser-DCP system for direct determination of metals in solid and powder samples. From Ref. [30], reprinted by permission of the Society for Applied Spectroscopy, Frederick, MD.

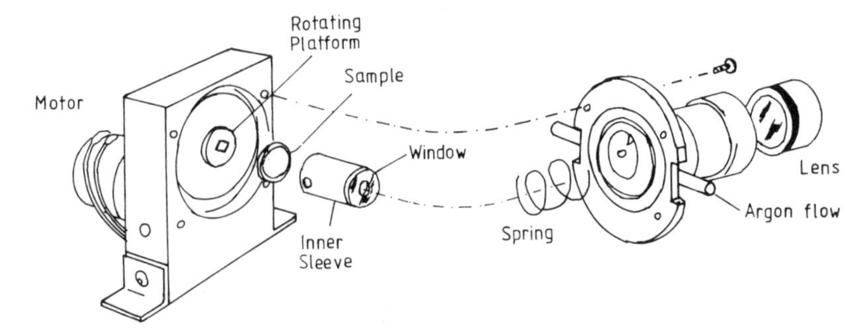


**Figure 71.** Schematic of ablation chamber for laser-DCP spectrometry. From Refs. [29, 30], reprinted by permission of the Society for Applied Spectroscopy, Frederick, MD.

chamber and an interface between the laser ablation chamber and the DCP were developed (Figs. 71 and 72). The effect on emission intensity with time for LM-DCP at a pulse rate of 20 Hz for a pelletized sample containing 1000 $\mu$g/g of copper in copper oxide in a cellulose matrix is shown in Fig. 73. The precision of determination of major components ranges between 2.8 and 9.9%.

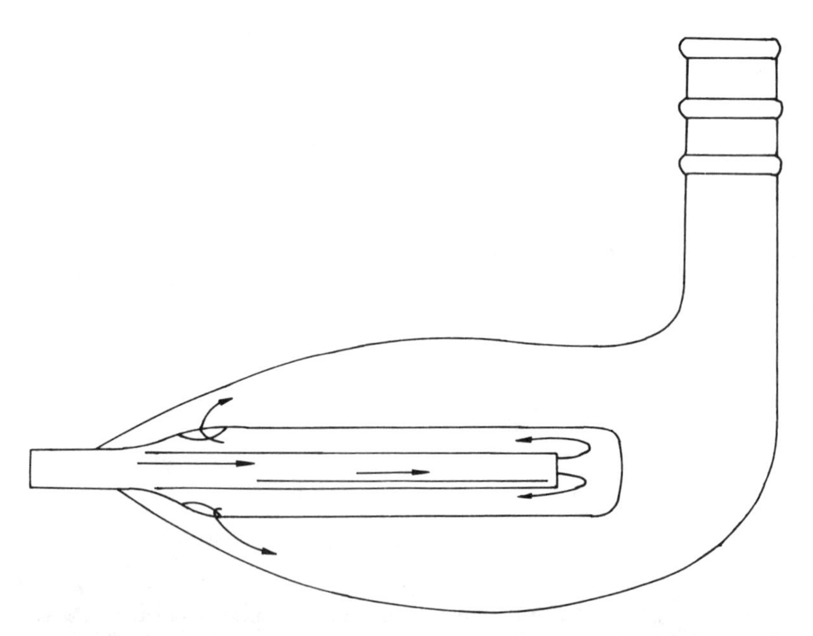

**Figure 72.** Schematic of interface between laser ablation chamber and DCP. From Ref. [29], reprinted by permission of the Society for Applied Spectroscopy, Frederick, MD.

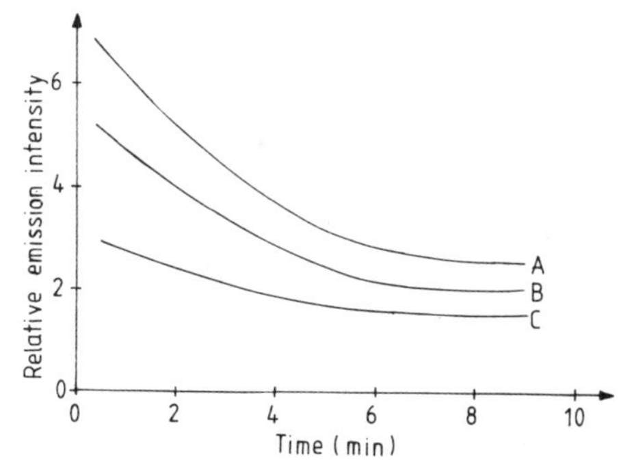


**Figure 73.** Effect of relative emission intensity of laser pulse rate with time: (*A*) 20 Hz; (*B*) 10 Hz; (*C*) 1 Hz. From Ref. [30], reprinted by permission of the Society for Applied Spectroscopy, Frederick, MD.

**TABLE 42. Detection Limits for Various Metals Using Plasma Emission Spectrometry with Sample Introduction by Laser Ablation**

| | | | Detection Limit (μg/g) | | | |
|---|---|---|---|---|---|---|
| | | | ICP[a] | | | |
| Metal | Wavelength (nm) | Sample | *Q*-Switched | Normal | DCP[b] | MIP[c] |
| Al | 396.2 | Steel | 20 | 2 | — | 9.3 |
| Co | 228.8 | Steel | 8 | 0.6 | — | — |
| | 324.7 | Steel | — | — | 48 | — |
| Cr | 267.7 | Steel | 10 | 1 | — | — |
| | 357.9 | Steel | — | — | 31 | 13 |
| | 357.9 | Al alloy | — | — | — | — |
| Cu | 324.7 | Steel | 9 | 0.3 | 80 | 2.4 |
| | 324.7 | Powder | — | — | 21 | — |
| Mn | 257.6 | Steel | 3 | 0.3 | 18 | — |
| Mo | 218.6 | Steel | 20 | 2 | — | — |
| | 379.8 | Steel | — | — | 30 | — |
| | 386.4 | Al alloy | — | — | — | 22 |
| Ni | 231.6 | Steel | 20 | 1 | — | — |
| | 341.5 | Steel | — | — | 32 | 12 |
| Ti | 334.4 | Steel | — | — | 50 | — |
| | 365.5 | Al alloy | — | — | — | 22 |
| V | 310.2 | Steel | 20 | 1 | — | — |
| | 437.9 | Steel | — | — | 131 | — |
| Zr | 339.2 | Steel | — | — | 50 | — |

*Source*: After Ref. 31; reprinted by permission.

[a] According to Ref. 9.

[b] According to Ref. 31.

[c] According to Ref. 28.

## 6.4. COMPARISON OF LM-ICP-OES, LM-MIP-OES, AND LM-DCP-OES

Sneddon and Mitchell [31] compared the above-named methods concerning detection limits (Table 42). In general, the MIP and DCP detection limits are comparable, with the ICP detection limits being generally superior.

## REFERENCES


1. J. A. C. Broekaert and P. W. J. M. Boumans, in *Inductively Coupled Plasma Emission, Spectroscopy*, Part I, *Methodology, Instrumentation and Performance*, P. W. J. M. Boumans, Ed., John Wiley & Sons, Inc., New York, 1987, Chap. 6, p. 296.
2. J. A. C. Broekaert, F. Leis, B. Raeymaekers, and G. Zaray, *Spectrochim. Acta*, **43B**, (1988).
3. F. N. Abercrombie, M. D. Silvester, and G. S. Stoute, *Proc. 28th Pittsburgh Conf. Analytical Chemistry and Applied Spectroscopy*, Cleveland, Ohio, 1977, Paper 406, *ICP Inf. Newsl.*, **2**, 309 (1977).
4. E. D. Salin, J. W. Carr, and G. Horlick, *Proc. 30th Pittsburgh Conf.*, Cleveland, Ohio, 1979, Paper 563.
5. J. W. Carr and G. Horlick, *Proc. 31st Pittsburgh Conf.*, Atlantic City, N.J., 1980, Paper 56.
6. M. Thompson, J. G. Goulter, and F. Sieper, *Jena Rev.*, 202 (1981); *Analyst*, **106**, 32 (1981).
7. H. Kawaguchi, I. Xu, T. Tanaka, and A. Mizuike, *Bunseki Kagaku*, **31**, 185 (1982).
8. J. W. Carr and G. Horlick, *Spectrochim. Acta*, **37B**, 1 (1982).
9. T. Ishizuka and Y. Uwamino, *Spectrochim. Acta*, **38B**, 519 (1983).
10. K. Dittrich, K. Niebergall, and R. Wennrich, *Fresenius Z. Anal. Chem.*, **328**, 330 (1987).
11. D. Günther and M. Gäckle, *Z. Chem.*, **28**, 227 and 258, 1988.
12. V. A. Fassel and R. N. Kniseley, *Anal. Chem.*, **46**, 1155A (1974).
13. G. Horlick, *Appl. Spectrosc.*, **30**, 113 (1976).
14. L. Moenke-Blankenburg, H. Moenke, J. Mohr, W. Quillfeldt, W. Grassme, and W. Schrön, *Spectrochim. Acta*, **30B**, 227 (1975).
15. W. Schrön, Patent DD 15 39 21 B1, 1980.
16. W. Schrön, G. Bombach, and P. Beuge, *Spectrochim. Acta*, **38B**, 1269 (1983).
17. *Spectroflame ICP*, Spectro Analytical Instruments GmbH, Kleve, FRG.
18. V. Karanassios and G. Horlick, *Appl. Spectrosc.*, **40**, 813 (1986).
19. G. M. Levy, A. Quaglia, R. E. Lazure, and S. W. McGeorge, *Spectrochim. Acta*, **42B**, 341 (1987).
20. S. W. McGeorge, J. B. Falconer, and R. L. Lyke, *Proc. 25th. CSI*, Toronto, Canada, 1987, Abstract H4.5.

21. D. Günther and M. Gäckle, Doctoral thesis, Martin–Luther–University, Halle–Wittenberg, Halle, German Democratic Republic, 1989–90 (in preparation).
22. A. U. Ramsing, J. Ruzicka, and E. H. Hansen, *Anal. Chim. Acta*, **129**, 1 (1981).
23. H. Matschiner, P. Sivers, and H. H. Rüttinger, *Wiss. Z. Univ. Halle*, **34**(H.5), 3 (1985).
24. O. Levenspiel and W. H. Smith, *Chem. Eng. Sci.*, **6**, 227 (1957).
25. L. Moenke-Blankenburg and U. Zander, Laboratory report, Martin–Luther–University, Halle, GDR, 1988.
26. Khaled, Al Hamadi and S. Hegewald, Diploma thesis, Martin–Luther–University, Halle, East Germany, 1988.
27. F. Leis and K. Laqua, *Spectrochim. Acta*, **33B**, 727 (1978); **34B**, 307 (1979).
28. T. Ishizuka and Y. Uwamino, *Anal. Chem.*, **52**, 125 (1980).
29. P. G. Mitchell, J. A. Ruggles, J. Sneddon, and L. J. Radziemski, *Anal. Lett.*, **18**(A14), 1723 (1985).
30. P. G. Mitchell, J. Sneddon, and L. J. Radziemski, *Appl. Spectrosc.*, **40**, 274 (1986).
31. L. J. Sneddon and P. G. Mitchell, *Int. Lab.*, **4**, 18 (1987).

CHAPTER

7

# LASER MICROANALYSIS BASED ON ATOMIC ABSORPTION SPECTROMETRY

In analytical atomic absorption spectrometry (AAS), one is largely dealing with the absorption of resonance radiation, which is defined as characteristic radiation of an element corresponding to the transfer of an electron from the ground state to a higher energy level. The analytical validity of making absorption measurements depends on the relationship between absorption and the concentration of partial pressure of the absorbing atoms. This relationship is, according to classical dispersion theory,

$$\int K_{v} d_{v} = \frac{\pi e^{2}}{mC} N_{v} f \tag{58}$$

where $K_v$ = absorption coefficient at frequency $v$
$e$ = electronic charge
$m$ = electric mass
$N_v$ = number of atoms per $cm^3$ capable of absorbing energy in the range $n$ to $d_v$

The $f$-value is the effective number of free electron oscillators per atom of the element in question responsible for the absorption effect produced by the incident radiation. (General biographies about AAS are given by Price [1], Welz [2], and Dittrich [3].)

This relationship is easily derived as follows. If the incident radiation of intensity $I_0$ enters a flame or a nonflame cell, emerging after attenuation by absorption with intensity $I$, the value log $I_0/I$ is defined as the absorbance $A$. The simplest calibration method is with the help of standards by $A = ac + b$ and determination of the unknown concentration $C_x = (A_x - b)/a$.

In the past three decades atomic absorption spectrophotometry has developed from a nearly unknown description of its principles into a well-established and universally employed method for the determination of a large number of elements in a variety of solutes. The principle of analyzing solids by atomizing analytes directly from the solid state offers definite advantages. The time-consuming decomposition step can be omitted, and the analysis can be carried out without the addition of reagents and without separation and/or preconcentration steps. The risks of introducing contaminants and of losing the elements to be determined are thus reduced considerably.

The disadvantages inherent to all AAS techniques are that they are destructive, that normally only one element can be determined at a time, and that the dynamic range is relatively small. Whereas interferences do not normally present any serious problems in flame AAS, they are frequently encountered in nonflame cells; with the latter type of cell, devices for background correction must be employed [4].

The laser-produced vapor cloud contains, depending on optimal parameters of the laser radiation, neutral ground-state atoms. Therefore, AAS methods can be used to identify the elements producing these atoms. Two advantages are to be expected from laser microanalysis in absorption compared with emission:

- Improvement in the accuracy of the determination of concentrations at high concentration of elements up to approximately 100% through an increase in the concentration sensitivity in comparison with the emission method. Whereas the emission calibration curves become flatter at concentrations of a few parts percent, the absorption calibration curves retain their steepness. By the increase in the sensitivity for high concentrations of the elements, in absorption it is possible to cover a concentration range of three to four orders of magnitude with a single linear calibration curve.
- The transition from the line-rich emission spectra to absorption spectra containing fewer lines simplifies evaluation.

The technique is applicable to the analysis of inorganic and organic materials in the form of solids (crystalline and amorphous), foils or sheets, cuttings of fibres, drillings, samples of soft or hard tissue, powders and powder pellets, as well as suspensions of solids in solid agents. Liquid samples, such as biological fluids, may be analyzed after being transformed into the solid state by drying, dry or plasma ashing, or lyophilization; these operations also serve the purpose of preconcentrating the analytes.

The method has been used to determine 47 elements present as trace or minor constituents. In some instance major elements have also been quantified. The analytes are usually present in the ppm and ppb range; extreme values are concentrations up to 50% and down to $10^{-8}$%.

The precision of the analytical data depends on a number of factors, the most important of which are the errors of sampling and measurement of absorbance. In general, a relative standard deviation of 5 to 10% for elements present at a higher concentration level, and one of 10 to 30% for analytes at a lower level, have to be considered as normal. The present technique is not to be recommended in instances where major and minor constituents are to be determined with a high degree of precision.

The accuracy depends on the care taken by the analyst; on the quality, state, and treatment of the apparatus and instrumentation; on the standard employed; and on the presence of systematic errors. In general, the present method compares favorably with other methods for the quantification of trace or minor elements in solids [4, 5].

Laser microanalysis based on atomic absorption spectrometry was employed by Piepmeier in 1966 [6] and by Mossotti, Laqua, and Hagenah between 1965 and 1967 [7]. Several authors followed [8–32]. An excellent review is given by Dittrich and Wennrich [33].

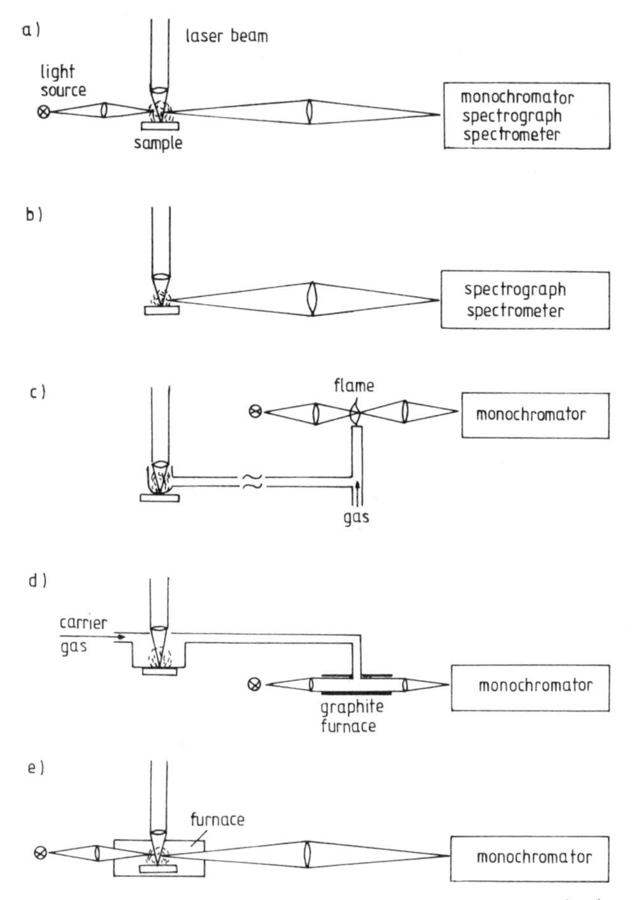


**Figure 74.** Five versions of atomic absorption with laser vaporization.

Mossotti et al. [7] and Piepmeier [6], using $Q$-switched lasers, have shown for samples in air at atmospheric pressure that the absorbing atoms outlast the brief period of light emission from the plasma, allowing atomic absorption measurements to be made well after the intense continuum emission pulse from the plasma. They used a pulsed continuum primary source with single-pass or multipass optics and a spectrograph or a spectrometer (the principle is shown in Fig. 74*a*). Several authors (e.g., [10, 17, 19, 26, 27, 32]) have used pulsed hollow cathode lamps as the primary light source, and results obtained indicate little additional line broadening due to pulsing, making them usable with low-resolution monochromators for atomic absorption measurements [9].

Karajakin and Kaigorodov [8], on the other hand, have used self-absorbed time-integrated spectra as an analytical tool by measuring self-absorption linewidths produced on film by a relatively high-resolution spectrograph. Moenke-Blankenburg et al. [14, 19] and Quillfeldt [20] have used the recombination continuum of the laser crater as a background radiator. A laser microanalyzer combined with a spectrograph served as atomizer and multielement analyzer, respectively (see the principle in Fig. 74*b*). Of the variants investigated [19], utilization of the recombination radiation of free electrons in the laser crater as background radiation showed the following favorable points in addition to those mentioned above:

- No separate background radiator, with supply and ignition unit, is required.
- Consequently, there is no synchronization problem as in the case of a separate background radiator.
- The second beam path for absorption that can be realized in the microscope of a laser microanalyzer, which was used in our experimental arrangement, permits simultaneous recording of the emission and absorption spectra of the atomic cloud. There is no loss of information in the emission spectrum compared with conventional laser microanalysis in emission. An increased amount of information is obtained from the absorption lines of the elements of high concentration in the absorption spectrum, which is recorded on the photographic plate below the emission spectrum [14, 19, 20] (Fig. 75).
- Since the recombination continuum radiates through the cloud of atomic vapor in the direction of propagation, a longer absorption pathway is achieved in comparison to the perpendicular passage of radiation with a separate background radiator. This leads, in some cases, to lower limits of detection for laser microatomic absorption spectral analysis than with the other variants.

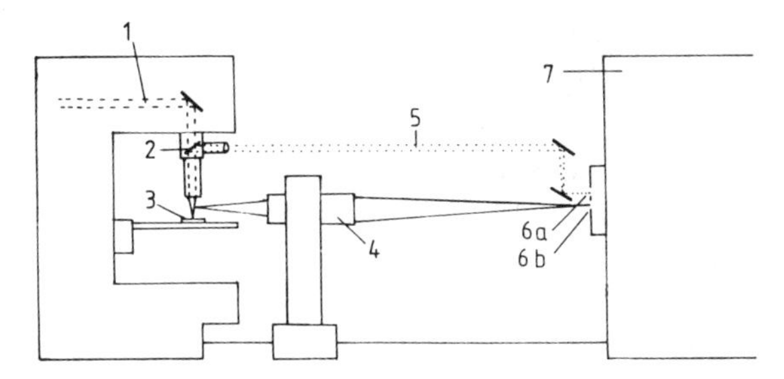


**Figure 75.** Simultaneous absorption and emission laser microanalysis. 1, Incident laserbeam; 2, dielectric long-pass filter (high transmission of laser wavelength; high reflection of UV–visible radiation up to about 500 nm); 3, sample; 4, condenser for the atomic emission pathway; 5, atomic absorption pathway; 6a and 6b, separate entrance slits for either pathway; 7, spectrograph.

- The use of this process is limited to materials that emit intense recombination radiation from the laser crater, such as metals, rocks, and minerals. Glasses and other transparent substances cannot be analyzed by this method, since the recombination radiation is too weak.
- The beam pathway for AAS requires special optical elements with a UV–visible laser achromat and a long-pass filter.

In a third variant (see Fig. 74*c*), a laser microanalyzer was, in accordance with a proposal of Kantor et al. [12, 13], tested through a coupling unit developed in combination with an atomic absorption spectrometer acquired commercially [14]. For special quantitative analytical problems (see Kantor et al. [22]), this method offers a considerable shortening of the time for analysis and the advantage of avoiding errors that can arise in the preparation of samples (contamination, loss of substance, etc.). In this case, the laser microanalyzer serves as an aerosol generator, the sample vapor produced by the laser evaporation being drawn off and fed into the AAS apparatus. Here the sample vapor is blown into the mixing chamber and excited in an air–acetylene flame.

The coupling of laser microanalyzer and atomic absorption spectrophotometer permits the immediate quantitative determination of one or a small number of elements in concentrations of more than 1 to 100 ppm. The well-known absorption-sensitive elements can be analyzed, and also those that form heat-resistant compounds, especially carbides, in the hot laser microplasma. Moenke-Blankenburg et al. [19] have studied the elements Mg, Cu, Ag, Pb, Zn, Sn, Fe, Cr, Ni, Mn, Co, Cd, Ti, Mo, and Si. The elements Ti, Mo, and Si could not be detected even at high concentrations. For all the other elements studied, the above-mentioned limits of detection apply. The reproducibility is $\leqslant 10\%$.

The coupling unit, consisting of a small suction chamber and a venturi nozzle operating on the principle of the water-jet pump, is suitable in principle for attachment to all types of AAS apparatus on the market.

Dittrich and Wennrich [28] determined trace elements in solid samples by the combination of laser ablation and vaporization and flameless AAS (for the principle, see Fig. 74*d*). A ruby laser with a passive *Q*-switch ablates and vaporizes micro amounts of solids in a chamber. The aerosol is transported by an Ar gas flow into a hot graphite furnace, where the solid particles and the vapor are atomized for the AA measurement. The relative standard deviation of 10 parallel analyses varied by [29] between 20 and 30%. Further investigations by Wennrich et al. [31] have shown that when a graphite furnace (AAS) is combined with laser ablation of solid samples, the results obtained are definitely influenced by the matrix composition. It seems

**TABLE 43. Some Results Obtained with Laser Microanalysis Based on Atomic Absorption**

| Elements | Matrices | Concentration Ranges | Reference |
|---|---|---|---|
| Cu | Powders | 0.3 to 10% | 8 |
| Cu, Mn | Al alloys | $10^{-3}$ to $10^{5}$% | 9 |
| Ag, Cu | Rocks | $10^{-10}$ g | 10 |
| Ag, Cu, Mn | Powders | | 11 |
| Fe | Alloys | ng | 12 |
| Ag | Alloys | % | 15 |
| Al, Cr, Cu, Mn, Mo, Ni, V | Steel | ppm | 17 |
| Ag, Cd, Co, Cr, Cu, Fe, Mg, Mn, Ni, Pb, Sn, Zn | Powders | ppm | 19 |
| Cr, Ni | Steel | 0.1 to 30% | 19 |
| Al, Cr, Cu, Mn, Mo, Ni | Steel | ppm | 20 |
| Mn | Steel | 0.3% | 21 |
| Ag, Au, Ni | Layers | ng | 22 |
| Cu | Al | $10^{-5}$ to $10^{-3}$ g/g | 23 |
| Cd | Biological tissues | ppm | 26 |
| Cu, Fe, Mg, Mn, Si | Al alloys | ppm | 27 |
| Ag | Cu, $Al_2O_3$, $Pb_3O_4$, pyrite | 10 to 200 pg/10 $\mu$l | 28 |
| Al, Ni | Brass | 800 to 2100 ppm | 29 |
| Ag, Sb | Minerals | ppm | 30 |
| Pb, Cd | Copper, aluminum oxide | ppm | 31 |
| Cr, Cu, Mn | Steel | $10^{-10}$ g | 32 |

*Source*: Reprinted with permission from *Prog. Analyt. Spectrosc.*, **9**, L. Moenke, copyright 1986, Pergamon Journals Ltd.

necessary that the chemical reactions taking place in the laser plume, the transport system, and the graphite furnace be investigated further. In addition to these effects, it is well known that variations on the interactions of laser radiation and target influence accuracy and precision. The authors emphasize that in quantitative analysis of trace elements by laser AAS the standard samples have to have the same composition in major elements as the unknown sample. Schrön [24, 30] built a specifically designed sample chamber from which the laser-produced aerosol is also transported into a graphite tube. The authors claim that this technique is suited for a large range of concentration and is relatively free of matrix interference.

Figure 74*e* shows a fifth variant, which deals with laser evaporation directly inside an absorption cell. Matousek and Orr [15], Ishizuka et al. [17], and Sumino et al. [26], using specially designed chambers, reported RSD values between 1 and 10%. Table 43 gives an overview of the applications of laser AAS obtained by a variety of authors.

## REFERENCES

1. W. J. Price, *Spectrochemical Analysis by Atomic Absorption*, Heyden & Son Ltd., Bristol, England, 1979.
2. B. Welz, *Atomabsorptionsspektrometrie*, Vol. 3, Verlag Chemie GmbH, Weinheim, FRG, 1983.
3. K. Dittrich, *Atomabsorptionsspektrometrie*, Akademie-Verlag, Berlin, GDR, 1982.
4. L. J. Langmyhr and G. Wibertoe, *Prog. Anal. At. Spectrosc.*, **8**, 193 (1985).
5. L. J. Langmyhr, *Fresenius Z. Anal. Chem.*, **322**, 654 (1985).
6. E. H. Piepmeier, Ph.D. thesis, University of Illinois, Urbana, Ill., 1966.
7. V. G. Mossotti, K. Laqua, and W.-D. Hagenah, *Spectrochim. Acta*, **23B**, 197 (1967); see also W.-D. Hagenah, L. Laqua, and V. Mossotti, *Proc. 12th Colloquium Spectroscopicum Internationale*, Exeter, England, 1965, p. 282.
8. A. V. Karjakin and W. A. Kaigorodov, *Zh. Anal. Khim.*, **23**, 930 (1968).
9. O. E. Osten and E. H. Piepmeier, *Appl. Spectrosc.*, **27**, 3, 165 (1973).
10. A. V. Karjakin, A. M. Pschelinzew, A. J. Schidlowski, E. K. Wulfson, and M. N. Zingarelli, *Zh. Prikl. Spektrosk.*, **18**, 4, 610 (1973).
11. E. K. Wulfson, A. V. Karjakin, and A. I. Schidlowski, *Zavod. Lab.*, **8**, 945 (1974); see also: *Zh. Prikl. Spektrosk.*, **22**, 14 (1975).
12. T. Kantor, L. Polos, P. Fodor, and E. Pungor, *Proc. 2nd EUROANALYSIS*, Budapest, Hungary, 1975, Abstracts 1–46, p. 68.
13. T. Kantor, L. Polos, P. Fodor, and E. Pungor, *Talanta*, **23**, 585 (1976).
14. L. Moenke-Blankenburg, W. Quillfeldt, and D. Böwe, *Proc. 7th National Conf. Atomic Spectroscopy*, Slanchev Bryag, Bulgaria, 1976.

15. J. P. Matousek and B. J. Orr, *Spectrochim. Acta*, **31B**, 475 (1976).
16. R. M. Manabe, Ph.D. thesis, Oregon State University, Corvallis, Oreg., 1977.
17. T. Ishizuka, Y. Uwamino, and H. Sunahara, *Anal. Chem.*, **49**, 1939 (1977).
18. A. Quentmeier, Doctoral thesis, University of Düsseldorf, Düsseldorf, FRG, 1977.
19. L. Moenke-Blankenburg, D. Böwe, and W. Quillfeldt, *Proc. 19th Colloquium Spectroscopicum Internationale*, Prague, Czechoslovakia, 1977.
20. W. Quillfeldt, *Jena Rev.*, 226 (1978).
21. R. M. Manabe and E. H. Piepmeier, *Anal. Chem.*, **51**, 2066 (1979).
22. T. Kantor, L. Bezür, E. Pungor, and P. Fodor, *Spectrochim. Acta*, **34B**, 341 (1979).
23. A. Quentmeier, K. Laqua, and W.-D. Hagenah, *Spectrochim. Acta*, **34B**, 117 (1979).
24. W. Schrön, DD-patent GO 1 N-224791, 1980.
25. M. L. Petuch, A. D. Shirokanov, and A. A. Jankovsky, *Zh. Prikl. Spektrosk.*, **32**, 414 (1980).
26. K. Sumino, R. Yamamoto, F. Hatayama, S. Kitamura, and H. Itoh, *Anal. Chem.*, **52**, 1064 (1980).
27. A. Quentmeier, K. Laqua, and W.-D. Hagenah, *Spectrochim. Acta*, **35B**, 139 (1980).
28. K. Dittrich and R. Wennrich, *Spectrochim. Acta*, **35B**, 731 (1980).
29. R. Wennrich and K. Dittrich, *Spectrochim. Acta*, **37B**, 913 (1982).
30. W. Schrön, G. Bombach, and P. Beuge, *Spectrochim. Acta*, **38B**, 1269 (1983).
31. R. Wennrich, K. Dittrich, and U. Bonitz, *Spectrochim. Acta*, **39B**, 657 (1984).
32. L. T. Suchov and G. E. Solotukhin, *Zh. Prikl. Spektrosk.*, **30**, 11 (1979).
33. K. Dittrich and R. Wennrich, *Prog. Anal. At. Spectrosc.*, **7**, 139 (1984).

CHAPTER

8

# LASER MICROANALYSIS BASED ON ATOMIC FLUORESCENCE SPECTROMETRY

Atomic fluorescence spectrometry as a method of chemical analysis was first proposed by Winefordner and Vickers [1] in 1964. Since the advent of the tunable dye laser [2, 3] spectroscopists have applied it to the generation of atomic fluorescence (Refs. 4 to 8 are several of the earlier references; Refs. 9 to 16 are some of most recent). In laser atomic fluorescence flame spectrometry the expected increase in sensitivity could not be reached to the full extent. Laser atomic fluorescence spectrometry has been shown to be more sensitive when combined with a graphite atomizer [9, 14–16]. This method, with electro-thermal atomization, has produced the lowest detection limits of spectroscopic methods for several elements [7, 14–19]. This method needs only a small analytical volume and can therefore be a micromethod for investigations of fluids, but it is not a method for laser microanalysis "in situ" of solid samples.

In 1979, Measures and Kwong [20–22] introduced the TABLASER, an acronym for Trace (Element) Analyzer Based on Laser Ablation and Selectivity Excited Radiation. This approach combines the microselectivity and in situ capability of the laser microprobe with the high sensitivity of laser selective excitation and saturation spectroscopy. It could, therefore, be the ideal ultramicro-ultratrace element analyzer, with the following characteristics:

- No sample preparation, allowing in situ measurement
- High sensitivity and low detection limit (i.e., ppb and attogram)
- Freedom from matrix effects (chemical and physical) so that calibration is independent of the substrate containing the element
- Linearity of response over a wide dynamic range of concentrations
- High selectivity for the element of interest
- Selective ultramicro sampling capability
- Depth profiling capability
- Real-time analysis
- Minimum variation in sensitivity between elements

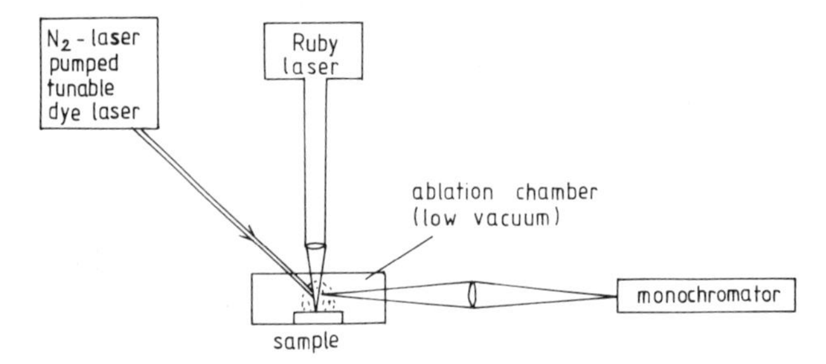


**Figure 76.** Principle of TABLASER. From Ref. [22], reprinted by permission of H. Kwong.

- Simultaneous multielement measurement possible
- Capability of isotope ratio measurements

The TABLASER has three additional important areas of application: the study of the expansion dynamics of neutral species resulting from laser-target interaction, measurement of the radiative lifetimes of atoms, and ultratrace element water analysis. Figure 76 shows the principles of a TABLASER.

## REFERENCES


1. J. D. Winefordner and T. J. Vickers, *Anal. Chem.*, **36**, 161 (1964).
2. P. D. Sorokin and I. R. Lankard, *IBM J. Res. Dev.*, **10**, 162 (1966).
3. F. P. Schäfer, W. Schmidt, and J. Volze, *J. Appl. Phys. Lett.*, **9**, 306 (1966).
4. L. M. Fraser and J. D. Winefordner, *Anal. Chem.*, **43**, 1693 (1971), and **44**, 444 (1972).
5. M. B. Denton and H. V. Malmstadt, *Appl. Phys. Lett.*, **18**, 485 (1971).
6. J. Kuhl and H. Spitschan, *Opt. Commun.*, **1**, 256 (1974).
7. S. Neumann and M. Kriese, *Spectrochim. Acta*, **29B**, 127 (1974).
8. J. D. Winefordner, *J. Chem. Educ.*, **55**, 72 (1978).
9. P. Wittman and J. D. Winefordner, *Can. J. Spectrosc.*, **29**, 75 (1984).
10. A. L. Lewis II and E. Piepmeier, *Appl. Spectrosc.*, **37**, 3, 523 (1983).
11. N. Omenetto and H. G. C. Human, *Spectrochim. Acta*, **39B**, 1333 (1984).
12. H. G. C. Human, N. Omenetto, P. Cavalli, and G. Rossi, *Spectrochim. Acta*, **39B**, 1345 (1984).
13. B. W. Smith, N. Omenetto, and J. D. Winefordner, *Spectrochim. Acta*, **39B**, 1389 (1984).
14. J. Tilch, H. Falk, H.-J. Paetzold, P. G. Mon, and K. P. Schmidt, *Proc. 24th CSI*, Garmisch-Partenkirchen, FRG, 1985, p. 67.

15. K. Dittrich and H.-J. Stärk, *Proc. 24th CSI*, Garmisch-Partenkirchen, FRG, 1985, p. 472.

16. H.-J. Stärk, Untersuchungen zur laserangeregten Atomfluoreszenzspektrometrie (LAFS) und zur laserangeregten Molekülfluoreszenzspektrometrie (LAMOFS) mit einem Graphtrohratomisator, Doctoral thesis, Karl Marx University, Leipzig, GDR, 1985.

17. M. A. Bolshov, A. V. Zybin, and I. I. Smirenkina, *Spectrochim. Acta*, **36B**, 1143 (1981).

18. M. A. Bolshov, A. V. Zybin, V. G. Koloshnikov, and M. V. Vasnetsov, *Spectrochim. Acta*, **36B**, 345 (1981).

19. N. Omenetto and J. D. Winefordner, in *Atomic Fluorescence Spectroscopy with Laser in Analytical Laser Spectroscopy*, N. Omenetto, Ed., Wiley-Interscience, New York, 1979.

20. R. M. Measures and H. S. Kwong, *Appl. Opt.*, **18**, 281 (1979).

21. H. S. Kwong and R. M. Measures, *Anal. Chem.*, **51**, 428 (1979).

22. H. S. Kwong, Laser ablation and selective excitation directed to trace element analysis, *UTIAS Report 234, CN ISSN 0082-5255*, Toronto, Canada, 1980.

CHAPTER

9

# LASER MICROANALYSIS BASED ON MASS SPECTROMETRY

## 9.1. FOUNDATION AND HISTORICAL DEVELOPMENT

Mass spectrometry characterizes ionized atoms, molecules, and fragments of molecules with respect to their mass/charge ratio after deviation in a magnetic or electric field or by reason of their translation energy.

The energy of generated ions of masses $m$ and velocity $v$ amounts to

$$E = \frac{m}{2} v^2 \tag{59}$$

In a homogeneous magnetic field of a static ion separation system, the charged particle describes a circle segment with the radius

$$r = \frac{1}{B} \sqrt{2U \frac{m}{e}} \tag{60}$$

where $B$ is the magnetic flow density and $U$ is the voltage.

Instruments that use magnetic *and* electric fields are said to be "double focusing." They reach resolutions between $M/\Delta M = 20{,}000$ and 200,000 (against resolutions of simple equipments of 200 to 2000).

Dynamic ion separation systems are based on several physical principles. Well known are the time-of-flight (TOF) mass spectrometer and the quadrupole mass spectrometer. In TOF instruments the ions are separated by their time of flight along a given distance $l$:

$$t = l \sqrt{\frac{1}{2U} \frac{m}{e}} \tag{61}$$

Taking $2Ue$ as a constant, the velocity $v$ of an ion of mass $m$ is proportional to the square root of the inverse of its mass:

$$v \simeq \sqrt{\frac{1}{m}} \tag{62}$$

This, in turn, means that the traveling time $t$ of an ion of mass $m$ for a given flight path $l$ is given by

$$t = l\sqrt{m} \tag{63}$$

The first use of a solid-state laser for laser micro mass spectrography was reported by Honig and Woolston as early as 1963 [1]. Since the demands made on both the performance of the laser and on the focusing optics were similar to those in laser micro emission analysis, it was natural to supplement a laser microanalyzer with an ion source and to couple it with a suitable mass spectrograph. The starting point for these experiments was the interest in extending local analysis in the micro region to the determination of isotopes in order, among other things, to open up new possibilities for tracer technique in solid-state research and in isotope analysis in geology and biology.

The first laser of the first-generation apparatus used delivered an output energy of 1 J and, with a pulse duration of 200 μs, a power of 5 kW [1]. It was therefore suitable for ionizing all elements with an ionization potential of

**TABLE 44. Ionization Potentials of Some Polyisotopic Elements**

| Element | Ionization Potential (eV) | Element | Ionization Potential (eV) |
|---|---|---|---|
| Li | 5.39 | In | 5.79 |
| B | 8.30 | Sn | 7.34 |
| Mg | 7.64 | Ba | 5.21 |
| Si | 8.15 | La | 5.61 |
| K | 4.34 | Ce | 6.91 |
| Ce | 6.11 | Pr | 5.76 |
| Ti | 6.82 | Nd | 6.31 |
| V | 6.74 | Sm | 5.63 |
| Cr | 6.76 | Eu | 5.67 |
| Fe | 7.87 | Gd | 6.16 |
| Ni | 7.63 | Yb | 6.22 |
| Cu | 7.72 | Dy | 6.82 |
| Ga | 6.00 | Er | 6.08 |
| Ge | 7.88 | Lu | 6.15 |
| Rb | 4.17 | Hf | 7.0 |
| Sr | 5.69 | Ta | 7.88 |
| Zr | 6.84 | W | 7.98 |
| Mo | 7.10 | Re | 7.87 |
| Ru | 7.36 | Tl | 6.11 |
| Pd | 8.33 | Pb | 7.42 |
| Ag | 7.57 | | |

$\leqslant 10$ eV in the microplasma (i.e., more than 50% of all elements of the periodic system) (Table 44). Trace analysis (ppm region) and isotope analysis can be performed in micro regions of solid surfaces down to $10^{-4}\ \mathrm{cm}^2$ (see Table 45). The effects of laser target interaction and micro plasma generation have been discussed in detail in Chapter 3.

The advantage of laser mass spectrographic local analysis of solids compared to other mass spectrographic methods is the substantially higher excitation temperature and the representative vaporization of the specimen. In addition, no special preparation of the sample is needed; that is, no transformation of the sample into measurable compounds and no separation of disturbing elements are necessary, in contrast to isotope analysis with ordinary thermionic sources. The experiments on laser micro mass spectrography have been carried out in three variants, with the following result:

- The thermionic spectrum obtained by laser radiation is completely free of background. Surface contaminants (hydrocarbons, etc.) are not detected.
- On the other hand, the low-voltage discharge mass spectra produced by laser radiation correspond to the well-known mass spectra obtained with externally ignited low-voltage ion sources or high-frequency spark ion sources; that is, they detect all the elements present in the vapor space, including gases and surface impurities.
- The electron-impact ionization of the neutral vapor produced by laser radiation is particularly suitable for analysis of organic solids or organic layers.

In the period 1963 to 1972, the development of laser ion mass spectrometers for the characterization of solids was proceeding simultaneously in several laboratories in different countries [3–56] (see Table 45). Some excellent review articles, doctoral thesis, and chapters of books provide a panoramic view of the broadening applications of the laser ion source in mass spectrometry between 1970 and 1980 [42, 57–66]. One hundred thirty-seven references are given by Dumas in 1970 [42], who described the effects of laser beam impact on different materials over laser power density ranges from $10^7$ to $10^9\ \mathrm{W\ cm^{-2}}$, and presented theoretical considerations for vapor and ion formation.

In 1971, Honig [58] described the mass analyzer systems available for laser ion sources and reviewed 61 publications. In 1971 and 1972 [59], Knox wrote chapters in two books which gave preferential treatment to results from free-running mode lasers to mass analysis by TOF mass analyzer, and discussed the interactions between laser radiation and solids. In 1973, Beahm [60] provided an interesting view of the limitations of laser ion sources for the characterization of solids. Eloy [61] gave detailed information on laser–solid interactions, with 76 references. In 1976 Cuna et al. [62] compared a laser ion

**TABLE 45. Survey of Some Mass Spectrometric Laser Studies between 1963 and 1972**

| References | Year | Laser Characteristic | | | | | Analyzer[a] |
|---|---|---|---|---|---|---|---|
| | | Energy (j) | Power (W) | Power Density ($W\,cm^{-2}$) | Beam Area ($cm^2$) | Pulse Length ($\mu s$) | |
| Honig and Woolston [1] | 1963–67 | 0.2–1 | $5 \times 10^3$ | $5 \times 10^7$ | $10^{-4}$ | 200 | DF |
| Giori et al. [2] | 1963 | 5–13 | $2 \times 10^4$ | $4 \times 10^7$ | $5 \times 10^4$ | 500 | Quad |
| Berkowitz and Chupka [3] | 1964 | 1 | $2 \times 10^3$ | $2 \times 10^7$ | $10^{-4}$ | 400 | SF |
| Lincoln et al. [4–6] | 1964–69 | 1 | $2 \times 10^3$ | $2 \times 10^6$ | $10^{-3}$ | 500 | TOF |
| Isenor [22] | 1964–65 | | $2.5 \times 10^7$ | $8 \times 10^{10}$ | $3 \times 10^{-4}$ | 0.02 | TOF |
| Knox et al. [7–9] | 1966–69 | 0.1 | | $10^6$ | | 200 | TOF |
| | | 5 | | $10^7$ | | 1000 | TOF |
| Bernal et al. [14] | 1966 | 0.1 | $3 \times 10^6$ | $6 \times 10^7$ | $5 \times 10^{-2}$ | 0.03 | TOF, Quad |
| Eloy et al. [16–18] | 1966–69 | <0.1 | $2 \times 10^4$ | $(2 \times 10^8)$ | $(10^{-4})$ | 5 | SF |
| Fenner et al. [19–21] | 1966–68 | 1 | $(5 \times 10^3)$ | | | (200) | SF |
| | | 0.01 | $3 \times 10^5$ | $1.5 \times 10^{10}$ | $2 \times 10^{-5}$ | 0.03 | TOF |
| Langer et al. [23, 24] | 1966–68 | 1 | $3 \times 10^7$ | $1 \times 10^{11}$ | $3 \times 10^{-4}$ | 0.03 | TOF |
| Namba et al. [26] | 1966 | 0.3 | $10^3$ | $10^6$ | $10^{-3}$ | 300 | TOF |
| | | | $10^7$ | $5 \times 10^9$ | $2 \times 10^{-3}$ | 0.05 | TOF |
| Bykovskii et al. [36–40] | 1969–70 | ≤0.5 | $10^7$ | $(10^{10})$ | $10^{-3}$ | 0.05 | TOF |
| | | | | | | | SF |
| Dietze and Zahn [54, 55]; Moenke and Moenke-Blankenburg [56] | 1970–72 | ≤1 | $2 \times 10^3$ | $10^6$–$10^8$ | $10^{-4}$ | 250 | DF–MH |

[a]DF double focusing; Quad, quadrupole; SF, single focus; TOF, time of flight; MH, Mattauch–Herzog.

source with RF spark ion source. Two reviews by Maksimov and Larin in 1976 [63] and Kovalev et al. in 1978 [64] are especially valuable for giving information on Soviet work. In 1980, Conzemius and Capellen [65] published a comprehensive overview with 287 references, and a bibliography in the same year [66].

The second generation of laser microprobe mass analyzers were developed from the middle of the 1970s to the beginning of the 1980s [67–83]. Two major areas of laser mass spectrometry for analytical purposes have evolved during the last decade: microprobe analysis, with a spatial resolution of approximately 0.5 $\mu$m and a sensitivity in the ppm to ppb range, and laser desorption of thermally labile, nonvolatile organic compounds.

All laser microprobe mass analyzers commercially available are based on essentially identical functional principles. They use short laser pulses ($\tau \approx 10$ ns) at wavelengths in the far UV, typically the quadrupled wavelength of $Q$-switched ND-YAG lasers at $\lambda = 265$ nm. Standard optical microscopes are used for sample observation and focusing of the laser beam with a spatial resolution of approximately 0.5 $\mu$m in the analysis of thin specimens and approximately 3 $\mu$m for surface analysis of bulk specimens. Typical irradiances in the laser focus on the sample surface range $10^7$ to $10^{11}$ W cm$^{-2}$. More recently, excimer lasers with 10-ns pulses at wavelengths of 308, 248, and 193 nm have been used.

Substantial effort has been directed toward understanding high-temperature phenomena. A particularly challenging problem has been to characterize the chemical and physical processes that occur at elevated temperatures and are not predictable by extrapolation from lower-temperature data. In recent years the study of high-temperature processes has been facilitated by developments in instrumentation and measurement techniques [84].

There has been increasing interest in the utilization of lasers to induce thermal description of ions and molecules from surfaces to study surface diffusion and reaction kinetics. A laser provides extremely rapid localized heating (see Chapter 3) of the surface in the small area of the incident beam while the rest of the surface is not appreciably perturbed [85–93].

Besides laser ionization mass spectrometry, laser resonance ionization spectrometry provides possibilities for ultrasensitive mass spectrometry of solids. The principle is based on the phenomenon of multiple photon ionization. By the absorption of the first photon, the material is evaporated and brought to an excited state, and absorption of the second photon brings the free atom to the region of the ionization continuum, producing an ion of a given kinetic energy. The entire process can take place using one laser if the energy of the two photons equals or exceeds the sum of the energy of evaporation and ionization. If this is not the case, two different laser quanta have to be used. The literature data show five cases for ionization by two-photon absorption. One of the lasers has to be a dye laser in order to be able to vary the energy of the light quantum that is necessary to locate the specific excited state. All the elements can be ionized except helium and neon [94–102]. This new field of mass spectrometry has been applied to measurements of radioactive isotopes which are useful as chronometers and tracers, and to ultrasensitive trace analysis in the sub-ppm range in solids [103].

A reference and abstract index of laser-induced mass spectroscopy has been given by Kaufmann in 1986 [104]. Of general interest are the *Proceedings of the 3rd International Laser Microprobe Mass Spectrometry Workshop*, 1986 [105].

## 9.2. ANALYTICAL FEATURES

The interest in microprobe instruments is characterized by different features: spatial resolution, capability of analytical information about all elements, high sensitivity, and sufficient mass resolution. For the choice of each component unit in the basic configuration of laser mass spectrometry, consisting of ion source, analyzer, and detector, these goals are being pursued.

Modern laser micro mass spectrometers utilize a Nd:YAG laser (*Q*-switched) which operates at four times the fundamental frequency ($4f$) at 265 nm. Therefore, from simple diffraction theory it is expected that this laser wavelength, together with a suitable objective, produces a smaller focused spot on the surface of the sample than it does with the fundamental frequency used in LM-OES (see Section 3.1). Theoretically, a spatial resolution of 0.5 $\mu$m should result at $4f$ operation [112]. Indeed, the lateral resolution can be below 1 $\mu$m in transmission geometry instruments. Perforation holes of 0.4-$\mu$m diameter have been demonstrated in thin samples of biological tissue [115]. In reflection geometry instruments, the minimum crater diameter is several micrometers [112, 116].

The capability of LM-MS is described by [113] as fast and as semiquantitative in the detection of all elements of the periodic table, of isotopes, and of molecules and molecule fragments. In principle, an unlimited $m/e$ range can be collected from a single pulsed laser ablation event. The reproducibility depends on a large number of factors, but standard deviations of better than 8% can be obtained [112]. LM-MS provides complete mass analysis at any single point of a solid sample surface and in addition, line and area scanes.

An attractive feature of the laser ion source is the high detection sensitivity. In MS several parameters have been used to describe sensitivity: for example, concentrational sensitivity attainable per laser pulse, and total ion charge accumulated at the detector per laser pulse. Tables 46 and 47 contain

**TABLE 46. Ionization Efficiency**

| Reference | Laser Energy (J) | Ion Production | | | |
|---|---|---|---|---|---|
| | | Atoms Eroded per Laser Shot | Ions Created per Laser Shot | Ions Created per Atoms Eroded | Ions Created per Laser Energy |
| Honig and Woolston [1] | 0.2 | $\sim 10^{17}$ | $5 \times 10^{12}$ to $2 \times 10^{14}$ | $3 \times 10^{-3}$ to $10^{-5}$ | $2 \times 10^{13}$ to $10^{15}$ |
| Fenner et al. [19–21] | 0.003 to 1 | $3 \times 10^{13}$ | $6 \times 10^{11}$ to $10^{12}$ | $3 \times 10^{-2}$ | $6 \times 10^{13}$ to $3 \times 10^{14}$ |
| Eloy et al. [16–18] | 0.005 to 0.01 | $2 \times 10^{14}$ to $5 \times 10^{14}$ | $2 \times 10^{11}$ | $4 \times 10^{-4}$ | $4 \times 10^{13}$ |
| Bykovskii et al. [36–40] | 0.02 to 0.5 | $10^{14}$ to $3 \times 10^{15}$ | $10^{14}$ | $\sim 1$ | $6 \times 10^{15}$ |
| Dietze and Zahn [54, 55] | 0.85 | $10^{17}$ | $3 \times 10^{13}$ to $3 \times 10^{14}$ | $3 \times 10^{-3}$ to $5 \times 10^{-5}$ | $4 \times 10^{13}$ to $4 \times 10^{14}$ |
| Zakharov and Protas [106] | 4 | $8 \times 10^{17}$ | $6 \times 10^{12}$ | $8 \times 10^{-6}$ | $1.5 \times 10^{12}$ |
| Hillenkamp et al. [67] | $10^{-7}$ | $\sim 10^{10}$ | — | — | — |
| Devyatykh et al. [107–109] | 0.4 to 1.6 | $3 \times 10^{16}$ | — | — | — |
| Bingham and Salter [110] | 0.005 | $10^{14}$ | $10^{14}$ | $\sim 1$ | $2 \times 10^{16}$ |
| Conzemius and Svec [111] | 0.001 | — | $6 \times 10^{11}$ | — | $6 \times 10^{14}$ |

*Source*: After Ref. 65; published with permission of Elsevier Scientific Publishing Company, Amsterdam.

**TABLE 47. Ion Collection Efficiency**

| References | Mass Resolution | Ion Detected, ID | | | | |
|---|---|---|---|---|---|---|
| | | ID per Laser Shot | ID per Ions Created | ID per J | ID Per J $s^{-1}$ | ID × Resolution per J $s^{-1}$ |
| Honig and Woolston [1] | Low | To $10^{10}$ | — | To $5 \times 10^{10}$ | — | — |
| Fenner et al. [19–21] | 6 | $6 \times 10^{6}$ | $10^{-5}$ | $6 \times 10^{8}$ | — | — |
| | 30 | $6 \times 10^{3}$ | $10^{-8}$ | $6 \times 10^{5}$ | — | — |
| | 30 | $10^{4}$ | $10^{-8}$ | $3 \times 10^{6}$ | — | — |
| Eloy et al. [16–18] | 200 | $2 \times 10^{8}$ | $10^{-3}$ | $4 \times 10^{10}$ | $3 \times 10^{9}$ | $6 \times 10^{11}$ |
| Bykovskii et al. [36–40] | 2000 to 3500 | $2 \times 10^{6}$ | $2 \times 10^{-8}$ | $2 \times 10^{8}$ | $2 \times 10^{10}$ | 3 to $5 \times 10^{13}$ |
| Dietze and Zahn [54, 55] | 1500 | $1 \times 10^{8}$ | $3 \times 10^{-7}$ | $1 \times 10^{8}$ | $5 \times 10^{6}$ | $8 \times 10^{9}$ |
| | 4000 | $3 \times 10^{8}$ | $1 \times 10^{-5}$ | $4 \times 10^{8}$ | $2 \times 10^{7}$ | $8 \times 10^{10}$ |
| Zakharov and Protas [106] | 1150 | $6 \times 10^{5}$ | $10^{-7}$ | $1.5 \times 10^{5}$ | — | — |
| Bingham and Salter [110] | 3000 | $10^{7}$ | $10^{-7}$ | $2 \times 10^{9}$ | $2 \times 10^{8}$ to $10^{11}$ | $6 \times 10^{11}$ to $3 \times 10^{14}$ |
| Conzemius and Svec [111] | 2000 | $6 \times 10^{5}$ | $10^{-6}$ | $6 \times 10^{8}$ | $6 \times 10^{12}$ | $1 \times 10^{15}$ |
| Busygin and Ulmasbaev [114] | 100 | $3 \times 10^{6}$ | — | $1 \times 10^{8}$ | $2 \times 10^{9}$ | $2 \times 10^{11}$ |

*Source*: After Ref. 65; reproduced with permission of Elsevier Scientific Publishing Company, Amsterdam.

information on ionization efficiency and ion collection efficiency. The best definition of ionization efficiency is: the quotient of ions created per total atoms eroded. It is of great significance. The best definition of ion collection efficiency is: the quotient of number of ions detected per number of ions created, which depends on the transmission of ions through the mass spectrometer, which is usually a compromise with mass resolution.

Element detection limits are in the ppm range and can be achieved with the consumption of only picograms of material. Detection limits below 1 ppm have been quoted in thin sections of quasi-biological tissues [117]. In glass microparticles detection limits are typically 10 ppm [118]. In bulk metal targets analyzed in the reflection mode, detection limits are in the range 1 to 10 ppm [112].

## 9.3. INSTRUMENTS

### 9.3.1. Ion Sources

In instrumental application of laser target interaction, it is necessary to distinguish between two types of processes: (1) photon absorption by transmission mode (Fig. 77*a*) and (2) photon absorption by reflection mode (Fig. 77*b* and *c*). In the first laser interaction configuration, the laser focusing beam induces and absorption zone behind the solid thin layer. The laser-generated microplasma emerges from the opposite side of the thin samples. In the second case the laser focusing beam induces an absorption zone and plasma creation before the solid surface as regards the laser incidence axis. Thus the expanding plasma spreads in the direction of the focusing lens. Among the main advantages of a laser transmission mode system is the fact that it is possible to reduce the distance from lens to sample to a minimum. In effect, the thin film of the specimen is located in an evacuated specimen chamber directly underneath a thin cover slide (quartz) which serves as both an optical window for the microscope and as a vacuum seal [71]. An additional advantage lies in the fact that the plasma expansion is possible only when the sample (facing the analyzer) is punctured by the laser beam. This gives a better definition of $t_0 = 0$, the zero point of the time of flight [119]. In this configuration the smaller laser spot sizes are $\leqslant 0.8\ \mu m$ in diameter and 2 to $5\ \mu m$ in depth vis-à-vis the sample.

A magnified view of the sample region for transmission geometry is depicted in Fig. 78. The target illustrated is a small particle supported by a thin film on a copper TEM grid.

Eloy [119] emphasized that the main disadvantages are that the sample requires difficult preparation and that molecular fragments in the mass spectra

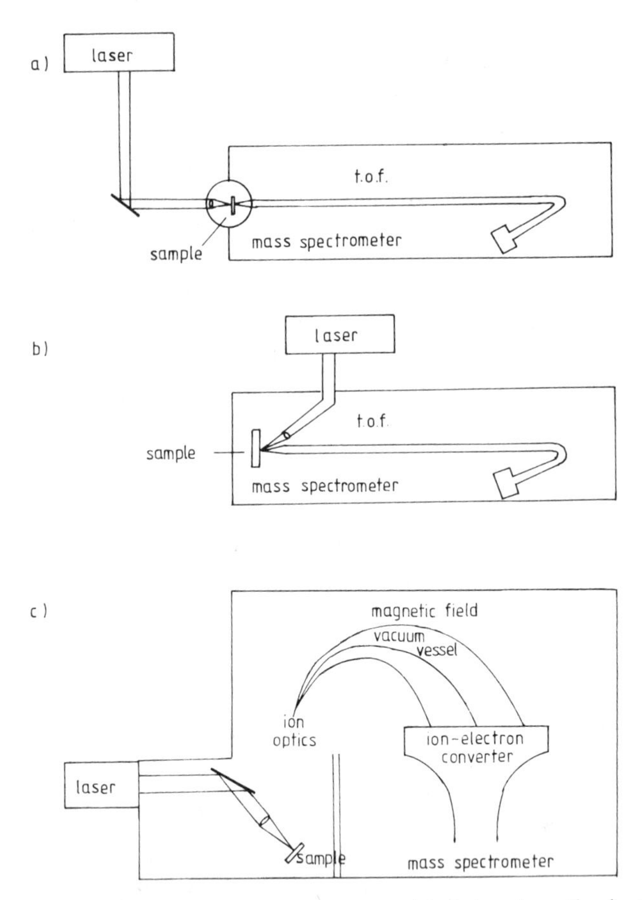


**Figure 77.** Principles of laser micro mass spectrometers: (*a*) photon absorption by transmission mode; (*b*) and (*c*) photon absorption by reflection mode. Types of instruments: (*a*) LAMMA 500, (*b*) LAMMA 1000 and LIMA, (*c*) LPMS. Reprinted by permission of Leybold-Heraeus GmbH, Köln, Hanau, FRG, Cambridge Mass Spectrometry Ltd., Cambridge, GB, and Eloy, Grenoble, France.

are relatively intense. Among the main advantages of laser reflection mode systems are the possibility of studying the bulky solid materials directly without preparation, the fact that the analytical useful laser power density is more easily defined when considering varied mmaterial, and the fact that the mass spectra are simplified because of small contaminations by organic fragments, polyatomic ions, and doubly charged ions. The interaction time can be reduced to reach a volatilized material thickness limit of 0.05 $\mu$m [119]. The geometrical configuration between the laser beam, sample, and ion optic axis

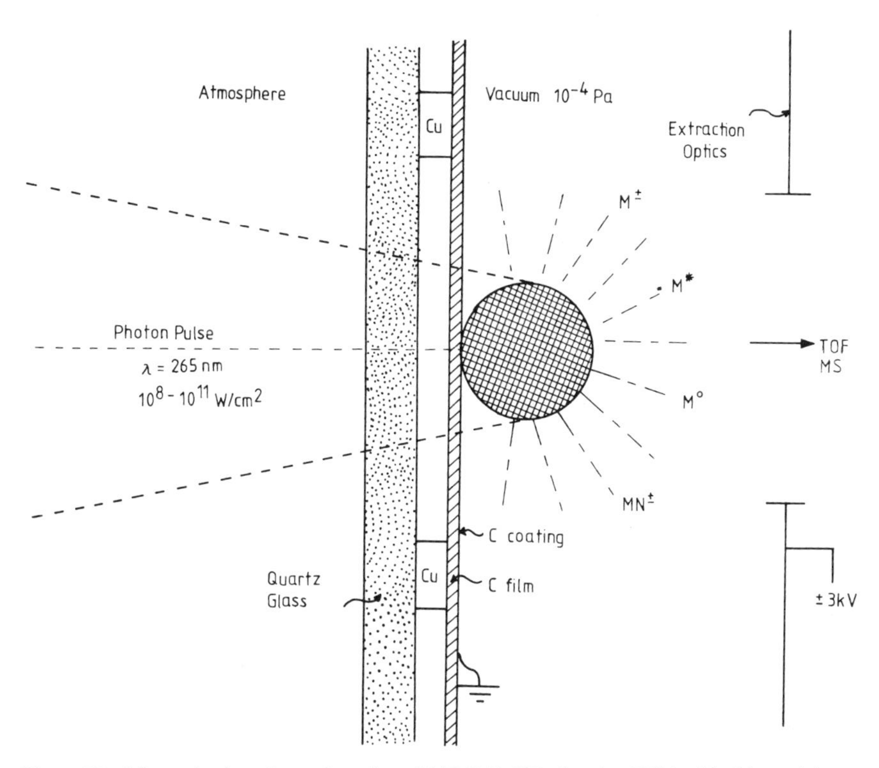


**Figure 78.** Schematic view of sample region of LAMMA 500, showing TEM grid with particle on carbon films. From Ref. [120] reprinted by permission of D. S. Simons, National Bureau of Standards, Gaithersburg, U.S.A.

varies within the angle ratios 45/90, 90/0, 45/45, 90/60, and 90/45 [65]. Typical parameters of solid-state laser systems used with laser mass spectrometry of solids are presented in Table 48.

The development of high-powered short-pulsed lasers and the availability of such lasers at shorter wavelengths have made possible their use for the analysis of more complex organic structures by a number of techniques, including multiphoton ionization, photodissociation, and laser desorption [121]. The latter method is considered a "soft" ionization technique, capable of producing intact molecular ions from complex, and sometimes rather large molecules. This ionization technique may be the method of choice for certain classes of compounds.

**TABLE 48. Solid-State Laser Systems Used with Laser Mass Spectrometry of Solids**

| Lasing Material | Wavelength (nm) | Operation Mode | Pulse Duration ($\mu$s) | Pulse Energy (J) | Pulse Power Density (W cm$^{-2}$) |
|---|---|---|---|---|---|
| Ruby | 694 | Normal | 500 to 1000 | 0.1 to 5 | $10^7$ |
| | 694 | *Q*-switched | 0.02 to 8 | $5 \times 10^{-4}$ to 1 | $10^7$ to $10^{11}$ |
| | 347 | *Q*-switched | 0.03 | 0.1 | $10^{10}$ |
| Nd-doped glass | 1060 | Normal | 500 | 3 | $10^7$ |
| | 1060 | *Q*-switched | 0.015 | 1 | $10^{10}$ |
| Nd-doped YAG | 1060 | *Q*-switched | 0.02 to 0.1 | 0.02 to $10^{-3}$ | $10^9$ |
| | 353 | | 0.015 | | $10^8$ to $10^{11}$ |
| | 265 | | 0.015 | | $10^8$ to $10^{11}$ |

*Source*: After Ref. 65; reproduced with permission of Elsevier Scientific Publishing Company, Amsterdam.

### 9.3.2. Analyzers

Mass spectrometry for inorganic analysis was originally developed with a RF spark ion source, a double-focusing Mattauch–Herzog mass spectrograph, and photographic plate detection. In the first years of its exploitation, spark source mass spectrometry (SSMS) was often considered to be a high-sensitivity substitute for the emission spectrograph, and was applied primarily as a semiquantitative survey method for multielement analysis. Electrical detection systems with an electron multiplier became available for quick real-time data collection; this detection method, largely as a result of the shortcomings of the ion formation process in the spark source, has not provided a real breakthrough. Hence current equipment is roughly equivalent to that which became available more than 30 years ago, but considerable technological and methodological progress has been achieved recently. A review of available mass analyzers suitable for laser coupling shows that the main advantage of the time of flight is the higher luminosity. The higher mass resolution powers are available on the double-focusing optics (mainly Mattauch–Herzog optics) but at the detriment of sensitivity.

A TOF spectrometer consists simply of a device for accelerating the ions to a uniform energy, a drift tube of lenth $l$ in which the ions are separated according to their mass $m$, and an ion detector at the end of the drift tube where the different ions arrive in discrete packs, one pack after the other. The detector, usually a secondary electron multiplier, delivers an analog signal at its output, the amplitude of which corresponds to the number of ions present in each pack. One of the advantages of the TOF mass spectrometer over other mass spectrometers is the fact that all sorts of ions are analyzed in one circle. That means that all analytical information about the area of interest is obtained and recorded simultaneously.

Eloy and Unsöld [70] developed a new panoramic electro-optic detector which permits a number of collected ions to be read directly. The purpose was to provide a new detector for instruments having a widespread focal plane. This panoramic detector is 100 times more sensitive than a photographic plate detector [80, 119].

Chamel and Eloy [82] published a comparison of some characteristics of laser mass spectrometers of the LAMMA and LPMS types (Table 49). It is completed by specifications of LIMA [123]. A Fourier transform mass spectrometer coupling with pulsed laser sources is ideal for high-resolution laser desorption and ionization analyses of labile, nonvolatile, high-molecular-weight compounds [124].

Nowadays, problems in surface and bulk analysis are solved by using a combination of techniques on the same spectrometer. If both bulk and surface analysis is required on the same material, the use of separate instruments can

**TABLE 49. Comparison of Three Types of Laser Mass Spectrometers**

| Characteristics | LAMMA 500 to 1000; Leybold-Heraeus GmbH, Köln, FRG | LPMS: CEA-CEN, Grenoble, France, and GSF, München-Neuherberg, FRG | LIMA: Cambridge Mass Spectrometry Ltd., Cambridge, England |
|---|---|---|---|
| Samples | Thin sections (0.1 to 2 μm), compact samples up to 200 mm in diameter | | 20 mm in diameter, 10 mm thick |
| Diameter of the analyzed zone | 0.3 to 1 μm | 3 to 100 μm | 1 to 3 μm in reflection, ⩽ 1 μm in transmission |
| Spatial resolution (lateral and in depth) | 0.3 to 1 μm | 1 to 3 μm | |
| Laser | Nd:YAG | Nd: YAG | Nd:YAG, *Q*-switched |
| Wavelength | 353 nm, 265 nm | 1.06 μm, 530 nm, 353 nm, 265 nm | 1.06 μm, 530 nm, 265 nm |
| Pulse duration | 15 ns | 3 to 4 ns | 4 to 5 ns |
| Power densities in desorption mode | $10^7$ to $10^9$ W cm$^{-2}$ | | 0 to $10^{11}$ W cm$^{-2}$ continuously variable |
| Power densities in pyrolysis mode | $10^9$ to $10^{11}$ W cm$^{-2}$ | | |

| | | | |
|---|---|---|---|
| Power densities necessary to create a cold plasma (with an equivalent ionic temperature) | $10^{11}$ to $10^{12}$ W cm$^{-2}$ | $10^{9}$ to $10^{10}$ W cm$^{-2}$ | |
| Ionization factor | $10^{-2}$ to $10^{-1}$ | $10^{-1}$ to $7 \times 10^{-1}$ | |
| Mass separation | TOF | Magnetic analyzer and TOF | TOF |
| Transmission factor | $10^{-1}$ to $5 \times 10^{-1}$ | $10^{-3}$ to $5 \times 10^{-2}$ | |
| Mass resolution ($M/\Delta M$) | 400 to 1000 | 100 to 600 | 500 FWHM at mass 208 |
| Dynamic range of signals | $10^{2}$ | $10^{3}$ to $10^{4}$ | |
| Detection limits | $10^{-18}$ to $10^{-20}$ g<br>0.1 to 2 ppm | $10^{-13}$ to $10^{-16}$ g,<br>1 to 100 ppm | 1 ppm |
| Reproducibility | 5%, 10 to 50% | 5 to 50% | |

*Source*: After Ref. 82; published with permission; completed by [122, 123].

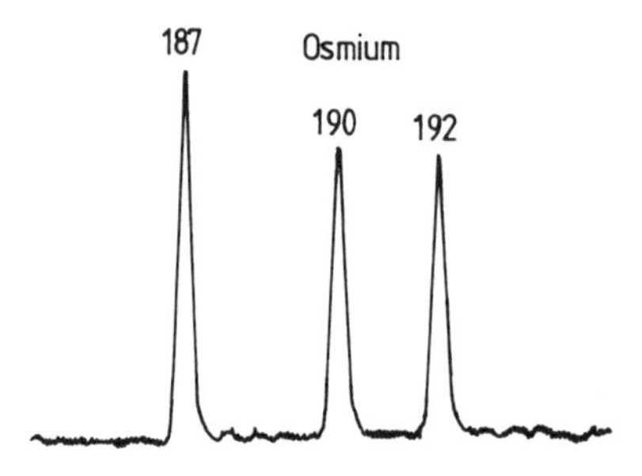


**Figure 79.** Positive ion mass spectrum of osmium isotopes from single laser pulse, showing radiogenig $^{187}$Os and control spikes of $^{190}$Os and $^{192}$Os. From Ref. [120], reprinted by permission of D. S. Simons.

not only be time consuming, but where the specimen is to undergo a series of successive treatments, such as gas reaction, corrosion, temperature cycling, and so on, with analyses to be conducted in between, regular removal into ambient atmosphere for the separate analyses could change the composition. There is therefore a growing tendency to combine two and more techniques. The IX23L Laser Microprobe (LAMMS) is constructed to ultrahigh vacuum and can be fitted with a range of primary ion sources for submicrometer-

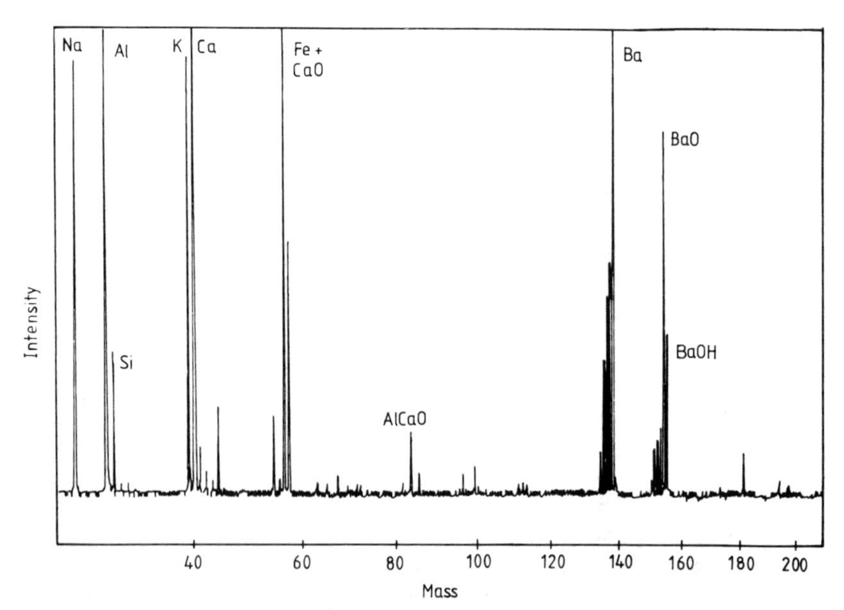


**Figure 80.** Portion of positive ion mass spectrum of single microsphere of K 309 glass. From Ref. [120], reprinted by permission of D. S. Simons.

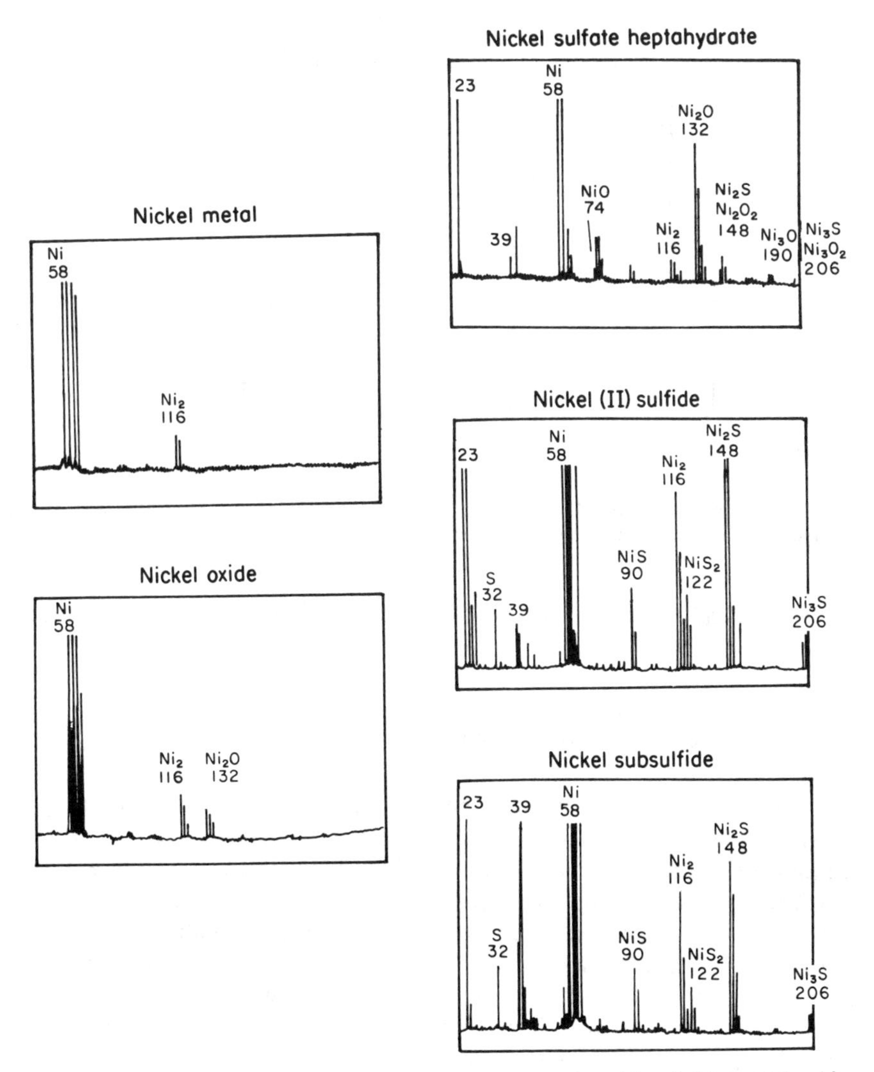


**Figure 81.** Positive ion mass spectra of single micrometer-sized particles of nickel metal and four nickel compounds. From Ref. [120], reprinted by permission of D. S. Simons.

resolution static TOF SIMS imaging, where high mass resolution ($M/\Delta M >$ 15000 at 100 amu) is guaranteed with a mass range greater than 10,000 amu [125].

A field-ion-atom probe is described as the most powerful microanalytical technique yet devised. It combines atomic spatial resolution with individual atom detection and analysis. The addition of pulsed laser-field evaporation

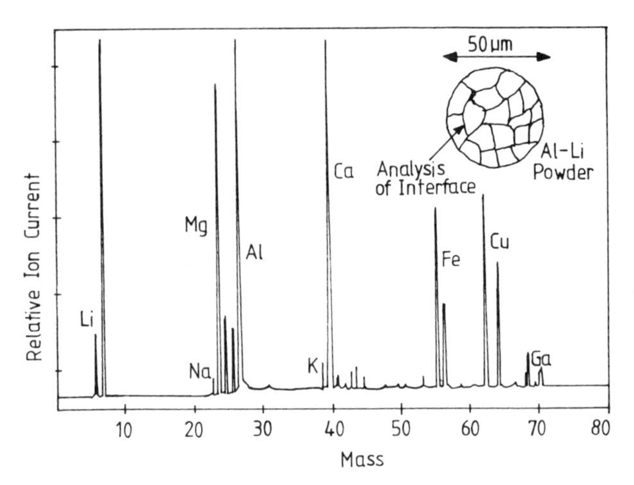


**Figure 82.** Positive ion mass spectrum from cell wall of aluminium–lithium alloy showing major impurities present. From Ref. [120], represented by permission of D. S. Simons.

extends applications of the technique beyond traditional metals and alloys into semiconductor materials [125].

## 9.4. FIELDS OF APPLICATION

Bibliographies of applications published through 1982 [66] and 1986 [104] are available, as are reports of three international workshops [77, 83, 105]. Quantitative analysis still has some limitations. Since the precision of every analytical device depends on the reproducibility of the processes involved, and since the interaction of laser light with a specimen is a highly nonlinear process, the system is rather sensitive to statistical fluctuations of the laser parameters and to inhomogeneity of the specimen. Presumptions for quantitative laser microanalysis of solids are described in Chapter 4. Quantitative analysis is possible only if the laser output is stabilized as far as possible and the specimen is relatively homogeneous. Relative standard deviation in such cases could be $< 5\%$ at higher concentrations ($> 10$ ppm) and $< 15\%$ at low concentration levels. Often, the determination of elemental concentration ratios is sufficient.

### 9.4.1. Chemical Analysis of Inorganic Materials

Examples of up-to-date (since 1980) applications in mineralogy/geology, inorganic chemistry, metallurgy, and related fields are given in Table 50.

**TABLE 50. Laser Micro Mass Analyses of Inorganic Materials**

| Elements | Matrices | Aims of Analyses | Results | References |
|---|---|---|---|---|
| Mg, Si, K, Ca, Fe | Mineral fibers, chrysotile, amphibole, croicidolite, glass fibers, nonfibrous zeolithes, chabazite, erionite | Identification and microanalysis of asbestos and glass fibers by environmental sampling and after deposition in human tissue | Qualitative analysis of fibers of 1 to 2 $\mu$mm diameter, studies of leaching effects | Spurny et al. [126] |
| Si, Al, Fe, Mg, Ca, Li, Na, K, Rb, Cs, Ti, Mn, Cr, Zr, F | Augite, hornblende, and lepidolite | Quantitative analysis of silicate minerals by secondary ion mass spectrometry and laser probe mass analysis | Comparison of results | Beusen et al. [127] |
| Ti, V, Cr, Mn, Co, Ni, Cu | Meteoritic troilite (FeS) | Mass spectra of positive and negative ions and quantitative studies in the range 23 to 80 amu | Trace element contents, comparison with pyrite of terrestrial origin | Weinke et al. [128] |
| $Mg^{24}/Si^{28}$ and adsorbed organic compounds | Asbestos | Study of asbestos using SIMS and LAMMA | Leaching experiments, changing of inorganic composition, surface adsorption of organic compounds | Van Espen et al. [129] |

**TABLE 50.** (*Contd.*)

| Elements | Matrices | Aims of Analyses | Results | References |
|---|---|---|---|---|
| Ca, Mg, | Dolomite | Geological application of ionization LTE model (with computer program ELOGET) | Stoichiometric composition | Eloy et al. [130, 131] |
| Hf, Si, Zr, Ge, U, | Zircon | | Stratigraphical variations of main components, concentration profiles of traces | |
| Na, K, Si | Quartz | Fluid inclusion (quartz); | Advantages of the LPMS instrument and the ELOGET program | |
| O, Na, Mg, S, Cl, K, Ca, Br, S | Halite | Studies in geological samples (halite) | | |
| Ca, Mg, Cl, S, Cr; Na, Mg, Al, Si, P, S, Cl, K, Ca, Sc, Ti, V, Cr, Mn, Fe, Ce, Ni, Cu, Zn, Ga, Ge, As, Br, Rb, Sr, Cs, Ba, Y, Zr, Mo, La, Ce | Dolomite; BM-standard of the ZGI, GDR (Central Geological Institute, Berlin, GDR) | Determination of concentrations with different methods: SSMS, RCS, LIMS, and RSC | Comparison of quantitative results | Dietze and Becker [103] |
| Na, Al, Si, S, K, Ca, Ti, Mn, Fe, Co, Cu, Zn, As, Zr, Ag, Pb, Bi | Pb–Zn oxide ores, Bolkardağ (Turkey) | Qualitative analyses | Spectra and their interpretation | Akyüz et al. [132] |
| Rare earth elements and Th, Y, Bi, Co | Malatya Th-REE deposits | | | |
| Th, Ce, La, Li, Na, Mg, Al, S, K, Ca, Bi | Monazite | | | |

| | | | | |
|---|---|---|---|---|
| $Ti^+$, $TiO^+$, $TiO^+$, $Ti_2O^+$, $Ti_2O_3^+$, $Li^+$, $Na^+$, $Al^+$, $K^+$, $Fe^+$ | Rutile | | | |
| $^{187}Re$, $^{187}Os$, $^{190}Os$, $^{192}Os$ | Perrhenic acid | Determination of the half-life of $^{187}Re$, important to geo- and cosmochemistry as a nucleochronometer | Direct determination of the $^{187}Re$ by observing the in-growth with time of the decay product $^{187}Os$ relative to $^{190}Os$ and $^{192}Os$ (see Fig. 79) | Simons et al. [120, 133–135] |
| Na, Mg, Al, Si, K, Ca, Ti, Fe | Coal fly ash | Quantitative analysis | Standard deviations of relative peak intensities from 5.1 to 17.6% | Yokozawa et al. [136] |
| As | Si | Study of recrystallization effects of amorphous Si with As implantation | | Eloy et al. [142] |
| Si | Si | Pulsed laser melting of amorphous silicon | Time-resolved measurements and model calculations | Lowndes et al. [143] |
| H | Si | Study of concentration of hydrogen in amorphous silicon thin samples (thickness $\sim 1\ \mu m$) | Different profiles of hydrogen, comparison of results with those of IR absorption and nuclear reactions | Eloy [119] |
| | $Ga_{1-x}Al_xAs$ | Effect of alloy composition on secondary yields | Semiconductor analysis; results are compared with data obtained using SIMS | Edge et al. [144] |

**TABLE 50.** *(Contd.)*

| Elements | Matrices | Aims of Analyses | Results | References |
|---|---|---|---|---|
| H, He, Li, Be | Thin films | Detecting the elements that are inaccessible to EPMA | Advantages of LIMA in thin-film analysis and depth profiling | Clarke et al. [145] |
| $Si_m^+$ and $Si_mO_n^+$ | Silica; powdered $SiO_2$, monodispersed silica spheres, quartz glass, crystalline mineral quartz | Occurrence of atomic and molecular ions, spectral interferences | Relative negative and positive ion yields and weighted average O/Si ratios for clusters with a given number of Si atoms | Michiels et al. [146] |
| O, Si, Na, Al, Ca, Cr, Mn, Fe, Ni, Zn, Sr, Zr, Mo | Glass | Quantitative microanalysis of insolators, studies of solid inclusions, multielemental profiles of concentration, superficial homogeneity of materials and phases | Use of computer program ELOGET for correction: reduction of error to 4 to 80% | Eloy [147] |
| | Alkali halides | Study of ion formation from polar and ionic compounds | Occurrence of $(Alk_{n-1}Hal_n)^- + Alk^+$ and $(Alk_nHal_{n-1})^+ + Hal^-$ | Jöst et al. [148] |
| | CsI and I | Multiphoton laser ionization mass spectrometry of cesium iodide and atomic iodine | Study of the mechanism | Balooch and Olander [149] |

| | | | | |
|---|---|---|---|---|
| $M_mO_n^{\pm}$, $M^+$, $M^-$ | $M_xO_y$: $Sc_2O_3$, $Ho_2O_3$, BeO, ZnO, CuO, $Ag_2O$, $As_2O_3$, $Sb_2O_4$, $Bi_2O_3$, $Nb_2O_5$, $Ta_2O_5$, etc. | Cluster ion distributions and correlation with fragment valence | Cluster ion distributions depend on the valence electron configuration of the metal in the oxide; correlation exists between $MO^+/M^+$ ratio and the bond dissociation energy of the $MO^+$ ion | Michiels and Gijbels [150] |
| $NH_4$, Na, K | Ammonium salts: $NH_4Cl$, $NH_4NO_3$, $(NH_4)_2SO_4$ | Alalysis of inorganic ammonium compounds in marine aerosol particles | Quantitative analysis of $NH_4$ of five samples given in mol % | Otten et al. [151] |
| Ga | GaAs | Nonresonant multiphoton ionization of neutrals | LM-MS combined with laser postionization enhances the absolute sensitivity by an order of magnitude | |
| Na, Al, Si, K, Ca, Fe, CaO, AlCaO, Ba, BaO, BaOH | NBS glass microspheres | Studying performance characteristics | Mass spectrum (see Fig. 80) | Simons [120] |
| $P/^{28}Si$, $(P + PO)/^{28}Si$ | Phosphosilicate glass | Quantitative analysis | Ion intensity ratios versus P concentration with RSD of 30 to 50% | Odom et al. [152] |
| $^{11}B/^{28}Si$ | Glass | Depth profile | Comparison with SIMS | |

**TABLE 50.** *(Contd.)*

| Elements | Matrices | Aims of Analyses | Results | References |
|---|---|---|---|---|
| Hg, Cd, Te | HgCdTe | Stoichiometric measurement | The sum of ion intensities of Hg and Cd is equal to Te ion intensities | |
| Na, AlO | HgCdTe | Surface analysis | Determination | |
| K, Ca, Sc, Ti, V, Cr, Mn, Fe, Ni, Co, Cu, Zn, Ga, Ge, As, Se, Rb, Si, Y, Zr, Nb, Mo, Ag, Cd, In, Sn, Sb, Te, Cs, Ba, etc. | NBS standard reference glass No. 611 | Application note | Information about analytical possibilities | Leybold-Heraeus [153] |
| $(Cs_{n+1}I_n)^+$ | Cs I | Kinetics of ion formation | Ion formation interpreted in terms of a nonequilibrium process | Schueler et al. [154] |
| $TiO_2$, $Ti_2O_3$, TiO | Powdered titanium oxides | Stoichiometric determination | O/Ti ratio differs enough to observe substantial differences | Michiels et al. [155] |
| $Ti^+$, TiO, $Ti_mO_n^+$, $TiO_2H^+$, $TiH^+$, $TiOH^+$ | $TiO_2$ | Ion kinetic energy measurement | Kinetic energy measurement has proven useful in developing a tentative model of the ion formation process for this specific sample | |
| Ca, CaF | $CaF_2$ | Applications of resonance ionization mass spectrometry | Performance in detectivity, dynamic range, study of desorption/ablation/ damage process, study of interfacial phenomena | Nogar et al. [156] |

| | | | | |
|---|---|---|---|---|
| W, Ta, Sb, Fe, Hf, Pd, Zn, Cu, U, Mn, Sn, Ti, Ni, Zr, Au, Ag, Sc, Al | Metals and alloys | Application of field desorption mass spectrometry | Qualitative pilot study, critical evaluation and outlook | Schulten and Muller [162] |
| H, B, C, O, CH, Na, Ti, Mg, Fe, Mn | Fe-Ti alloys | Comparison of LAMMA and SIMS/Auger measurements | Lateral resolution (with LAMMA) of about 0.5 $\mu$m and several 1000 Å depth resolution; SIMS of about 0.2 $cm^2$ and 3 to 10 Å in depth | Hammer et al. [163] |
| Fe, Cr, Ni, Mn, Si, Ti, V, Co, Cu, Zr, Mo, Nb, Sn, Ta, W, Pb | Steel | Application example | Crater diameter, 3 to 5 $\mu$m; crater depth, 0.1 to 0.3 $\mu$m; detection limit, 13 ppm for Ta | Leybold-Heraeus [164] |
| Ag, Au | Ag/Au-sandwich foils | Investigation of laser-induced damage, evaporation, and ionization with homogeneous inorganic target foils | Equilibrium ionization models and experimental values are in reasonable agreement | Fürstenau [165] |
| $^{54}Fe$, $^{56}Fe$, $^{63}Cu$, $^{65}Cu$, $^{64}Zn$, $^{66}Zn$, $^{67}Zn$, $^{68}Zn$ | Steel, brass | Quantitative analysis of isotopes with LMA 10-MS | RSD, 0.02 to 0.07%; for $\bar{x}$, 0.1 to 0.9% | Nýari et al. [102] |
| 23 elements | Ta | Quantitative determination | 0.1 to 4880 ppm; 0.25 to 7.71 ppm | Dietze and Becker [103] |
| Ni, NiO, $NiSO_4 \cdot 7H_2O$, NiS, $Ni_3Si_2$ | Ni particles from pollution sources | Particle analysis | Using molecular ions as a fingerprint for identification of compound stoichiometry (see Fig. 81) | Simons et al. [120, 166] |

**TABLE 50.** (*Contd.*)

| Elements | Matrices | Aims of Analyses | Results | References |
| --- | --- | --- | --- | --- |
| Mg, Ca, Fe, Cu, Ga | Al-Li alloys | Surface microanalysis | Determination of the impurities at intercellular boundaries (see Fig. 82) | |
| $CrO_n^-$ ($n = 1$ to 4) | $Fe_2O_3$, $SiO_2$, $Al_2O_3$ | In situ identification of chromium oxidation states to study dust formed during stainless steel machining | Cr(III) and Cr(VI) were estimated with a given methodology | Gondouin and Muller [167] |
| Cu, Ni<br>Ga, As, In | Cu-Ni alloy<br>$GaInAs_2$ | Application of statistics to analysis of binary and ternary compounds to estimate accuracy and reproducibility | Ratios of the elements can be measured with accuracies of 0.5%; results are in agreement with LTE model | Harris and Wallach [168] |
| $Ca^+$, $CaO^+$, CaOH, $Ca_2O^+$, Na, Al, K, Fe, and organic compounds | Thin metal sheet | Corrosion phenomena and metallic surface defects | Qualitative information is higher in comparison with other techniques (Raman, Auger, glow discharge), quantitative information for the major compounds are available | Krier and Muller [169] |

Application notes are given by Cambridge Mass Spectrometry Ltd. regarding calcite, C and N in silicate glasses, mineral grains, red mud, coal dust, explosives, contamination on microelectronic devices, photoresist on device, buried contamination in Al run, sidewall contact window, coatings, Al and Al alloys, superconducting wire, turbine blades, carbon fiber, and so on.

For further literature on geological applications: see Refs. 65, 66, 137–141.

For further literature about mass analysis of inorganic materials: see Refs. 65, 66, 157–161.

TABLE 51. **Laser Micro Mass Analyses of Organic Materials**

| Elements | Matrices | Aims of Analyses | Results | References |
|---|---|---|---|---|
| $^{44}Ca$ | Retina of crayfish | Quantitative LAMMA analysis of biological samples (standards and isotope labeling) | Reproducibility, $\leqslant 5\%$; lateral resolution, 2 to 3 $\mu$m; use of isotopes as a label for kinetic studies is tempting | Schröder [170] |
| $Li^+$, $Na^+$, $K^+$, $Rb^+$, $Cs^+$, $TL^+$ | Muscle cells | Localization of alkalimetal ions in living muscle cells | Accumulation of alkalimetal ions in sections of freeze-dried and plastic embedded biological material can be detected simultaneously with high sensitivity | Edelmann [171] |
| K/Na ratio | Cochlea wall | While in the perilymphatic space the K/Na ratio concentration corresponds to the usual composition of extracellular fluids (1:30); the K/Na ratio in the endolymphatic fluid is nearly reversed, to 30:1 | Profile of K/Na ratio across spiral ligament and stria vascularis before and after 3 min of anoxia | Orzulakova et al. [172] |
| $^{23}Na$, $^{39}K$, $^{41}K$, $^{40}Ca$, | Heart and skeletal | Test of optimal | K/Na-ratio determination | Hirche et al. [173] |

**TABLE 51.** *(Contd.)*

| Elements | Matrices | Aims of Analyses | Results | References |
|---|---|---|---|---|
| $^{24}Mg$, $^{20}Mg$, $^{26}Mg$ | muscle | preparation techniques, such as shock-freezing and freeze-drying to assure that the "natural" in situ state is preserved | wthin $\pm 20\%$; Na/Mg ratio cannot be interpreted properly; $^{40}Ca$ can be clearly discriminated | |
| Ca, Na, K, diethyl-2-hexylphthalate | Human skin | Studies of disturbances of $Ca^{2+}$ and $PO_4{}^{3}$ metabolism and accumulation of $CaHPO_4$ in the skin of dialysis patients; the verification of loci for Al deposits in tissues | Wide applicability of laser-induced microanalysis is demonstrated | Kupka et al. [174] |
| F | Dental hard tissue | Preparation methods and LAMMA analysis | Comparison of different analytical methods for determination of fluorine | Gabriel et al. [175] |
| U | Alga dunaliella | Localization and determination of the internal cellular uranium content in *Dunaliella* following addition of uranium carbonate to seawater | Confirmation of the presence of Si, P, K, Ca, Na, and U; one cell contained an average of $1.5 \times 10^{15}$ g U; on a cell diameter of 5 $\mu$m, this is a concentration of 24 ppm | Sprey and Bochem [176] |

| | | | | |
|---|---|---|---|---|
| Pb | Cells of chlorophyta | Lead accumulation of diverse algal species | Comparison with ECX (x-ray microanalysis) indicates the superior sensitivity of the LAMMA analysis | Lorch and Schäfer [177] |
| Lichexanthone | Laurea benguelensis | Studies of pigments in situ; distribution within thallus and fruit bodies; lichenology | Xanthones, anthraquinones, biosynthraquinones, usnic acid, pulvic acid derivates, and other substances (e.g., depsides and depsidones) may be identified | Mathey [178] |
| Na, K (ratio) | Macobacteria cells | Quantitation of Na and K and the Na/K ratio | Comparison with results obtained with AAS and NAA; standard deviation of the mean value) of the resulting $Na^+$ or $K^+$ contents was less than 5% | Seydel and Lindner [179] |
| | Vegetative bacteria cells of the genus Bacillus | Sample preparation technique for analysis; distinguishing between B. cereus and B. anthracis | The standard eviation of only ±10% observed for several of the signals seems to justify further efforts in this direction | Böhm [180] |

**TABLE 51.** *(Contd.)*

| Elements | Matrices | Aims of Analyses | Results | References |
|---|---|---|---|---|
| Ions of phospholipids | Escherichia coli bacteria | Information on molecular weight and chemical structure | Positive ion LD-mass spectra of three phospholipids (DPPE, DPPC, DPPG) and of dipalmitoylphosphatidic acid as well as negative ion LD-mass spectra of the fatty acid composition | Seydel and Lindner [181] |
| Al, Pb | Bone cells, bone | Localization of Al in bone of patients with dialysis osteomalacia; Pb distribution in cellular elements of bone marrow | Quantitative local and distribution analysis with precision of about 10 to 20% | Schmidt [182] |
| $^{23}Na^{+}$, $^{39}K^{+}$, $^{41}K^{+}$ (ratios) | E. coli treated with drug HN 32; single cells | Information on inorganic and organic composition of single bacterial cells differentiation (taxonomy) between various bacterial strains | Cation content (Na/K); fingerprint evaluation of molecular fragments; distinction between bacteria of different genera | Lindner and Seydel [183] |
| $Ca^{2+}$, $Na^{+}$, $K^{+}$, $Mg^{2+}$ | Phosphate and oxalate | Calcium cytochemistry, ultrastructural localization | Combination of OPA-LAMMA can localize Ca in subcellular compartments | Jacob et al. [184] |

| | | | | |
|---|---|---|---|---|
| Pt | Kidney tissue | Nephrotoxicity studies, platinum complexes, and antitumor activity; aluminum toxicity of drugs for patients with chronic renal failure | Pt concentration of 20 to 30 $\mu$g/g, present in the cells and not in the extracellular renal material; ultrastructural localization of aluminum in fresh liver and bone biopsies | Verbueken et al. [185] |
| Al, Fe, Na, K, Ca, Os, Pb | Liver and bone biopsies | | | |
| $^{127}I$ | Amiodarone drug administered to dogs and rats; analysis of lungs and lymph nodes, prepared using conventional methods | Subcellular localization of a drug by two independent methods: laser microprobe mass analysis and autoradiography | Iodine can be detected; matrix effects prevent the detection of heavy fragment ions of the drug | De Nollin et al. [186] |
| $^{23}Na^+$, $^{39}K^+$ (ratio) | Mycobacterium leprae | In vivo studies of drug kinetics; in virto studies of drug action on M. leprae | Single-cell mass analysis is possible; the influence of the drug on the $Na^+/K^+$ ratio is shown | Seydel et al. [187] |
| Cd | Kidney cortex of rats | Quantititative investigations on the distribution of elements and chemical compounds | Calibration by $I_{Cd}/I_{org\,mass}$ v. Cd concentration | Schmidt [188] |
| P | Histological sections | | P distribution by referring the mass line of $I_{PO}$ | |
| $CO_3^{2-}$ | Human urinary calculi | | Differences between struvite stones and apatite stones | |

**TABLE 51.** (*Contd.*)

| Elements | Matrices | Aims of Analyses | Results | References |
|---|---|---|---|---|
| Pb | Histological sections of tissues of poisoned rats | Localization in tissues | Better understanding of the relationships between total and subcellular tissue concentrations of toxic metals | Vandeputte et al. [189] |
| Elements of inorganic and organic microliths | Kidney sections | Identification | Calcium phosphate and calcium oxalate, organic moieties of amorphous and crystalline intrarenal deposits have been identified | Verbueken et al. [190] |
| Mass range 400 to 800 $m/e$; (M—H)anions | Rat bile, human bile | Comparison of LIMS spectra of those of fast atom bombardment (FAB) or liquid SIMS technique | Negative ion LIMS spectra and discussion | Hitzman and Odom [191] |
| K, Ca, Mg, Na | Fungal vesicles of Trifolium pratense L. endomycorrhizas | Elemental composition was to be determined | Results obtained with fixed and unfixed samples were compared | Strullu et al. [192] |
| Be | Epithelioid cell granulomas in affected tissue | Diagnosing chronic beryllium disease (CBD) | Detection of Be in 14 cases | Williams and Kelland [193] |
| Herbicide | Soyabean leaf | Observation of the location of herbicides on leaves | Spectra | Cambridge Mass Spectrometry [194] |

| | | | | |
|---|---|---|---|---|
| Mg, Al, Ca, Fe, Mn, Ba, Pb | Spruce needles | Investigation on healthy and acid rain–affected spruce needles | Developing of the preparation method; collapsed cells show high concentrations of Mn and Fe | Goossenaerts et al. [195] |
| $^{287}(M+H)^+$, $^{257}(M—H)^-$, $^{285}(M+H)^+$, $^{353}(M+H)^+$, $^{271}(M—H)^-$ | Cryosections of lichens | In situ investigation of morphology and chemistry of plants | Results confirmed the hypothesis | Mathey et al. [196] |
| $C_n^+$, $C_nH^+$, $C_nH_2^+$, $C_nH_3^+$, Na, K, AL, Co, $C_nH_m^{\pm}$, $C_nF_m^{\pm}$ | Polyethylene, polyvinyl-chloride, polytetra-fluoroethylene, polyphenyl-methacrylate, polybenzyl-methycrylate | Mass spectrometry of molecular solids | Characterization of both the hydrocarbon backbone of simple straight-chain polymers and of slight changes in side chain functionality | Gardella et al. [198] |
| Ionic compounds, nonionic compounds | Organic salts, alkali halides, inorganic complex salts | Ion formation in laser desorption mass spectrometry | Negative ions are desorbed with the same efficiency as positive ions; quasi-molecular ions are present | Heinen et al. [199, 200] |
| ACl components | Amino acids, peptides | Possibility to obtain spectra even of nonvolatile and/or thermally labile organic molecules | Classification scheme for spectra of these molecules as a function of their structural properties | Schiller et al. [201] |

**TABLE 51.** (*Contd.*)

| Elements | Matrices | Aims of Analyses | Results | References |
|---|---|---|---|---|
| | Leucine, tetracene | Studies about fragmentation and cross-linking processes of organic molecules as secondary processes of radiation damage | Analysis of crystalline organic films damaged by 100-keV electron irradiation | Bernsen et al. [202] |
| | Bis(1-phenyl-1,3-butanedionate)-oxovanadium; TTP, TTPNi, TTPCo, TTPCu, TTRRh$Cl_2$, TTP + $ZnCl_2$ | Analysis of organo-metallic compounds | Characterization of small amounts of organo-metallic nonvolatile compounds | Müller et al. [203] |
| | | Laser desorption techniques of nonvolatile organic substances | A survey of the state of development is given | Hillenkamp [204] |
| Diverse | Diverse | Organic analysis; analysis of polymers; analysis of organic layers on metal support | Application note for LAMMA 1000 | Leybold-Heraeus [205] |
| | Bis(dimethyl-amino)benzophenone, bianthrone, asparagine, glyco-cholic acid, stachyose, | Application of structural analysis | Spectra of organic com-pounds, general characteristics, organic salts, high-molecular-weight compounds, polymers; quantita- | Hercules et al. [206] |

| | | | | |
|---|---|---|---|---|
| | Triphenylarsen, triphenylantimon, triphenylwismut, triphenylgermanium, and cadmiumcyclo-hexabutyrat | Detection of metal-organic complexes out of organic matter | Positive and negative ion spectra were interpreted | Ollmann et al. [207] |
| Mg, Al, Si, K, Ca, Fe | Surfaces of asbestos fibers | Analysis of organic impurities at the surfaces of asbestos fibers | Adsorption of benzo[*a*]-pyrene on crocidolite from the gas phase and from a benzene solution, benzidine from aqueous solution and *N*, *N*-dimethylaniline | De Waele et al. [208] |
| | Tetramethylammonium chloride, leu-enkephalin, cyclodextrin | Laser desorption MS for nonvolatile organic molecules | Peptides, oligosaccharides, nucleotides, and other biologically important compounds, which are difficult to analyze by traditional methods, seem to be particularly amenable to laser desorption | Tabet and Cotter [209] |
| $M_mO_n^{\pm}$, $M^+$, $M^-$ | $M_xO_y$ | Cluster ion distributions and correlation with fragment valence in laser-induced mass spectra of oxides | The positive and negative cluster ion distribution show a strong correlation with the valence electron configuration of the metals in the oxide | Michiels and Gijbels [210] |

**TABLE 51.** (*Contd.*)

| Elements | Matrices | Aims of Analyses | Results | References |
|---|---|---|---|---|
| Fe<br>Co<br>Ni<br>Sn<br>TL<br>Pb<br>and other 20 elements | $C_{10}H_{14}FeO_4$<br>$Br_3Co(C_6H_5PF_2)_3$<br>$(C_5H_5)_2Ni$<br>$(C_8H_{17})_2SnCl_2$<br>$C_5H_5TL$<br>$Pb(CH_3COO)_2$<br>and 20 other inorganic compounds; hemoglobin; chlorophyll | Laser-assisted field desorption mass spectrometry of inorganic and organometallic compounds | Demonstration for molecular weight determination structural elucidation and purity control | Schulten et al. [211] |
| Co, Cs | Polyester resin | Quantitative analysis capabilities | Using the correction program ELOGET for calculation of electronic temperature and plasma density | Eloy [212] |
| $Al^-$, $Cl^-$, $SiO_2^-$, $SiO_3^-$, $^{166}(M—H)^-$, $^{122}(M—H—CO_2)^-$, $^{46}(NO_2)^-$ | Paranitrobenzoate | Example for the capability to ionize organic molecules with little fragmentation | Influence of the nature of the silver film for pNBA desorption (see Figs. 83 and 84) | Simons [120] |
| $^{30}CH_2{=}NH_2$<br>$^{44}H_2C—CH_2$ (with $NH_2$ on C)<br>$^{57}CH_2—CH(CH_3)_2$ | Leucine | Information on damage effects | | Bernsen et al. [213] |

| | | | | |
|---|---|---|---|---|
| $^{74}CH—COOH$ (with $NH_2$ on CH) <br> $^{86}R—CH{=}NH_2$ | | | | |
| $(M+H)^+$, $(M—H)^-$, $(M+A)^+$, $(M—A)^-$, $(C_4HO_4)^-$ | Oxocarbon squaric acid (3,4-dihydroxy-3-cyclobutene-1,2-dione) and its salts; $A_2C_4O_4$ (A = cation) | Laser desorption mass spectrometry | Positive and negative ion spectra, molecular weight information | Byrd et al. [214] |
| $(M+CH_3{}^+)$, $(M+CH_3{}^-)$ | Quaternary ammonium salts | Analysis of zwitterionic quaternary ammonium salts of different structural classes | A dominant intermolecular alkyl transfer reaction leading to "pair production" has been observed | Balasanmugam and Hercules [215] |
| $ML_3{}^+$, $ML_2X^+$, $MLX^+$, $ML^+$, $(L+H)^+$, $Ag(NO_3)_2{}^-$, $NO_3{}^-$, $OCN^-$, $CN^-$ et al., (M = metal cation, L = bpy or *o*-phen) | 2,2′-Bipyridine and *o*-phenanthroline complexes | Spectra | Structurally significant fragment ions are observed; the negative ion spectra provide information about the anion and the formal oxidation state of the central metal atom | Balasanmugam et al. [216] |
| $M^+$, $(M+H)^+$, $(M—H)^+$, $(M—H_2)^+$, $(M—H_3)^+$ | Polycyclic aromatic hydrocarbons (PAHs) | Positive and negative ion spectra; comparison between LAMMA 500 and LAMMA 1000 | Discussion of the spectra and the comparison of two instruments | Balasanmugam et al. [217] |

**TABLE 51.** *(Contd.)*

| Elements | Matrices | Aims of Analyses | Results | References |
|---|---|---|---|---|
| | Polyglucols with average molecular weights $\leqslant 1000$ | Determining molecular weight distribution of low-molecular-weight polymers by measuring the relative intensity of each oligomer | $\bar{M}_n$ values derived were reproducible to $\pm 2\%$ RSD | Mattern and Hercules [218] |
| Fragments | Polyfluoroethylenes | Analysis of polymers | The polymers investigated ranged from poly(vinyl-fluoride) to polytetra-fluoroethylene: fragmentation mechanism, quantitative study of structural fragments, comparison with $^{19}F$ NMR spectro-scopy | Mattern et al. [219] |
| Quasi-molecular ions | Aliphatic and aromatic amino acids | Mass spectra | In $\alpha$-$\beta$ fission, the phenyl-alanine quasi-molecular ion gives the $\alpha$-fragment, but no $\alpha$-fragment is observed for tryptophan | Parker and Hercules [220] |
| Four triphenylmethane dyes as contaminants | Polymer surface | Molecular map | 900 mass spectra; LM-MS is capable of providing both the chemical identity and the location of trace components | Novak et al. [221] |

| | | | | |
|---|---|---|---|---|
| Carbon cluster-type ions $C_n^-$ and $C_nH^-$ | Polycyclic aromatic hydrocarbons, the corresponding aza heterocyclic and oxygenated analogs | Approach for structural interpretation | A model for interpretation has been developed | Van Vaeck et al. [222] |
| Ions and neutral species | Phthalic and benzoic acid and alkaline salts | Metastable decay of laser-desorbed ions from aromatic organic compounds | Some new information about fragmentation and metastable decay after short-pulse UV laser desorption | Rosmarinowsky et al. [223] |
| Compounds | Gramicidin S (cyclic decapeptide) with molecular weight of 1141.49; and aluminum phthalocyanine chloride (APC) and oxidized insulin (mass $3496 \pm 1$) | Examples for mass spectrometry at high masses | Unlimited mass range | Dingle et al. [224] |

Further literature references on analyses of organic materials using beam techniques can be found in the *Proceedings of the 3rd International Laser Microprobe Mass Spectrometry Workshop* [105].

Further literature about application in biology and botany: see Ref. 197.

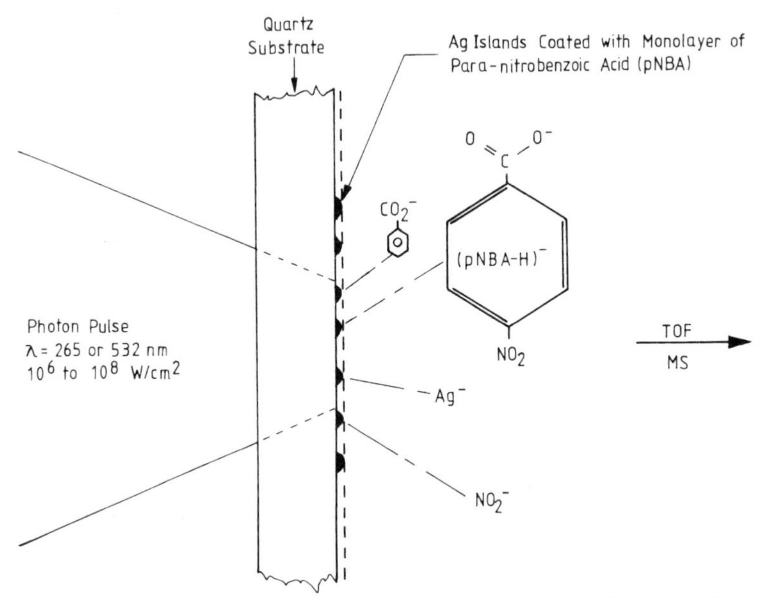


**Figure 83.** Schematic view of sample geometry for pNBA desorption from silver island film. From Ref. [120], reprinted by permission of D. S. Simons.

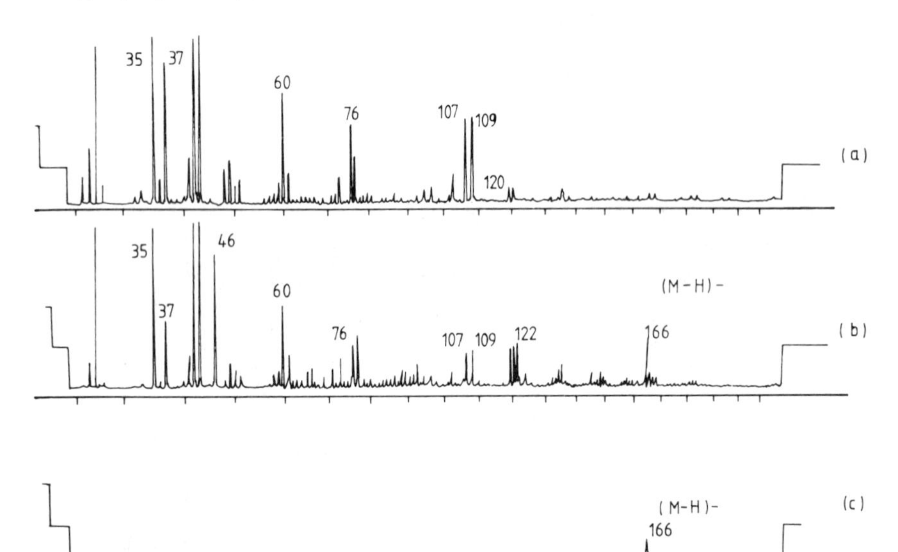


**Figure 84.** Negative ion mass spectra of (*a*) uncoated silver island film, (*b*) film coated with monolayer of *para*-nitrobenzoate, laser energy density about 400 mJ cm$^{-2}$, and (*c*) coated film as in (*b*), laser energy density about 100 mJ cm$^{-2}$. From Ref. [120], reprinted by permission of D. S. Simons.

### 9.4.2. Chemical Analysis of Organic Materials

Up-to-date applications in the fields of biology, medicine, and organic chemistry are described in Table 51.

## REFERENCES


1. R. E. Honig and J. R. Woolston, *Appl. Phys. Lett.*, **2**, 138 (1963); *Proc. IUPAC Meeting*, London, 1963, Paper Cl-17; R. E. Honig, *Appl. Phys. Lett.*, **3**, 8 (1963); *Proc. 12th Annual Conf. Mass Spectrometry*, Montreal, 1964, Paper 38; *Ann. N.Y. Acad. Sci.*, **137**, 262 (1966); *Proc. 15th Annual Conf. Mass Spectrometry*, Denver, 1967, subcommittee VII, part 2.
2. F. A. Giori, L. A. McKenzie, and E. J. McKinney, *Appl. Phys. Lett.*, **3**, 25 (1963).
3. J. Berkowitz and W. A. Chupka, *J. Chem. Phys.*, **40**, 2735 (1964).
4. K. A. Lincoln, *Proc. 12th Annual Conf. Mass Spectrometry*, Montreal, 1964, Paper 79; *Anal. Chem.*, **37**, 541 (1965); *Pure Appl. Chem. Suppl.*, 323 (1969); *Int. J. Mass Spectrom. Ion Phys.*, **2**, 75 (1969).
5. K. A. Lincoln and D. Werner, *15th Annual Conf. Mass Spectrometry*, Denver, Colo., 1967.
6. K. A. Lincoln and F. A. Wodley, *17th Annual Conf. Mass Spectrometry*, Dallas, Tex., 1969.
7. B. E. Knox and F. J. Vastola, *Chem. Eng. News*, **44**, 48 (1966).
8. B. E. Knox, *15th Annual Conf. Mass Spectrometry*, Denver, Colo., 1967, Paper 9; *Mater. Res. Bull.*, **3**, 329 (1968).
9. B. E. Knox and V. S. Ban, *Mater. Res. Bull.*, **3**, 337, 885 (1968); *16th Annual Conf. Mass Spectrometry*, Pittsburgh, Pa., 1968, Paper 127; *Bull. Am. Phys. Soc.*, **14**, 419 (1969).
10. S. I. Anisimov, A. M. Bonch-Bruevich, M. A. Elyashevich, Y. A. Imas, N. A. Pavlenko, and G. S. Romanov, *Sov. Phys.-Tech. Phys.*, **11**, 945 (1967), transl. from *Zh. Tekh. Fiz.*, **36**, 1273 (1966).
11. G. A. Askaryan and M. S. Rabinovich, Ionization and plasma generation by laser radiation, in *Proc. Lebedev Phys. Inst.*, Vol. 32, *Plasma Physics*, D. V. Skobeltsyn, Ed., Nauka Press, Moscow, 1966, pp. 76–83; transl. for Consultants Bureau, New York, 1968.
12. L. A. Arbuzova, V. A. Danilkin, Ya. A. Imas, V. A. Molchanov, and A. G. Mileshkin, *Ind. Lab. (USSR)*, **34**, 1440 (1968), transl. from *Zavod. Lab.*, **34**, 1199 (1968).
13. L. A. Arbuzova, A. M. Bonch-Bruevich, V. A. Danilkin, Y. A. Imas, V. A. Molchanov, and A. G. Mileshkin, *Metody Opred. Issled. Sostoyaniya Gazov Metal*, 269 (1968).
14. E. Bernal, L. P. Levine, and J. F. Ready, *Rev. Sci. Instrum.*, **37**, 938 (1966).
15. D. Damoth, *Proc. 2nd NATO Advanced Study Institute on Mass Spectrometry*, 1966, p. 49.

16. J. F. Eloy and J. L. Dumas, *Rev. Method. Phys. Anal.*, **2**, 251 (1966).
17. J. F. Eloy, P. Contamin, R. Stefani, and A. Cornu, *CEA-Conf.-1111*, Commissariat à l'Energie Atomique, Grenoble, C.E.N.G., 1968; *16th Annual Conf. Mass Spectrometry*, Pittsburgh, Pa., 1968.
18. J. F. Eloy, *Rev. Method. Phys. Anal.*, **4**, 161 (1968); **5**, 157 (1969); *Proc. 1st. Conf. International. Sourc. Ions*, Saclay, France, 1969, pp. 617–631.
19. N. C. Fenner and N. R. Daly, *Rev. Sci. Instrum.*, **37**, 1068 (1966); *Lasers Opt. Non Conv.*, **8**, 87 (1967); *J. Mater. Sci.*, **3**, 259 (1968).
20. N. C. Fenner, *Phys. Lett.*, **22**, 421 (1966).
21. N. C. Fenner, R. G. Ridley, R. W. D. Hardy, and N. R. Daly, *14th Annual Conf. Mass Spectrometry*, Dallas, Tex., 1966, Paper 25.
22. N. R. Isenor, *Appl. Phys. Lett.*, **4**, 152 (1963); *Can. J. Chem.*, **42**, 1413 (1964); *J. Appl. Phys.*, **36**, 316 (1965).
23. P. Langer, G. Tonon, F. Floux, and A. Ducauze, *IEEE J. Quantum Electron.*, **2**, 499 (1966).
24. P. Langer, B. Pin, and G. Tonon, *Rev. Phys. Appl.*, **3**, 405 (1968).
25. G. H. Megrue, *Science*, **157**, 1555 (1967); see also in *Recent Developments in Mass Spectroscopy*, K. Ogata and T. Hayakawa, Eds., University of Tokyo Press, Tokyo, 1970, pp. 654–656; *Proc. International Conf. Mass Spectroscopy*, Kyoto, 1969.
26. S. Namba, P. H. Kim, and A. Mitsuyama, *J. Appl. Phys.*, **37**, 3330 (1966).
27. J. F. Ready, E. Bernal, G. Levine, and L. P. Levine, *AD-477 231, Semiannual Report*, Nov. 1965; *AD-636 680*, Honeywell Research Center, 1966, NTIS.
28. J. F. Ready, *Ind. Res.*, **7**, 44 (1966); *Effects of High-Power Laser Radiation*, Academic Press, Inc., New York, 1971.
29. F. J. Vastola, A. J. Pirone, and B. E. Knox, *14th Annual Conf. Mass Spectrometry*, Dallas, Tex., 1966.
30. F. J. Vastola, A. J. Pirone, P. H. Given, and R. R. Dutcher, *Proc. 154th Meeting American Chemical Society*, Chicago, 1967, abstract L-59.
31. F. J. Vastola and A. J. Pirone, Ionization of Organic Solids by Laser Irradiation, in *Advances in Mass Spectrometry*, Vol. 4, E. Kendrick, Ed., Institute of Petroleum, London, 1968, pp. 107–111, presented at *4th International Mass Spectrometry Conf.*, Berlin, 1967.
32. F. J. Vastola, A. J. Pirone, and R. O. Mumma, *16th Annual Conf. Mass Spectrometry*, Pittsburgh, Pa., 1968, Paper 105; *Org. Mass Spectrom.*, **3**, 101 (1970).
33. F. J. Vastola, *Appl. Spectrosc.*, **22**, 374 (1968); and *7th Natl. Meeting Society for Applied Spectroscopy*, Chicago, 1968, Paper 172; *18th Annual Conf. Mass Spectrometry*, San Francisco, 1970, Paper J.9.
34. V. S. Ban and B. E. Knox, *Int. J. Mass Spectrom. Ion Phys.*, **3**, 131 (1969); *J. Chem. Phys.*, **51**, 524 (1969); *17th Annual Conf. Mass Spectrometry*, Dallas, Tex., 1969, Paper 22; *J. Chem. Phys.*, **52**, 243, 248 (1970).
35. V. S. Ban, Mass spectrometric studies of laser induced vaporization of solids,

Thesis, Pennsylvania State University, University Park, Pa., 1969, University Microfilms 70–655.

36. Yu. A. Bykovskii, V. I. Dorofeev, V. I. Dymovich, B. I. Nikolaev, S. V. Ryzhikh, and S. M. Silnov, *Sov. Phys.-Tech. Phys.*, **13**, 986 (1969), transl. from *Zh. Tekh. Fiz.*, **38**, 1194 (1968); *Sov. Phys.-Tech. Phys.*, **14**, 955 (1970), transl. from *Zh. Tekh. Fiz.*, **39**, 1272 (1969).
37. Yu. A. Bykovskii, A. G. Duboladov, N. N. Degtyarenko, V. F. Elesin, Yu. P. Kozyrev, and I. N. Nikolaev, *Zh. Eksp. Teor. Fiz.*, **56**, 1819 (1969).
38. Yu. A. Bykovskii, N. N. Degtyarenko, V. I. Dymovich, V. F. Elesin, Yu. P. Kozyrev, B. I. Nikolaev, S. V. Ryzhikh, and S. M. Silnov, *Sov. Phys.-Tech. Phys.*, **14**, 1269 (1970), transl. from *Zh. Tekh. Fiz.*, **39**, 1694 (1969).
39. Yu. A. Bykovskii, N. N. Degtyarenko, V. F. Elesin, Yu. P. Kozyrev, and S. M. Silnov, *Radiophys. Quantum Electron.*, **13**, 703 (1970), transl. from *Izv. Vyssh. Uchebn. Zaved. Radiofiz.*, **13**, 891 (1970); *Sov. Phys.-Tech. Phys.*, **15**, 2020 (1971), transl. from *Zh. Tekh. Fiz.*, **40**, 2578 (1970).
40. Yu. A. Bykovskii, V. I. Dymovich, Yu. P. Kozyrev, V. N. Nevolin, and S. M. Silnov, *Sov. Phys.-Tech. Phys.*, **15**, 1877 (1971), transl. from *Zh. Tekh. Fiz.*, **40**, 2401 (1970).
41. P. Contamin, A. Cornu, J. F. Eloy, and R. Stefani, *16th Annual Conf. Mass Spectrometry*, Pittsburgh, Pa., 1968.
42. J. L. Dumas, *Rev. Method. Phys. Anal.*, **1**, 47 (1967); *Rev. Phys. Appl.*, **5**, 795 (1970); Étude de la Photoionisation des Cibles Métalliques en Vue d'Application à la Spectrométrie de Masse, Thesis, University of Grenoble, Grenoble, France, 1970.
43. W. Demtröder and W. Jantz, *Plasma Phys.*, **12**, 691 (1970).
44. M. O. Hobbs, A. J. Getzkin, and R. A. Meyer, *16th Annual Conf. Mass Spectrometry*, Pittsburgh, Pa., 1968, Paper 134.
45. A. F. Haught, D. H. Polk, and W. J. Fader, *Phys. Fluids*, **13**, 2842 (1970).
46. F. S. Karn and J. Singer, *Fuel*, **47**, 235 (1968).
47. F. S. Karn, R. A. Friedel, and A. G. Sharkey, *Carbon*, **5**, 25 (1967); *Fuel*, **47**, 193 (1968); **48**, 297 (1969); *Chem. Ind.*, (*London*), 239 (1970).
48. F. S. Karn, A. G. Sharkey, A. F. Logar, and R. A. Friedel, Coal investigations using laser irradiation, *PB-190 031*, U.S. Dept. of Interior, Bureau of Mines, 1970, N 70-15659, BM-RI-7328.
49. G. F. Ivanovskii, L. M. Blyumkin, S. V. Varnakov, and L. P. Lisovskii, *Ind. Lab.* (*USSR*), **34**, 1524 (1968), transl. from *Zavod. Lab.*, **34**, 1263 (1968).
50. G. F. Ivanovskii and S. V. Varnakov, *Ind. Lab.* (*USSR*), **35**, 1151 (1969), transl. from *Zavod. Lab.*, **35**, 959 (1969).
51. W. K. Joy, W. R. Ladner, and E. Pritchard, *Nature* (*London*), **217**, 640 (1968); *Fuel*, **49**, 26 (1970).
52. S. H. Kahn, F. A. Richards, and D. Walsh, *IEEE J. Quantum Electron.*, **1**, 359 (1965).
53. V. G. Mossotti, D. W. Golightly, and W. C. Phillips, *Proc. 15th CSI*, Madrid, 1969.

54. H.-J. Dietze and H. Zahn, *Exp. Tech. Phys.*, **20**, 389 (1972).
55. H. Zahn and H.-J. Dietze, *Exp. Tech. Phys.*, **20**, 401 (1972).
56. H. Moenke and L. Moenke-Blankenburg, *Mikrochim. Acta (Wien), Suppl.*, *5*, 377 (1974), presented at *6th Kolloquium Metallkundliche Analyse*, Vienna, Austria, 1972; L. Moenke-Blankenburg, *Nouv. Rev. Opt. Appl.*, **3**, 243 (1972).
57. C. DeMichelis, *IEEE J. Quantum Electron.*, **6**, 630 (1970).
58. R. E. Honig, in *Laser Interaction and Related Plasma Phenomena*, Vol. 1, H. J. Schwarz and H. Hora, Eds., Plenum Press, New York, 1971, pp. 85–108.
59. B. E. Knox, in *Dynamic Mass Spectrometry*, Vol. 2, D. Price, Ed., Heyden & Son Ltd., London, 1971, Chap. 2, pp. 61–96; Laser ion source analysis of solids, in *Trace Analysis by Mass Spectrometry*, A. J. Ahearn, Ed., Academic Press, Inc., New York, 1972, Chap. 14, pp. 423–444.
60. E. C. Beahm, An investigation of the laser source mass spectrometer, Thesis, Pennsylvania State University, University Park, Pa., 1973, University Microfilms 74–4215.
61. J. F. Eloy, Applications des Sources d'Ions à Laser à l'Analyse des Solides par Spectrométrie de Masse, in *Précis de Spectrométrie de Masse Analytique*, A. Cornu, Ed., Presses Universitaires de Grenoble, St. Martin d'Hères, France 1975, ISBN 2706100516, Chap. 9, pp. 381–409.
62. C. Cuna, D. Ioanoviciu, and S. Cuna, *Stud. Cercet. Fiz.*, **28**, 481 (1976).
63. G. A. Maksimov and N. V. Larin, *Russ. Chem. Rev.*, **45**, 1091 (1976), transl. from *Usp. Khim.*, **45**, 2121 (1976).
64. I. D. Kovalev, G. A. Maksimov, A. I. Suchkov, and N. V. Larin, *Int. J. Mass Spectrom. Ion Phys.*, **27**, 101 (1978).
65. R. J. Conzemius and J. M. Capellen, *Int. J. Mass Spectrom. Ion Phys.*, **34**, 197 (1980).
66. J. M. Capellen and R. J. Conzemius, *Bibliography of Publications on Laser Interaction with Solids and Laser Mass Spectrometry of Solids*, IS-4715, Ames Laboratory, Iowa State University, Ames, Iowa, 1980; R. J. Conzemius, D. S. Simons, Zhao Shankai, and G. D. Byrd, in *Microbeam Analysis—1983*, Ron Gooley, Ed., San Francisco Press, Inc., San Francisco, 1983, pp. 301–328.
67. F. Hillenkamp, E. Unsöld, R. Kaufmann, and R. Nitsche, *Appl. Phys.*, **8**, 341 (1975).
68. R. Wechsung, F. Hillenkamp, R. Kaufmann, R. Nitsche, and H. Vogt, *Mikroskopie (Wien)*, **34**, 47 (1978).
69. R. Wechsung, German patent P 2540 505.4-33.
70. J. F. Eloy and E. Unsöld, European patent 78.400.174.5; see also R. Stefani, *Analusis*, **16**, 147 (1988).
71. R. Kaufmann, F. Hillenkamp, and R. Wechsung, *Eur. Spectrosc. News*, **20**, 41 (1978).
72. J. F. Eloy, P. Curie, and M. Curie, *Proc. Seminary Méthodes nouvelles d'etudes des solides minéraux*, Paris, 1979.
73. R. Kaufmann and F. Hillenkamp, *Ind. Res. Dev.*, 145 (1979).

74. R. Kaufmann, F. Hillenkamp, R. Wechsung, H. J. Heinen, and M. Schürmann, *Scanning Electron Microsc.*, **2**, 279 (1979).

75. N. Fürstenau, F. Hillenkamp, and P. Nitsche, *Int. J. Mass Spectrom. Ion Phys.*, **31**, 85 (1979).

76. T. Floren, Diploma thesis, Frankfurt, FRG, 1980.

77. *Proc. Workshop on Ion Formation from Organic Solids*, Münster, FRG, 1980.

78. *Proceedings of the First LAMMA-Symposium*, Düsseldorf, FRG, 1980, *Fresenius Z. Anal. Chem.*, **308**(H. 3), 193–320 (1981).

79. N. Fürstenau, Doctoral thesis, Frankfurt, FRG, 1981.

80. E. Deloule and J. F. Eloy, *Chem. Geol.*, **37**, 191 (1982).

81. K. F. J. Heinrich, Ed., *Microbeam Analysis—1982*, San Francisco Press, Inc., San Francisco, 1982.

82. A. Chamel and J. F. Eloy, *Scanning Electron Microsc.*, 841 (1983).

83. *Proceedings of the 2nd LAMMA-Symposium*, Borstel, FRG, 1983, Leybold-Heraeus GmbH, Köln, FRG, 1984.

84. A. Hartford, Jr., *Pure Appl. Chem.*, **56**, 1555 (1984).

85. D. A. McCrery, E. B. Ledford, Jr., and M. L. Gross, *Anal. Chem.*, **54**, 1435 (1982).

86. R. Viswanathan, D. R. Burgess, Jr., P. C. Stair, and E. Weitz, *J. Vac. Sci. Technol.*, **20**, 605 (1982).

87. G. Wedler and H. Ruhmann, *Surf. Sci.*, **121**, 464 (1982).

88. E. Onyiriuka, R. L. White, D. A. McCrery, M. L. Gross, and C. L. Wilkins, *Int. J. Mass Spectrom. Ion Phys.*, **46**, 135 (1983).

89. R. B. Van Breeman, M. Snow, and R. J. Cotter, *Int. J. Mass Spectrom. Ion Phys.*, **49**, 35 (1983).

90. G. J. Q. van der Peyl, W. J. van der Zande, V. Bederski, A. J. H. Boerboom, and P. G. Kistemaker, *Int. J. Mass Spectrom. Ion Phys.*, **47**, 7 (1983).

91. R. J. Cotter, *Anal. Chem.*, **56**, 455 A (1984).

92. R. B. Hall and A. M. DeSantalo, *Surf. Sci.*, **137**, 421 (1984).

93. M. G. Sherman, J. R. Kingsley, D. A. Dahlgren, J. C. Hemminger, and R. T. McIver, Jr., *Surf. Sci.*, **148**, 125 (1985).

94. D. W. Beekman, T. A. Callcott, S. D. Kramer, E. T. Arakuwa, and G. S. Hurst, *Int. J. Mass Spectrom. Ion Phys.*, **34**, 39 (1980).

95. V. I. Balikin, G. J. Bekov, V. S. Letokhov, and V. I. Mishin, *Usp. Fiz. Nauk*, **132**, 293 (1980).

96. C. M. Miller, N. S. Nogar, A. J. Ganzarz, and W. R. Shield, *Ann. Chem.*, **54**, 2377 (1982).

97. J. P. Young and D. L. Donohme, *Anal. Chem.*, **55**, 88 (1983).

98. D. L. Donohne and J. P. Young, *Anal. Chem.*, **55**, 378 (1983).

99. N. S. Nogar, R. K. Sandler, and S. W. Downey, *Proc. SPIE, Los Alamos Conf. Optics*, 1983, p. 192.

100. J. E. Parks, H. W. Schmitt, G. S. Hurst, and W. M. Fairbank, *Thin Solids Films*, **108**, 69 (1983).
101. T. B. Lucatorto, C. W. Clark, and L. I. Moore, *Opt. Commun.*, **48**, 406 (1984).
102. I. Nyári, J. Frecska, L. Matus, and I. Opauszky, *Proc. 3rd Conf. SSMS and SIMS*, Soũs, Jizerské Hora, Czechoslovakia, 1984, p. 22.
103. H.-J. Dietze and S. Becker, *ZFI-Mitt.*, **101**, 73 (1985).
104. R. Kaufmann, *LIMS, Reference and Abstract Index*, University of Düsseldorf, Düsseldorf, FRG, 1986.
105. *Proc. 3rd International Laser Microprobe Mass Spectrometry Workshop*, University of Antwerp, Belgium, 1986.
106. V. P. Zakharov and I. M. Protas, *Instrum. Exp. Tech.* (*USSR*), **16**, 846 (1973), transl. from *Prib. Tekh. Eksp.*, **3**, 162 (1973).
107. G. G. Devyatykh, N. V. Larin, G. A. Maksimov, and A. I. Suchkov, *Russ. J. Phys. Chem.*, **47**, 1638 (1973), transl. from *Zh. Fiz. Khim.*, **47**, 2917 (1973); *J. Anal. Chem. USSR*, **29**, 1313 (1974), transl. from *Zh. Anal. Khim.*, **29**, 1516 (1974); *J. Anal. Chem. USSR*, **30**, 560 (1975), transl. from *Zh. Anal. Khim.*, **30**, 664 (1975).
108. G. G. Devyatykh, B. A. Nesterov, G. A. Maksimov, and N. V. Larin, *Sov. Tech. Phys. Lett.*, **1**, 152 (1975), transl. from *Pis'ma Zh. Tekh. Fiz.*, **1**, 318 (1975).
109. G. G. Devyatykh, S. V. Gaponov, I. D. Kovalev, N. V. Larin, W. I. Luchin, G. A. Maksimov, L. I. Pontus, and A. I. Suchkov, *Sov. Tech. Phys. Lett.*, **2**, 356 (1976), transl. from *Pis'ma Zh. Tekh. Fiz.*, **2**, 906 (1976).
110. R. A. Bingham and P. L. Salter, *Int. J. Mass Spectrom. Ion Phys.*, **21**, 133 (1976); *Anal. Chem.*, **48**, 1735 (1976).
111. R. J. Conzemius and H. J. Svec, *Proc. 26th Annual Conf. Mass Spectrometry*, St. Louis, M. 1978, Paper RB-13; *Anal. Chem.*, **50**, 1854 (1978).
112. T. Dingle and B. W. Griffiths, in *Microbeam Analysis—1984*, A. D. Romig, Jr., and J. I. Goldstein, Eds., San Francisco Press, Inc., San Francisco, 1984, pp. 23–26.
113. T. Dingle, B. W. Griffith, and J. C. Ruckman, *Vacuum*, **31**, 571 (1981).
114. A. I. Busygin and B. Sh. Ulmasbaev, *Instrum. Exp. Tech.* (*USSR*), **21**, 171 (1978).
115. R. Kaufmann, in *Microbeam Analysis—1982*, K. F. J. Heinrich, Ed., San Francisco Press, Inc., San Francisco, 1982, p. 341.
116. H. J. Heinen, S. Meier, H. Vogt, and R. Wechsung, *Int. J. Mass Spectrom. Ion Phys.*, **47**, 19 (1983).
117. R. Kaufmann, F. Hillenkamp, R. Wechsung, H. J. Heinen, and M. Schürmann, *Scanning Electron Microsc.*, 279 (1979).
118. P. Surkyn and F. Adams, *J. Trace Microprobe Tech.*, **1**, 79 (1982).
119. J. F. Eloy, *Proc. 5th International Symposium on High Purity Material in Science and Technology*, Dresden, GDR, 1980, Vol. II, p. 96.
120. D. S. Simons, *Appl. Surf. Sci.* (1987).
121. R. J. Cotter, *Proc. 3rd International Laser Microprobe Mass Spectrometry Workshop*, Antwerp, Belgium, 1986, p. 49.

122. *LAMMA, Laser Micro Mass Analyser*, Leybold-Heraeus GmbH, Köln, FRG.
123. *LIMA, Laser Ionisation Mass Analyser*, Cambridge Mass Spectrometry Ltd., Cambridge, England.
124. *FTMS-2000 LC, Fourier Transform Mass Spectrometer Laser Desorption*, Nicolet Analytical Instruments, Madison, Wis.
125. *LAMMS, IX 23L Laser Microprobe, IX 23S Time-of-Flight SIMS, IX 23F Field-Ion Atom Probe Microanalyzer*, VG Ionex Ltd., Burgess Hill, West Sussex, England.
126. K. R. Spurny, J. Schörmann, and R. Kaufmann, *Fresenius Z. Anal. Chem.*, **308**, 274 (1981).
127. J. M. Beusen, P. Surkyn, R. Gijbels, and F. Adams, *Spectrochim. Acta*, **38B**, 843 (1983).
128. H. H. Weinke, E. Michiels, and R. Gijbels, *Int. J. Mass Spectrom. Ion Phys.*, **47**, 43 (1983).
129. P. van Espen, J. de Waele, E. Vasant, and F. Adams, *Int. J. Mass Spectrom. Ion Phys.*, **46**, 515 (1983).
130. J. F. Eloy, *Scanning Electron Microsc.*, **11**, 1915 (1984).
131. J. F. Eloy, M. Lebu, and E. Unsöld, *Int. J. Mass Spectrom. Ion Phys.*, **47**, 19 (1983).
132. S. Akyüz, T. Akyüz, J. K. De Waele, and F. C. Adams, *Proc. 3rd International Laser Microprobe Mass Spectrometry Workshop*, Antwerp, Belgium, 1986, p. 11.
133. D. S. Simons, M. Lindner, and D. A. Leich, *Proc. 3rd International Laser Microprobe Mass Spectrometry Workshop*, Antwerp, Belgium, 1986, p. 183.
134. M. Lindner, D. A. Leich, R. J. Borg, G. P. Russ, J. M. Bazan, D. S. Simons, and A. R. Date, *Nature (London)*, **320**, 246 (1986).
135. D. S. Simons, *Int. J. Mass Spectrom. Ion Phys.*, **55**, 15 (1983/84).
136. H. Yokozawa, T. Kikuchi, K. Furuya, S. Ando, and K. Hoshino, *Proc. 3rd International Laser Microprobe Mass Spectrometry Workshop*, Antwerp, Belgium, 1986, p. 221.
137. F. Freund, H. Kathrein, H. Wengeler, R. Knobel, and H. J. Heinen, *Geochim. Cosmochim. Acta*, **44**, 1319 (1980).
138. S. Henstra, E. B. A. Bisdom, A. Jongerius, H. J. Heinen, and S. Meier, *Beitr. Elektronenmikrosk. Direktabb. Oberfl.*, **13**, 63 (1980); see also *Fresenius Z. Anal. Chem.*, **308**, 280 (1981).
139. W. Herr, P. Englert, U. Herpers, E. A. Watts, and G. A. Whittaker, *Meteoritics*, **15**, 300 (1980).
140. E. B. A. Bisdom, S. Henstra, A. Jongerius, H. J. Heinen, and S. Meier, *Neth. J. Agric. Sci.*, **29**, 23 (1981).
141. P. K. Dutta and Y. Talmi, *Fuel*, **61**, 1241 (1982).
142. J. F. Eloy, J. C. Brunn, P. Baker, J. P. Alibert, and M. Dipuy, paper given at *Journées d'Étude des Traitements per Faisceau d'Énergie*, Grenoble, France, May 27–28, 1982.

143. D. L. Lowndes, R. F. Wood, and J. Narayan, *Phys. Rev. Lett.*, **52**, 561 (1984).
144. G. J. Edge, D. D. Hall, and D. E. Sykes, *Proc. ECASIA 85*, Veldhoven, The Netherlands, 1985, p. 93.
145. N. S. Clarke, J. C. Ruckman, and R. R. Davey, *Proc. ECASIA 85*, Veldhoven, The Netherland,s 1985, p. 23; *Surf. Interface Anal.*, **9**, 31 (1986).
146. E. Michiels, A. Celis, and R. Gijbels, in *Microbeam Analysis 1982*, San Francisco Press, Inc., San Francisco, 1982, p. 383; see also *Int. J. Mass Spectrom. Ion Phys.*, **47**, 23 (1983).
147. J. F. Eloy, *J. Phys.* (*Suppl. 2*), **45**, C2-265 (1984).
148. B. Jöst, B. Schueler, and F. R. Krueger, *Z. Naturforsch.*, **37A**, 18 (1982).
149. M. Balooch and D. R. Olander, *Int. J. Mass Spectrom. Ion Phys.*, **51**, 155 (1983).
150. E. Michiels and R. Gijbels, *Anal. Chem.*, **56**, 1115 (1984).
151. Ph. Otten, F. Bruynseels, and R. Van Grieken, *Proc. 3rd International Laser Microprobe Mass Spectrometry Workshop*, Antwerp, Belgium, 1986, p. 159.
152. R. W. Odom, C. J. Hitzman, and B. W. Schueler, *Mat. Res. Soc. Symp. Proc.*, **69**, 265 (1986).
153. Leybold-Heraeus GmbH, Köln, FRG, *Application 11*, 1982.
154. B. Schueler, F. R. Kruger, and P. Feigl, *Int. J. Mass Spectrom. Ion Phys.*, **47**, 3 (1983).
155. E. Michiels and R. Gijbels, *Spectrochim. Acta*, **38B**, 1347 (1983); E. Michiels, T. Mauney, F. Adams, and R. Gijbels, in *Laser Microprobe Mass Analysis* (*LAMMA*), Part 2: *Titanium Dioxide*, manuscript 1985.
156. N. S. Nogar, E. C. Apel, C. M. Miller, and R. C. Estler, *Proc. Conf. Laser Applications to Chemical Analysis*, Lake Tahoe, Nev., 1987, p. 96.
157. F. J. Bruynseels and R. E. van Grieken, *Anal. Chem.*, **56**, 871 (1984); *Spectrochim. Acta*, **38B**, 853 (1983).
158. P. Wieser, R. Wurster, and U. Hass, *Fresenius Z. Anal. Chem.*, **308**, 260 (1981).
159. U. Haas, P. Wieser, and R. Wurster, *Fresenius Z. Anal. Chem.*, **308**, 270 (1981).
160. T. Dingle and B. W. Griffiths, *Proc. ECASIA 85*, Veldhoven, The Netherlands, 1985, p. 86; in *Microbeam Analysis—1985*, J. T. Armstrong, Ed., San Francisco Press, Inc., San Francisco, 1985, p. 315.
161. M. J. Southon, A. Harris, V. Kohler, S. J. Mullock, and E. R. Wallach, *Proc. ECASIA 85*, Veldhoven, The Netherlands, 1985, p. 92.
162. H. R. Schulten and R. Muller, *Fresenius Z. Anal. Chem.*, **304**, 15 (1980).
163. E. Hamer, W. Gerhard, C. Plog, and R. Kaufmann, *Fresenius Z. Anal. Chem.*, **308**, 287 (1981).
164. Leybold-Heraeus GmbH, Köln, FRG, *Application 10*, 1982.
165. N. Fürstenau, *Fresenius Z. Anal. Chem.*, **308**, 201 (1981).
166. I. H. Musselman, R. W. Linton, and D. S. Simons, in *Microbeam Analysis—1985*, J. T. Armstrong, Ed., San Francisco Press, Inc., San Francisco, 1985, p. 337.

167. S. Gondouin and J. F. Muller, *Proc. 3rd International Laser Microprobe Mass Spectrometry Workshop*, Antwerp, Belgium, 1986, p. 87.

168. A. Harris and E. R. Wallach, *Proc. 3rd International Laser Microprobe Mass Spectrometry Workshop*, Antwerp, Belgium, 1986, p. 95.

169. G. Krier and J. F. Muller, *Proc. 3rd International Laser Microprobe Mass Spectrometry Workshop*, Antwerp, Belgium, 1986, p. 133.

170. W. H. Schröder, *Fresenius Z. Anal. Chem.*, **308**, 212 (1981).

171. L. Edelmann, *Fresenius Z. Anal. Chem.*, **308**, 218 (1981); *Scanning Electron Microsc.*, 875 (1984); *Physiol. Chem. Phys. Med. NMR*, **16**, 499 (1984).

172. A. Orzulakova, R. Kaumann, C. Morgenstern, and M. D'Haese, *Fresenius Z. Anal. Chem.*, **308**, 224 (1981).

173. H. Hirche, J. Heinrichs, H. E. Schaefer, and M. Schramm, *Fresenius Z. Anal. Chem.*, **308**, 224 (1981).

174. K. D. Kupka, W. W. Schropp, C. Schiller, and F. Hillenkamp, *Fresenius Z. Anal. Chem.*, **308**, 229 (1981).

175. E. Gabriel, Y. Kato, and P. J. Rech, *Fresenius Z. Anal. Chem.*, **308**, 234 (1981).

176. B. Sprey and H. P. Bochem, *Fresenius Z. Anal. Chem.*, **308**, 239 (1981).

177. D. W. Lorch and H. Schäfer, *Fresenius Z. Anal. Chem.*, **308**, 246 (1981).

178. A. Mathey, *Fresenius Z. Anal. Chem.*, **308**, 249 (1981).

179. U. Seydel and B. Lindner, *Fresenius Z. Anal. Chem.*, **308**, 253 (1981); *Int. J. Quantum Chem.*, **XX**, 505 (1981).

180. R. Böhm, *Fresenius Z. Anal. Chem.*, **308**, 258 (1981).

181. U. Seydel and B. Lindner, *Proc. 2nd LAMMA Symposium*, Borstel, FRG, 1983, p. 91; U. Seydel, B. Lindner, H.-W. Wollenweber, and E. T. Rietschel, *Eur. J. Biochem.*, **145**, 505 (1984).

182. P. F. Schmidt, *Proc. 2nd LAMMA Symposium*, Borstel, FRG, 1983, p. 99.

183. B. Lindner and U. Seydel, *Proc. 2nd LAMMA Symposium*, Borstel, FRG, 1983, p. 111.

184. W. A. Jacob, M. De Smedt, M. Borgers, and S. De Nollin, *Proc. 2nd LAMMA Symposium*, Borstel, FRG, 1983, p. 119.

185. A. H. Verbueken, G. J. Paulus, F. L. Van de Vyver, G. A. Verpooten, W. J. Visser, M. E. De Broe, and R. E. Van Grieken, *Proc. 2nd LAMMA Symposium*, Borstel, FRG, 1983, p. 129.

186. S. De Nollin, W. Jacob, L. Van Vaeck, Y. Berger, W. Cautreels, B. Remandet, and P. Vic, *Proc. 3rd International Laser Microprobe Mass Spectrometry Workshop*, Antwerp, Belgium, 1986, p. 53.

187. U. Seydel, B. Lindner, and A. M. Dhople, *Int. J. Lepr.*, **53**, 365 (1985); B. Lindner, G. Dethlefs-Bubritzki, and U. Seydel, *Proc. 3rd International Laser Microprobe Mass Spectrometry Workshop*, Antwerp, Belgium, 1986, p. 139.

188. P. F. Schmidt, *Proc. 3rd International Laser Microprobe Mass Spectrometry Workshop*, Antwerp, Belgium, 1986, p. 169; P. F. Schmidt, R. Barckhaus, and W.

Kleimeier, *Trace Elem. Med.*, **3**, 19 (1986); P. F. Schmidt, B. Hagen, and D. B. Leusmann, in *Microbeam Analysis—1985*, J. T. Armstrong, Ed., San Francisco Press, Inc., San Francisco, 1985, p. 331.

189. D. F. Vandeputte, A. H. Verbueken, W. A. Jacob, M. E. De Broe, and R. E. Van Grieken, *Proc. 3rd International Laser Microprobe Mass Spectrometry Workshop*, Antwerp, Belgium, 1986, p. 191.
190. A. H. Verbueken, G. A. Verpooten, E. J. Nouwen, M. E. De Broe, and R. E. Van Grieken, *Proc. 3rd International Laser Microprobe Mass Spectrometry Workshop*, Antwerp, Belgium, 1986, p. 207.
191. C. J. Hitzman and R. W. Odom, *Proc. International Conf. Lasers '84*, 1984, p. 89.
192. D. G. Strullu, A. Chamel, J. F. Eloy, and J. P. Gourret, *New Phytol.*, **94**, 81 (1983).
193. W. J. Williams and D. Kelland, *J. Clin. Pathol.*, **39**, 900 (1986).
194. Cambridge Mass Spectrometry Ltd., *Application Notes.*, 1987.
195. C. H. Goossenaerts, A. H. Verbueken, R. E. Van Grieken, W. A. Jacob, and H. J. Van Praag, *Proc. 3rd International Laser Microprobe Mass Spectrometry Workshop*, Antwerp, Belgium, 1986, p. 91.
196. A. Mathey, L. Van Vaeck, and W. Steglich, *Proc. 3rd International Laser Microprobe Mass Spectrometry Workshop*, Antwerp, Belgium, 1986, p. 145.
197. L. Moenke-Blankenburg, *Prog. Anal. Spectrosc.*, **9**, 335 (1986), Refs. 552–573.
198. J. A. Gardella, Jr., D. M. Hercules, and H. J. Heinen, *Spectrosc. Lett.*, **13**, 347 (1980).
199. H. J. Heinen, H. Vogt, and R. Wechsung, *Fresenius Z. Anal. Chem.*, **308**, 290 (1981).
200. H. J. Heinen, *Int. J. Mass Spectrom. Ion Phys.*, **38**, 309 (1981).
201. C. Schiller, K. G. Kupke, and F. Hillenkamp, *Fresenius Z. Anal. Chem.*, **308**, 304 (1981).
202. P. Bernsen, P. F. Schmidt, and L. Reimer, *Fresenius Z. Anal. Chem.*, **308**, 309 (1981).
203. J. F. Müller, C. Berthé, and J. M. Magar, *Fresenius Z. Anal. Chem.*, **308**, 312 (1981).
204. F. Hillenkamp, *Int. J. Mass Spectrom. Ion Phys.*, **45**, 305 (1982).
205. Leybold-Heraeus GmbH, Köln, FRG. *Application 12–14*, 1982.
206. D. M. Hercules, R. J. Day, K. Balasanmugam, T. A. Dang, and C. P. Li, *Anal. Chem.*, **54**, 280A (1982).
207. B. Ollmann, K. D. Kupka, and F. Hillenkamp, *Int. J. Mass Spectrom. Ion Phys.*, **47**, 31 (1983).
208. J. K. De Waele, E. F. Vansant, P. van Espen, and F. C. Adams, *Anal. Chem.*, **55**, 671 (1983).
209. J. C. Tabet and R. J. Cotter, *Anal. Chem.*, **56**, 1662 (1984).
210. E. Michiels and R. Gijbels, *Anal. Chem.*, **56**, 1115 (1984).
211. H. R. Schulten, P. B. Monkhouse, and R. Mueller, *Anal. Chem.*, **54**, 654 (1982).

212. J. F. Eloy, *J. Phys.* (*Suppl. 2*), **45**, C2-265 (1984).

213. P. Bernsen, L. Reimer, and P. F. Schmidt, *Ultramicroscopy*, **7**, 197 (1981).

214. G. A. Byrd, A. J. Fatiadi, D. S. Simons, and E. White V, *Org. Mass Spectrom.*, **21**, 63 (1986).

215. K. Balasanmugam and D. M. Hercules, in *Microbeam Analysis—1982*, K. F. J. Heinrich, Ed., San Francisco Press, Inc., San Francisco, 1982, p. 389; *Anal. Chim. Acta*, **16**, 1 (1984).

216. K. Balasanmugam, R. J. Day, and D. M. Hercules, *Inorg. Chem.*, **24**, 4477 (1985).

217. K. Balasanmugam, S. K. Viswanadham, and D. M. Hercules, *Anal. Chem.*, **58**, 1102 (1986).

218. D. E. Mattern and D. M. Hercules, *Anal. Chem.*, **57**, 2041 (1985).

219. D. E. Mattern, F.-T. Lin, and D. M. Hercules, *Anal. Chem.*, **56**, 2762 (1984).

220. C. D. Parker and M. Hercules, *Anal. Chem.*, **57**, 698 (1985); *Anal. Chem.*, **58**, 25 (1986).

221. F. P. Novak, Z. A. Wilk, and D. M. Hercules, *J. Trace Microprobe Tech.*, **3**, 149 (1985).

222. L. Van Vaeck, J. Claereboudt, J. De Waele, E. Esmans, and R. Gijbels, *Anal. Chem.*, **57**, 2944 (1985).

223. J. Rosmarinowsky, M. Karas, and F. Hillenkamp, *Int. J. Mass Spectrom. Ion Processes*, **67**, 109 (1985).

224. T. Dingle, B. W. Griffith, and S. J. Mullock, in *Microbeam Analysis—1986*, A. D. Romig, Jr., and W. F. Chambers, Eds., San Francisco Press, Inc., San Francisco, 1986, p. 475.

CHAPTER

10

# LASER MICRO ICP MASS SPECTROMETRY

Considerable progress has been achieved during the past decade in the development and application of inductively coupled plasma (ICP) sources for spectrochemical analyses. Of the various combinations of the ICP source the greatest popularity of the ICP is found as a photon or ion source for optical emission spectrometry and mass spectrometry. ICP-OES is a mature technology with numerous manufacturers, and laboratory instruments are located throughout the world. ICP-MS has grown considerably since the first commercial instrument was announced in 1983. The growth of ICP-MS has been documented by Douglas [1].

The main features of the ICP-MS systems, which have been developed by Gray [2] and Douglas [3] consist of a conventional ICP system, an interface between the plasma and special vacuum systems, an ion lens assembly for the formation of an ion beam, a quadrupole mass analyzer, and a manual or computer-based control and data-handling system, as shown in Fig. 85. Technical data for two commercial instruments are given in Table 52.

Laser ablation of solid samples for introduction into an ICP-MS is done in the same manner as used in LM-OES and LM-ICP-OES (see Chapters 3, 5, and 6). An advantage of this technique is the separation of sample ablation and ionization; the ablated material may be in any chemical state since vaporization, atomization, and ionization for analytical purpose are performed subsequently in ICP.

Gray [7] used a ruby laser with 1.5 J energy output and accumulated in practice 5 to 10 shots of about 0.5 J. The burn diameters were approximately 0.5 mm (e.g., no real microanalysis were done). For bulk analysis homogeneity of samples plays an important role. Gray [8] milled such samples and compacted them with a binder (20% w/w Elvacite) into a flat disk.

Detection limits in the solid are typically 0.1 $\mu g\, g^{-1}$ for the full mass range scan. The detection limit is one to three orders of magnitude lower than in LM-ICP-OES, although the equipment is more complex and sophisticated than corresponding LM-ICP-OES systems and does not yet approach the precision and reliability of the best of these [8, 9].

Arrowsmith [6] used a solid-state laser with typical energy of approximately 200 mJ in a $Q$-switched mode and realized a spot diameter of 20 $\mu$m.

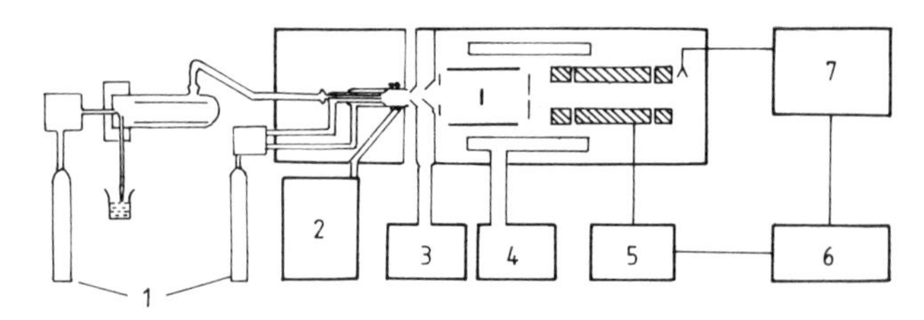


**Figure 85.** Schematic of an ICP-MS instrument with quadrupole mass spectrometer also suitable for LM-ICP-MS. 1, Argon; 2, RF generator; 3, mechanical pump; 4, cryo pump; 5, quadrupole mass spectrometer; 6, data-handling system; 7, computer. From Ref. 5 reprinted by permission of Perkin-Elmer & Co. GmbH, Überlingen, FRG.

**TABLE 52. Commercially Available ICP Mass Spectrometers**

| VG Plasma Quad; VG Laser Lab: VG Elemental, Winsford, Cheshire, England | ELAN[a]: Perkin-Elmer, Norwalk, CT |
|---|---|
| *Laser* | |
| Nd: YAG, *Q*-switched (Pockels cell)<br>1064 nm<br>1000 to 250 mJ (*Q*-switched)<br>1 to 15 Hz<br>8 to 140 ms pulse length | A laser ablation sample introduction system has been developed (see [6]) |
| *Sample cell* | |
| 60 mm high × 40 mm diameter<br>Quartz window | |
| *Spot size* | |
| 10 to 300 μm | |
| *ICP* | |
| Cool gas, 13 liters/min<br>Auxiliary gas, 0.5 liter/min;<br>Carrier gas, 1.0 liter/min | |
| Ar/Ar<br>27.2 MHz<br>1.35 kW | Ar/Ar<br>27.2 MHz<br>2.5 kW |
| *MS* | |
| Quadrupole<br>1 to 300 amu | Quadrupole<br>1 to 300 amu,<br>Resolution, 1 to 0.6 ± 0.1 amu<br>Stability, ±0.05 amu per 8 h |
| *Detector* | |
| MCA channels: 512 to 4096 | Channel electron multiplier (CEM) |

[a] Formerly Sciex Ltd., Thornhill, Ontario, Canada, ELAN 250 ICP-MS (see [6]).
*Source*: Refs. 4 and 5.

**TABLE 53. Application of LM-ICP-MS[a]**

| Matrix | Elements | Aims and Results |
|---|---|---|
| Ceramics | Be, Mg, Fe, Cd, Pr, Sm, Gd, Dy, Er, Yb, Th, B, Al, Mo, Ce, Nd, Eu, Tb, Ho, Tm, Lu, U | Demonstration of the full mass range of $^{9}$Be to $^{238}$U; semiquantitative analysis of silicon nitride; full multielement analysis for each sample in 60 s |
| Glasses | Pb, Rb, Ag, Sr, Th, U | Semiquantitative analyses with RSD of 3 to 9%; estimation of detection limits of 0.1 ppb on a full range scan |
| Geochemical samples | Co, Zn, Cu, Fe | Calibration curves |
| Standards of NBS semiconductors | Ni, Zn, Sn, Er, Hf, Au, Th, Cu, Ag, Dy, Yb, W, Pb, U | Rapid method for surveying solid materials for contamination |
| Metal alloys: Au and steel | Ag, Pd, Pt in Au, B in steel | Quantitative analyses |

*Source*: After Ref. 10; reprinted by permission of VG Elemental.
[a] Further literature can be found in [11, 12].

The sample of a 10 × 10 cm area, together with an XYZ table, was housed in a large gastight box which was filled with ICP carrier gas (normally argon) at atmospheric pressure. The design of the sample chamber was found to be critical. Ideally, it should be a small volume to minimize dilution and to lose ablated material to a large-volume outer box, with consequent problems of memory effects and the like. The laser is run at 10 to 20 Hz to produce a nearly continuous signal.

Examples for application are given by [10] and summarized in Table 53.

## REFERENCES

1. D. J. Douglas, *ICP Inf. Newsl.*, **12**, 100 (1986).
2. A. L. Gray and A. R. Date, *Analyst*, **108**, 1033 (1983); see also A. L. Gray, in *Advances in Mass Spectrometry 1985*, J. F. J. Todd, Ed., John Wiley & Sons, Inc., New York, 1986, p. 243.
3. D. J. Douglas, *Can. Res.*, **16**, 55 (1983); see also D. J. Douglas and R. S. Houk, *Prog. Anal. At. Spectrosc.*, **8**, 1 (1985).

4. *VG Plasmaquad* and *VG Laserlab*, VG Elemental, a division of VG Isotopes, Ion Path, Road Three, Winsford, Cheshire, England.
5. *Elan*, Perkin-Elmer Corp., Norwalk, Conn.
6. P. Arrowsmith, *Anal. Chem.*, **59**, 1437 (1987); *Proc. Plasma Winter Conf.*, R. Barnes, Ed., San Diego, Calif., 1988, p. 54, paper M17.
7. A. L. Gray, *Analyst*, **110**, 551 (1985).
8. A. L. Gray, *Proc. Analytiktreffen 1986*, Neubrandenburg, Karl-Marx-Universität, Leipzig, GDR, 1987, Vol. II, p. 311.
9. A. L. Gray, *Fresenius Z. Anal. Chem.*, **324**, 561 (1986); *J. Anal. Atom. Spectrom.*, **1**, 403 (1986); *Spectrochim. Acta*, **40B**, 1525 (1985).
10. Technical information, *PQ 709 A-E*, VG Elemental, Winsford, Cheshire, England.
11. H. Budzikiewicz, *Mass Spectrom. Rev.*, **7**, 359 (1988).
12. A. L. Gray, *Inductively Coupled Plasma Source Mass Spectrometry*, John Wiley & Sons, New York, 1988, pp. 257–300.

CHAPTER

11

# COMPARISON WITH OTHER METHODS OF SOLID-STATE ANALYSIS

Methods of in situ microanalysis have to permit qualitative and quantitative analytical characterization of microregions of solids. Full analytical characterization requires information about type and quantity of localized and distributed elements (respectively, isotopes) and molecules, their state of chemical bonding, and their structure.

Usually, physical techniques based on photon, electron, neutron, and ion beams are applied. Due to the different physical properties and features of these analytical tools, the information content varies from technique to technique (Table 54 and Fig. 86).

In situ microanalysis is used in routine analytical work as well as in research laboratories. The field is developing rapidly due to the ever-increasing requirements for analytical characterization of materials and substances. Strong efforts are being made to develop or improve methods in order to increase spatial resolution, detection power, precision, and accuracy of analyses.

The main goal of laser microanalytical methods is elemental analysis for obtaining information about the type and quantity of elements in microregions of solids and about distribution of these elements. The major technique in use in this field is electron probe microanalysis (EPMA) [1–3]. Figures of merit are shown in Table 54 and Fig. 86. Development efforts in EPMA concern the quantitative analysis of second-period elements (B, C, N, O), including efforts to improve correction procedures [4–7]. Quantitative analysis of these elements can now be performed with an accuracy of a few percent. Simultaneous registration of the complete x-ray spectrum, including light elements, is made possible by the use of windowless energy-dispersive x-ray detectors. Detection limits of approximately 1% are achieved for C, N, and O even with energy-dispersive systems [8].

Progress in detection power for light-element analysis can also be achieved with secondary ion mass spectrometry (SIMS) [9–12]. All elements, including hydrogen, can be analyzed, which is not the case with all the other methods discussed in this chapter. For quantitative analysis of light elements with SIMS, relative sensitivity factors have to be determined. Well-defined samples

**TABLE 54. Comparison of Various Methods for Surface Analysis**

| | Method | | | | | |
|---|---|---|---|---|---|---|
| Property | LM-OES | LM-MS | EPMA | SIMS | AES | ESCA |
| Principle | | | | | | |
| Excitation | Photons | Photons | Electrons | Ions | Electrons | Photons |
| Emission | Photons | Ions | Photons | Ions | Electrons | Electrons |
| Lateral resolution ($\mu m$) | 10 | 0.5 | 1 | 0.5–1 | 0.05 –0.1 | 1000 |
| Depth resolution ($\mu m$) | 1 | 0.5 | 1 | 1 | 0.002–0.005 | 0.5–10 |
| Detection limit (ppm) | 10–1000 | 10–1000 | 10–1000 | 1 | 1000 | 1000 |
| Detection elements | All, without gases | All | $Z > 3$ | All | $Z > 2$ | $Z > 1$ |
| Isotopes | No | All | No | All | No | No |
| Chemical bonds | No | Special cases | No | Yes | Special cases | Yes |
| Element distribution | Yes | Yes | Yes | Yes | Yes | No |
| Depth profiles | Yes | Yes | Yes | Yes | Yes | No |
| Molecules | No | Yes | No | Yes | No | No |

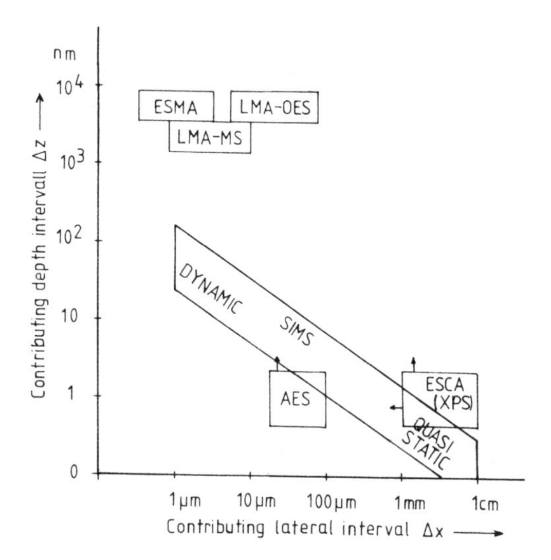


**Figure 86.** Spatial information domains of some methods for in situ microanalysis.

based on comparable matrices have to the used after characterization with a technique yielding quantitative results of high accuracy, where EPMA is well suited due to a similar lateral resolution [13]. An accuracy of 25% can be obtained with SIMS if the matrix effects are carefully controlled [14].

In fulfilling the increasing demands for materials analysis characterization of submicrometer regions of a solid, laser analytical methods are limited to $\geqslant 0.5\,\mu m$ (LM-MS). Submicrometer (or nanometer) analysis is possible with Auger electron spectrometry (AES) by measuring an electron signal originating from the outermost surface layers of a solid. AES offers good lateral resolution (50 to 100 nm) and excellent depth resolution (2 to 5 nm). These advantages and the possibility of quantification make AES the best technique for micro (and nano) distribution analysis at surfaces.

Submicrometer distribution analysis with SIMS using finely focused ion beams is developing rapidly. Focusing of high-energy ions (approximately 60 keV) to a spot of 20 nm has been possible since 1985 [15]. Practical distribution analysis is at present performed with approximately 0.5 $\mu m$ lateral resolution. Strong efforts to develop this area in the direction of quantitative three-dimensional distribution analysis are presently under way [16–18].

In situ microanalysis of trace elements could be obtained by generation of mass spectra by a photon beam with high-energy density as in LM-MS or by an ion beam as in SIMS. All techniques based on electron excitation of an analytical signal suffer from a high electron excitation or bremsstrahlung background [19].

For many scientific and technical questions, isotopic specificity of in situ microanalysis is required. Mass spectrometric methods are isotope specific. The greatest progress in the past few years has been made by SIMS and LM-MS. Isotopic ratio measurements can be made for small regions approximately 10 $\mu$m in diameter [20].

For speciation in microdistribution analysis, LM-MS, SIMS, and XPS offer great potential. Laser-excited mass spectra enable speciation by evaluation of molecular peaks and fragments, as does SIMS with liquid metal ion sources employing very low current densities. Identification is performed by evaluation of the secondary ion molecular fragmentation pattern [21, 22].

The determination of structural features is usually combined with other analytical information, such as chemical composition. Beyond other interesting new developments for gaining microstructural information, LM-MS is well suited for the in situ structural characterization of solids.

The major trends that can be seen in the progress of in situ microanalysis concern the development of methods for the following:

- Quantitative local and distribution analysis of elements in micrometer and nanometer regions
- Sensitive trace analysis in microregions
- Highly accurate isotope analysis in microregions
- Speciation of compounds in microanalysis
- Structural characterization of microregions

New perspectives are opened up by instrumental refinement of existing methods and the development of new ones, by incorporation of mathematical procedures for signal evaluation, and by the use of a combination of various analytical methods.

## REFERENCES

1. H. Malisa, *Elektronenstrahlmikroanalyse*, Springer-Verlag, Vienna, Austria, 1966.
2. O. Brümmer, J. Heydenreich, K. H. Krebs, and H. P. Schneider, *Festkörperanalyse mit Elektronen, Ionen und Röntgenstrahlen*, VEB Deutscher Verlag der Wissenschaften, Berlin, GDR, 1980.
3. K. F. J. Heinrich, *Electron Beam X-Ray Microanalysis*, Van Nostrand Reinhold, Company, Inc., New York, 1981.
4. J. Ruste, *J. Microsc. Spectrosc. Electron*, **4**, 123 (1979).
5. J. D. Brown and R. H. Packwood, *X-Ray Spectrom.*, **11**, 187 (1982).
6. A. P. Van Rosenstiel, P. Schwaab, and J. D. Brown, *Mikrochim. Acta Suppl.*, **10**, 199 (1983).

7. H. L. Pouchou and F. Pichoir, *Rech. Aerosp.*, **3**, 13 (1984).
8. W. Braue, H. J. Dudek, and G. Ziegler, *Mikrochim. Acta Suppl.*, **11**, 1 (1985).
9. C. A. Anderson, Ed., *Microprobe Analysis*, Wiley-Interscience, New York, 1973.
10. K. F. J. Heinrich, and D. E. Newbury, *Secondary Ion Mass Spectrometry*, NBS SP 427, National Bureau of Standards, Washington, D.C., 1975.
11. M. Grasserbauer, G. Stingeder, and M. Pimminger, *Fresenius Z. Anal. Chem.*, **315**, 575 (1983).
12. M. Grasserbauer, H. J. Dudek, and M. F. Ebel, *Angewandte Oberflächenanalyse mit SIMS, AES, XPS*, Springer-Verlag, Berlin, 1986.
13. M. Grasserbauer, *Fresenius Z. Anal. Chem.*, **324**, 544 (1986).
14. M. Grasserbauer, G. Stingeder, H. Ortner, W. Schintlmeister, and W. Wallgram, *Fresenius Z. Anal. Chem.*, **314**, 340 (1983).
15. R. Levi-Setti, *Annual Summer Symposium on Analytical Chemistry*, Clarkson University, Potsdam, N.Y., 1985.
16. W. F. Steiger, H. Rüdenauer, H. Gnaser, P. Pollinger, and H. Studnicka, *Mikrochim. Acta Suppl.*, **11**, 111 (1983).
17. G. H. Morrison and M. G. Moran, *Springer Ser. Chem. Phys.*, **36**, 178 (1984).
18. A. Benninghoven, F. Rüdenauer, and H. W. Werner, *Secondary Ion Mass Spectrometry*, John Wiley & Sons, Inc., New York, 1985.
19. H. W. Werner and R. P. H. Garten, *Rep. Prog. Phys.*, **47**, 221 (1984).
20. E. Zinner and M. Grasserbauer, *Springer Ser. Chem. Phys.*, **19**, 292 (1982).
21. J. A. Gardella, *Trends Anal. Chem.*, **3**, 129 (1984).
22. D. Briggs and A. B. Wootton, *Surf. Interface Anal.*, **4**, 109 (1982).

# INDEX


AAS (atomic absorption spectroscopy), 2, 24, 40, 207–214
Absolute detection limits, 126–127, 129, 133, 137, 138, 144, 145, 147, 148, 150, 151, 158, 160, 167–169
Absorbance, 207
Absorber, saturable, 14–15, 30–35
Absorption, 44
Absorption coefficient, 15, 207
Accuracy, 75–77
Active medium, 5–10
Additional excitation, 48–57
Aerosol, 182–183, 186–191
AES (Auger electron spectroscopy), 276–277
AFS (atomic fluorescence spectroscopy), 2, 215–216
Alloys, analysis:
  aluminum, 117, 142–152, 192, 194, 198, 203, 212, 244
  antimony, 147
  beryllium, 153
  brass, 148, 149, 194, 198, 212, 243
  bronze, 144, 163, 166, 196
  copper, 117, 144, 145, 149–153, 161, 162, 164, 165, 244
  ferrous, 117, 142–153, 160–162, 198, 203, 212, 243
  gold, 117, 142, 149, 152, 160, 165, 166, 243, 273
  lead, 147, 165
  magnesium, 145
  nickel, 142, 147, 244
  rhodium, 160
  silver, 164, 165, 243
  steel, *see* Alloys, analysis, ferrous
  titanium, 194, 243
  uranium, 143
  zirconium, 143
Analysis of variance, 80–83, 86–89
Analytical procedural error, 80, 83
Analyzer:
  LM-AAS, 209–212
  LM-AFS, 215–216
  LM-DCP-OES, 201–202
  LM-ICP-MS, 271–272
  LM-ICP-OES, 183–188
  LM-MIP-OES, 199
  LM-MS, 227–236
  LM-OES, 93–104, 104–120, 140
Aperture, 21
Application, 3, 120–171, 191–196, 199–203, 208–213, 216, 236–259
Atmosphere:
  argon, 57, 59, 61, 62, 64, 183–187
  helium, 57, 63
  hydrogen, 63
  nitrogen, 57, 64
  oxygen, 57, 64
Auxiliary:
  discharge, 50–57
  spark gap, 95

B, Ba, Be, Bi, *see* Elements
Bacteria, analysis of, 247–249
Balance of energy, 43–48
Biology, 167–170, 246–251
Bipyridine, 255
Bis(dimethylamino)benzophenone, 252
Bis(1-ethylquinolin-4-yl)trimethinecyanine-iodide, 31
Bis(4-di-methylaminodithiobenzil)nickel, 15
Blackening of lines, 18, 83
Bleachable dye, 32–35
Blood, 168
Boltzmann:
  constant, 41
  relation, 39
Bone, 170, 196
Borax melting, 82
Borosilicate glass, 155
Brain, 167, 168
Brass, *see* Alloys, analysis
Bronze, *see* Alloys, analysis
Bulk analysis, 3, 102, 135, 136, 143–145, 147, 149, 152, 271
Butterfly wings, 168

C, Ca, Cd, Ce, Co, Cr, Cu, *see* Elements

Calcified tissues, 16
Calcium cytochemistry, 248
Calibration, 84, 85, 134, 140, 207, 273
Capacity:
  of spark discharge, 53, 95
  specific heat, 27, 38
Carbonates, 135
Carrier gas, 183, 184, 186, 209, 272, 273
Catoptric objective, 96
CCD (charge-coupled device), linear and matrix, 110, 118
Celestite, 132
Cells:
  of bacteria, 247
  of blood, 168
  of bone, 247
  of chlorophyta, 247
  for passive $Q$-switch, 15
  Pockels and Kerr, 7, 12, 13, 30
  for samples, *see* Chamber for analytical samples
Ceramics, analysis of, 156–158, 171, 196, 273
Chalcopyrite, 133
Chalcosite, 133
Chamber for analytical samples, 59, 183–187, 201, 202, 216, 273
Chemistry as application field, 141, 236
Chromite, 129
Chrom spinel, 154
Cluster analysis, 193
Coal, analysis of, 138, 239
Cochlea wall, 245
Coins, 163–165
Collection efficiency of ions, 226
Comparison, 28, 232, 250, 255, 275–278
Computer, 118, 140, 187
Concentration of analyte, 41, 128–170, 190, 191, 194, 196, 198, 207, 211, 212
Conductivity, thermal, 26–28, 37, 38, 45
Confidence interval, 75
Cool gas, 272
Correlation, linear, 86, 139
Counting of spikes, 16
Crater:
  depth, 3, 27, 28, 34, 227
  diameter, 3, 34, 227
  radius, 21–23
  shape, 30, 31, 35, 36
Crime detection, 141
Cross excitation, 49, 51, 54, 56
$Cr^{3+}$ in ruby, 6, 7
Cryptocyanine, 31
Czerny–Turner system, 114, 186

DCP (direct current plasma), 200–204
Degrees of freedom, 75, 79, 87, 88
Density:
  of electrons, 42–44, 151
  of energy, 37
  of ions, 43, 44
  of magnetic flow, 219
  of mass, 25, 26, 78
  of power, 23, 26, 30, 35, 36, 39, 41, 222, 230, 232, 233
  of spectrum lines, 50, 51
  of vapor, 49
Dental hard tissue, 246
Detection:
  photoelectric, 106–120
  photographic, 105–106
Detection limit, *see* Absolute detection limits; Relative detection limits
Detectors:
  channel electron multiplier, 272
  charge-coupled device, *see* CCD
  diode array, 108–111
  OID (optoelectronic image device), 107, 112
  PMT (photoelectron multiplier tube), 106, 112
  SIT (silicon intensified target) vidicon, 108, 113–117
  SPD (self-scanned photodiode arrays), 111–114, 119
Depth:
  of crater, *see* Crater
  by layer analysis, 3, 117, 135, 142
  profile, 117, 157, 215, 276
  resolution, 276, 277
Deviation, *see* RSD (relative standard deviation)
Dielectrically coated mirror, 11
Diffraction grating:
  diffraction angle, 105
  dispersion, 104, 105
Diode array, *see* Detectors
Direct analysis of nonconducting specimens, 100
Discharge:
  between electrodes, *see* Auxiliary, discharge

in the flashtube, 10
in hollow cathode, 93
Disk-type sample chamber, 185
Dispersion in flow injection systems, 190
Distribution analysis, 3, 133, 134, 136, 137, 139, 141, 153, 159, 163, 276
Dolomite, 130, 238
Doppler broadening, 49
Double focusing spectrometer, 222
Dried samples, 167, 169, 170
DSID (direct sample insertion device), 182
Duration:
of spark discharge, 52
of spikes, 10, 23, 94, 232
Dust, 163
Dy, *see* Elements
Dye laser, 216

Effect(s):
of argon flow rate, 200
caused by absorption of focused laser radiation, 24, 35
of emission intensity, 203
of laser impact on a steel sample, 30, 31
in laser micro plume, 44
of microwave power, 200, 201
Effective diameter of the resonator rod, 22
Efficiency:
ratio between input and output energy of laser radiation, 10
high quantum, 112
of ion collection, 226
of ionization, 225
Einstein coefficient, 40
Ejected material, 46
Elastic scattering, 43
Electrodes, *see also* Auxiliary, discharge
gap separation, 51
position of, 54
electrical conditions:
capacity, 95
for electrode discharge, 95
inductance, 95
resistance, 95
triggering, 95
voltage, 95
Electroerosion, 182, 183
Electron:
density, *see* Density
mass, 41
Electronic charge, 207
Electro-optical switches, 12, 13
Elements:
aluminum, 38, 39, 43, 52, 121, 126, 128, 130, 133, 135–138, 142, 154–170, 192–194, 199, 212, 236, 238
antimony, 62, 124, 127, 147, 164, 196
argon, *see* Atmosphere
arsenic, 123, 126, 238, 239, 244
barium, 124, 127, 129, 131, 132, 134, 139, 154, 163, 196
beryllium, 121, 126, 129, 133, 137, 143, 153, 156, 159, 161, 166, 250, 273
bismuth, 28, 125, 127, 129, 130, 132, 135, 166, 238
boron, 121, 126, 129, 130, 132, 136, 143, 154, 168, 169, 196, 220, 241, 273
bromide, 58, 238
cadmium, 28, 51, 123, 127–129, 132, 164, 197, 211, 212, 245
caesium, 40, 124, 237, 238, 245
calcium, 40, 61, 121, 126, 130–136, 138, 142, 154–170, 196
carbon, 39, 52, 58, 148, 229, 251
chloride, 58
chromium, 43, 115, 117, 122, 126, 130–136, 138, 142–153, 154–170, 191, 192, 194, 211, 212, 220, 244
cobalt, 65, 82, 122, 126, 131–133, 136, 143, 148, 191, 194, 211, 238
copper, 28, 38, 39, 43, 52, 56, 64, 113, 117, 122, 126, 142–153, 154–170, 191, 192, 199, 211, 220
gallium, 117, 122, 126, 132, 135, 220, 241, 244
germanium, 52, 122, 126, 153, 161, 220
gold, 113, 125, 127, 148, 149, 160, 165, 273
hafnium, 125, 127, 238
hydrogen, 58, 239, 240, 243
indium, 123, 127, 142
iodide, 58
iron, 28, 52, 56, 65, 122, 126, 130–138, 143–170, 192, 196, 198, 211, 220
lead, 28, 52, 62, 117, 125, 127, 130, 132, 133, 150, 161–166, 196, 198, 211, 212, 250, 254, 273
lithium, 64, 121, 126, 132, 169, 220, 236, 245
magnesium, 28, 61, 121, 126, 130–138, 154–170, 192, 193, 211, 220, 246
manganese, 51, 61, 65, 117, 122, 126, 130–139, 142–153, 154–170, 188, 191, 194, 198, 211, 212

Elements (*Continued*)
mercury, 125, 127, 138, 162, 163, 165, 241
molybdenum, 28, 116, 123, 127, 132, 142, 143, 148, 162, 191, 198, 211, 212, 220
nickel, 28, 38, 56, 122, 126, 131, 133, 135, 137, 142–153, 192–194, 196, 199, 211, 212, 220, 235
niobium, 123, 127, 131, 148
nitrogen, 58. *See also* Atmosphere
osmium, 125, 234, 239, 249
oxygen, 58. *See also* Atmosphere
palladium, 113, 117, 123, 127, 132, 166, 220, 273
phosphorus, 58, 121, 126, 148, 153, 167, 168, 191, 241, 249
platinum, 35, 50, 51, 61, 125, 127, 166, 249, 273
potassium, 121, 132, 137, 170, 220, 245, 247
rare earth (lanthanides), 124, 127, 131, 132, 134, 161, 196, 220, 238, 273
rhenium, 239
rhodium, 123, 160
rubidium, 123, 220, 237, 245, 273
scandium, 121, 126, 131
selenium, 129, 153, 161
silicon, 61, 121, 126, 128–138, 142–153, 192, 193, 211, 220, 240
silver, 28, 38, 113, 117, 123, 127, 152, 211, 212, 220
sodium, 40, 114, 164, 170, 245, 247
strontium, 123, 127, 129, 134, 161, 220, 238
sulphur, 58, 121, 148, 191
tantalum, 125, 127, 129
tellurium, 124, 127, 242
thallium, 125, 164, 245, 254
thorium, 125, 238, 273
tin, 28, 39, 52, 117, 123, 127, 129, 135, 143, 149, 196, 211, 212, 254
titanium, 28, 65, 122, 126, 145, 155–170, 198, 211, 220, 238, 242
tungsten, 38, 51, 52, 65, 125, 127, 146, 243
uranium, 65, 125, 143, 243, 246, 273
vanadium, 65, 122, 126, 144, 149, 161, 171, 194, 212, 220
yttrium, 123, 127, 161, 238, 242
zinc, 40, 52, 122, 126, 128–136, 148–150, 154, 157, 166, 196, 198, 211, 212
zirconium, 123, 127, 129, 153, 156, 158, 161, 220, 237, 240

Emission:
behavior of the laser, 11
duration of cross-excited microplasma, 51
factor of radiation, 48
of energy by hollow radiation, 47
intensity, 115–117
signals, 185
spectrum, 114
spontaneous, 5, 40
stimulated, 5
Energy:
balance, *see* Balance of energy
density, *see* Density
of discharge between electrodes, 53
of incident laser radiation, 46–48
kinetic, 46, 47
of laser output, 10, 23, 33, 43–48, 222, 225, 230
level, 8, 9, 40
losses, 38, 43–48
measuring, 15–18
remainder, 48
EPMA (electron probe micro analysis), or ESMA, 275, 277
ESCA (electron spectroscopy for chemical analysis), 277
ETA (electrothermal analysis), 182, 183
Etendue, 105
Evacuation, 55
Excitation:
additional, *see* Additional excitation
of atoms, 24
collisional, 42
conditions, 56
cross, *see* Cross excitation
gas media, 57
potentials, 37
under variable pressure, 57
Expanding radiant plasma, 21
Exposure time, 105
Eyepiece, 97

Fe, *see* Elements
Feldspar, 128
FET (field effect transistor), 112
Fiber optic, 185, 187
Fibers, 159, 237
Fibromas, 169
Field–ion–atom probe, 235
Film, 159, 240
Filter, 16, 159

Fingerprint, 114
Fisher test, 79, 83, 88, 89
Flashover voltage, 60, 95
Flashtube, 6–10, 15, 44–45, 53, 94, 100
Flow:
  injection analysis, 190
  meter, 199
  rate, 184, 200
  velocity, 190
Fluctuation, 10, 77
Fluid inclusion, 136
Fluorescence spectroscopy, *see* AFS
Flux, 12, 26
Focal:
  length of a lens, 21, 22, 99, 118, 184
  plane, 111
  spot, 23, 24
Focus, 21
Focusability, 9
Focusing of laser radiation, 21, 24
Forensic science, 141
Four-level laser, 9
Free diameter of a rod, 22
Frequency, 40, 54, 55, 207
Free-running mode, 193
Fused sample, 76, 140

Ga, Ge, *see* Elements
GaAs, 147, 241
Galena, 129, 131
Gap, 49, 51
Garnets, 6, 8, 130
Gas, *see* Atmosphere
Gating, 108–110
Geochemistry, 128–138, 273
Giant pulse, 13
Glan–Thompson types, 13
Glass, 153–155, 161, 162, 166, 196, 240, 241, 273
Glow discharge, 60
Gold, *see* Elements
  bronze, 166
  leaf, 165
  wire, 113, 117, 152
Gramicidin S, 257
Graphite, 63, 158, 159
Graphite furnace, 209
Grating, 104, 105
Gray cat iron, 146
Growth ring, 139
Guldberg–Waage, 42

H, Hg, halogens, *see* Elements
Halite, 238
Heat capacity, 26, 38
Heating effects, 24, 28, 38
Hemoglobin, 254
HgCdTe, 242
Histological sections, 249, 250
Hollow radiation, 47
Homogeneity, 11, 76–84, 86–88
Human:
  skin, 246
  urinary calculi, 249
Hydrocarbons, 254–257

In, Iodide, Iron, *see* Elements
ICP (inductively coupled plasma), 181–197, 271–273
Image:
  detector, 117
  recording technique, 107
Inclusion, 134, 136
Influence of sample parameters on vaporization, 37, 38
Inhomogeneity, 37
Instruments, 93–119, 231–233
Insulator, 41
Intensity of spectral lines, 17, 18, 56, 57, 61, 115–119
Interface, liquid, 43
Interaction, 21 ff., 45, 47, 80, 227
Internal standard, 196
Ion:
  collection efficiency, 226
  formation, 251
  kinetic energy measurement, 242
  sources, 227 ff.
  technique, "soft", 229
Ionization, 24, 28, 42, 48, 54, 223
  energy, 4
  potential, 220
Irradiation, 36
Isotope, analysis of, 220, 234, 276

Jet, water-jet pump, 212

K, *see* Elements
Kerr cell, 7, 30

La (rare earth), Li, *see* Elements
Lamberts law, 15

LAMMA, 228, 229, 232, 233
Lanthanide, *see* Elements (rare earth)
Laser:
  acronym, 5
  characteristic, 222
  energy, 23
  energy density, *see* Density
  non-switched (normal mode), 9–11, 23
  power, 23
  power density, *see* Density
  principle, 5–9
  *Q*-switching, *see* *Q*-switching
Laser microanalysis, methods:
  LM-AAS, 209–216
  LM-AFS, 217–220
  LM-DCP-OES, 202–205
  LM-ICP-MS, 271–274
  LM-ICP-OES, 183–197
  LM-MIP-OES, 198–201
  LM-MS, 221–270
  LM-OES, 93–182
LIMA, 228, 232, 233
Limits of detection, *see* Absolute detection limits; Relative detection limits
Local analysis, 3 ff.
LPMS, 228, 232, 233
LTE (local thermodynamic equilibrium), 40–43
Luminosity, 105

Macroanalysis, 3. *See also* Bulk analysis
Magnetic field, 56
Magnetite, 129
Malachite, 132
Manganese–iron accumulates, 137, 139
Mass:
  separation, 2, 233
  spectrometry, *see* LM-MS; LM-ICP-MS
Maser, optical, 5
Matrix effects, 64, 65
Medicine, 141, 167–170
Metallurgy, 140, 141
Metastable level, 8
Meteorites, 128, 131, 132, 136
Mg, Mn, Mo, *see* Elements
Microscope, 96–100
Microwave, 54
Mineralogy, 120
Minerals, 128–140
MIP (microwave-induced plasma), 196, 198, 200, 201, 203

N, Na, Nb, Ni, *see* Elements
Nd–glass laser, 8, 9, 230
Nd–YAG laser, 8, 9, 230
Nebulization, 191
Niccolite, 129
Nicol, prism, 13
Nitrates, 137
Nonconducting materials, analysis of, 154–159
Nylon, 159

Ores, *see* Minerals
Organic samples, 245–258
Orthopyroxene, 128
Os, *see* Elements
Oscillation, 11

P, Pd, Pt, *see* Elements
Paintings, 162, 164
Paints, 160, 162, 163, 165
Pancreas, 169, 170
Passive *Q*-switching, *see* *Q*-switching
Peak:
  area, 189, 198
  height, 189, 198
Photodiode array, *see* Detectors
Photographic and photoelectric detectors, *see* Detectors
Plagioclase, 128
Plasmaarray spectrometer, 188
PMT, *see* Detectors
Pockels cell, 12
Population, 9
Power, 10
Power density, *see* Density
Precision, 75, 194, 208
Pressings, 134, 135
Pressure, gas media, 57–64. *See also* Vapor pressure
PVC-tubing, 185–191
Pyrite, 128, 133, 136
Pyrrhotite, 129, 136

*Q*-switching:
  active, 7, 12–14
  passive, 7, 14–15, 29–35
Quadrupole mass spectrometer, 222, 272
Quantitative analysis, presumption for, 75–84
Quartz, 131, 238

Radio frequency, 54, 56

Rat:
kidney, 168
lumbar vertebrae, 170
pancreas, 170
sperm, 168
Rb, Re, R.E., Rh, *see* Elements
Reference materials, 77
Reflectivity, reflection of laser radiation on sample surface, 38, 43–46
Refractory materials, 156
Regression, linear, 84, 85
Relative detection limits, 126, 127, 195, 203
Reliability of signal measurements and calculations, 76
Reproducibility:
laser ICP systems, 192
laser mass spectrometry, 233
laser output, 10
sample removal processes, 76
spectral line intensities, 98
Resolving power, 104, 105
Resonator of solid-state lasers:
degree of reflection of the mirror coating, 10
diameter, 22
divergence, 22
oscillation and emission behavior, 11
refraction index, 22
stability of the components, 11
temperature of the resonator rod, 10
Resonance lines, 40
Restoration of art objects, 141
Rod-type sample chamber, 184
RSD (relative standard deviation), 75–79, 107, 194, 196
Ruby laser, 6, 7, 230
Rutile, 238

S, Sb, Sc, Se, Si, silver, Sr, *see* Elements
Sample chamber, *see* Chamber for analytic samples
Saturable absorber, *see* Absorber, saturable
Scanning laser microanalysis, 102
Scattering of laser radiation on sample surface, 44–46
Semiconductor, analysis of, 147, 153
Semi *Q*-switch, *see* *Q*-switching
Si, $SiO_2$, 239, 240
Siegenite, 129
Silicates, 132, 137
SIMS (secondary ion mass spectrometer), 275–278
Skin, 168, 169
Skutterudite, 129
Slurry atomization, 182
Spark gap, *see* Auxiliary, discharge, spark gap
Spatial resolving power, 3
Spectrographs, spectrometers, *see* Instruments and analyzer
Speed:
of analysis, 107
of ejected splashes, 47
Sphalerite, 128
Spikes, 10 ff.
number of, 23
number-measuring accessory, 15–18
time duration, 23
variation of parameters, 31
Standard rock samples, 135
Statistical weight, 39
Statistics in analytical chemistry, 75–89
Steel, *see* Alloys, ferrous
Stimulated emission, *see* Emission

Ta, Te, Th, Ti, Tl, *see* Elements
TABLASER (trace element analyzer based on laser ablation), 215, 216
Teeth, 167, 170
Temperature:
resonator rod, 10
sample, 24–28, 37–48
vapor, 35–37
Time behavior of ICP emission signals, 192
Timer, 49
$TiO_2$, 242
TOF (time-of-flight), 219–233
Transmission of the dye stuff, 14, 15
of a sample, 44
factor of mass spectrometer, 233
Transport of analyte by tubing, 182, 185–191
Trigger, 49

V, *see* Elements
Vapor pressure, 35–37
Vaporization:
of solids, 24
by normal laser pulses, 25–29
by *Q*-switched pulses, 29–35
energy, 47
volume of vapor, 47
Variance analysis, 86–89
Velocity:
of expansion of laser-induced plume, 51, 52
flow, 190

Vidicon, 107, 108, 111, 113, 117, 118

W, *see* Elements
Wavefront, 5
Wavelength:
of the laser radiation, 8, 9, 21
of spectral lines, 121–125
Welding seam, 147
Wollaston prism, 13

Y, *see* Elements
YAG (yttrium aluminum garnet), 6, 8, 230

Zr, *see* Elements
Zircon, 131

(*continued from front*)

Vol. 63. **Applied Electron Spectroscopy for Chemical Analysis.** Edited by Hassan Windawi and Floyd Ho

Vol. 64. **Analytical Aspects of Environmental Chemistry.** Edited by David F. S. Natusch and Philip K. Hopke

Vol. 65. **The Interpretation of Analytical Chemical Data by the Use of Cluster Analysis.** By

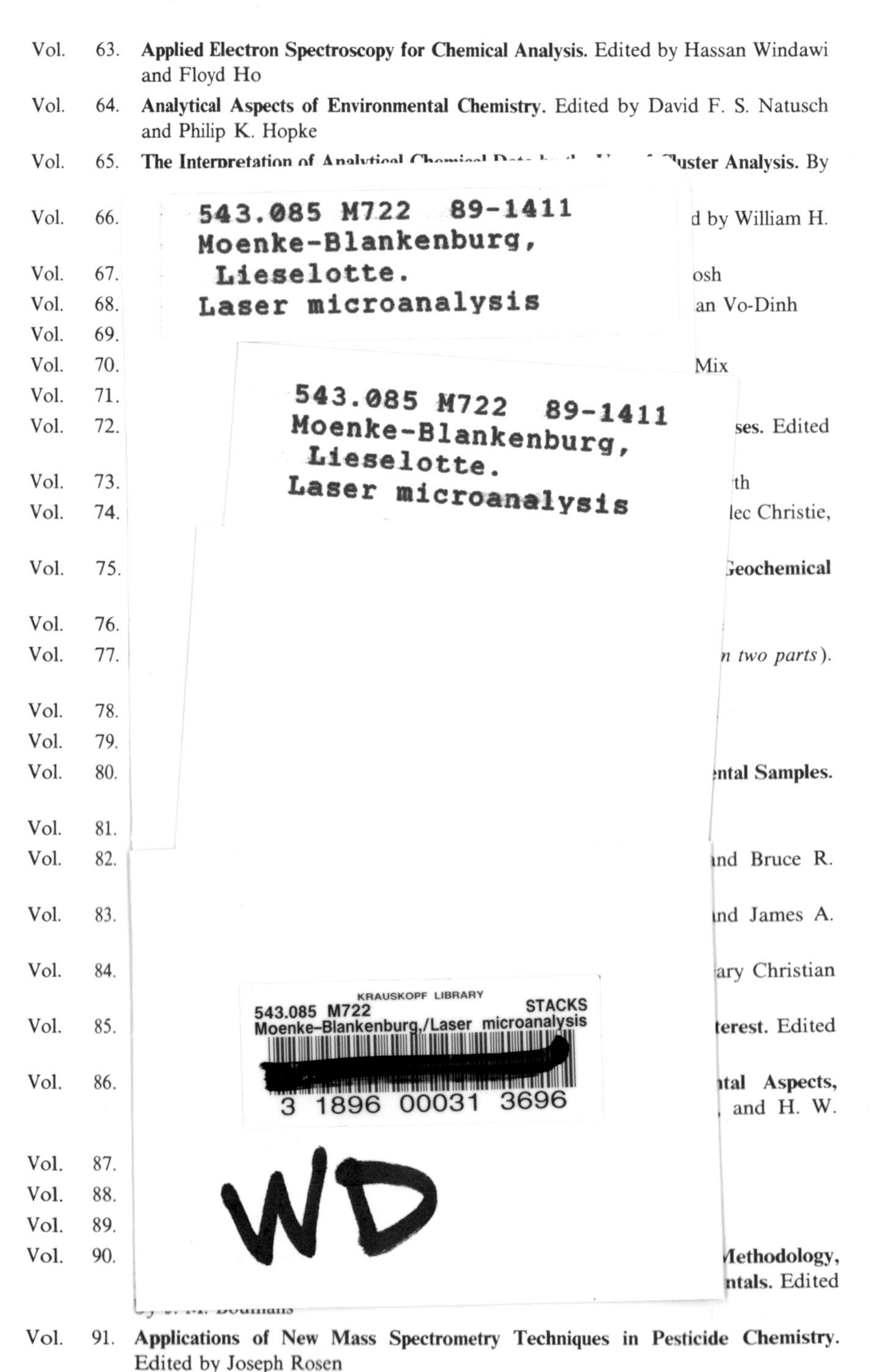


543.085 M722 89-1411
Moenke-Blankenburg,
Lieselotte.
Laser microanalysis



Vol. 66. ... d by William H.

Vol. 67. ... osh

Vol. 68. ... an Vo-Dinh

Vol. 69.

Vol. 70. ... Mix

Vol. 71.

Vol. 72. ... ses. Edited

Vol. 73. ... th

Vol. 74. ... lec Christie,

Vol. 75. ... Geochemical

Vol. 76.

Vol. 77. ... *n two parts*).

Vol. 78.

Vol. 79.

Vol. 80. ... ntal Samples.

Vol. 81.

Vol. 82. ... and Bruce R.

Vol. 83. ... and James A.

Vol. 84. ... ary Christian

Vol. 85. ... terest. Edited

Vol. 86. ... ntal Aspects, ... and H. W.

Vol. 87.

Vol. 88.

Vol. 89.

Vol. 90. ... Methodology, ... ntals. Edited by J. M. Boumans

Vol. 91. **Applications of New Mass Spectrometry Techniques in Pesticide Chemistry.** Edited by Joseph Rosen